A to Z of Thermodynamics

Oxford University Press, Great Clarendon Street, Oxford OX2 6DP

Oxford New York

Athens Auckland Bangkok Bogota Bombay
Buenos Aires Calcutta Cape Town Dar es Salaam
Delhi Florence Hong Kong Istanbul Karachi
Kuala Lumpur Madras Madrid Melbourne
Mexico City Nairobi Paris Singapore
Taipei Tokyo Toronto Warsaw
and associated companies in
Berlin Ibadan

Oxford is a trade mark of Oxford University Press

Published in the United States
by Oxford University Press, Inc., New York

Original French edition 'Dictionnaire de thermodynamique'
© InterEditions, Paris 1994

This English translation © Pierre Perrot, 1998

Published with the help of the Ministère de la Culture

A catalogue record for this book is available from the British Library

Library of Congress Cataloging in Publication Data

Perrot, Pierre.
    [Dictionnaire de thermodynamiques. English]
    A to Z thermodynamics/ Pierre Perrot.
    1. Thermodynamics—Dictionaries. I. Title.
    QC310.3.P47 1998
    621.402'1'03–dc21   97-51538   CIP

ISBN 0 19 856556 9 (Hbk)
ISBN 0 19 856552 6 (Pbk)

Typeset by Newgen Imaging Systems (P) Ltd., Chennai, India

Printed in Great Britain
by Biddles Ltd.,
Guildford, Surrey

Preface

To present Thermodynamics in the form of a dictionary may seem somewhat surprising. However, it allows readers to wander as in a garden, at their fancy or according to their need for explanation. In order that the walk be pleasant, the landscape must be attractive and varied; from that point of view, Thermodynamics fears no competition, but there are obstacles in the way, stones in the path; isn't Thermodynamics considered a fine intellectual structure, bequeathed by past decades, whose every subtlety only experts in the art of handling hamiltonians would be able to appreciate?

Successive generations of students always encounter the same difficulties in understanding, make the same mistakes in interpretation, fall into the same traps. In short, ramblers always stumble against the same stones! Relying on twenty years' experience in the teaching of this science dear to my heart, I have not hesitated to speak plainly when necessary, at the risk of seeming to breakdown open doors. I wanted to produce a work of demythification, to show that it is possible to go far with the minimum mathematical and physical baggage presented in the various entries, baggage whose weight does not exceed that which honest scientists can support when life is so burdened as to make them forget everything.

A dictionary needs to be rigorous; this one makes that claim, mainly in describing the principles. But Thermodynamics, not being an end in itself, has to be studied with a view to its applications. They are numerous and their development, which can extend to infinity, must necessarily be limited. Their selection reflects care to reveal the universality of reasonings in thermodynamics, capable of achievements other than the mere management of relationships between pressures, concentrations, and temperatures, to which academic teaching is often reduced, due to lack of time!

A complete and rigorous presentation must not leave any point in obscurity, which explains the length of the entries devoted to fundamental concepts, principles and functions. At this level, everything is proved and no difficulties are avoided. The applications spill over extensively on to what is taught in engineering courses. However, the whole remains accessible to first-year undergraduates. The treatment of the Thermodynamics of irreversible processes, limited to basic concepts, aims to give the reader the desire to look further in more specialized works.

Each entry, when possible, represents a whole and this is why repetition is not always avoided. Where useful, terms that are subjects of a particular entry in the dictionary are marked in *italic*. Finally, IUPAC recommendations have been followed, except in rare cases, exceptions justified by the need to avoid unnecessary heaviness. Superscripts $^\circ$ are used for standard states, and $^\ominus$ for reference states.

Simplicity and rigour are not necessarily synonymous with austerity, hence the small space allotted to matter that will prevent anyone dying a lunatic. Readers I hope will experience the pleasure I have felt when writing. Besides useful physico-mathematical developments, they will find short etymological considerations, historical notes, and discussion of classical paradoxes or current topics, together with personal thoughts that sometimes go beyond the framework of the discipline.

I cannot conclude without recognizing my debts: successive generations of students who have taught me that there is no shame in not admitting directly what becomes clear only after long years of maturing; my colleagues, who with their criticisms and suggestions have shown me that the availability of abundant literature does not make me secure from mistakes; my publisher, finally, knew how to persuade me that Nicolas Boileau, the author of *l'Art poétique* did not speak only to his contemporaries of the seventeenth century in wisely recommending:

Add sometimes, but oft delete

July 1997   Pierre Perrot

A

Absolute—Without reference to any particular system, as opposed to 'relative'. Among the physical quantities characterizing a system, some, such as volume or *entropy*, are defined in absolute value and are directly accessible to experiment. More often, thermodynamic quantities (energies, potentials) can only be defined in relative value, experiment giving access only to their changes.

An absolute scale of temperature is a scale defined independently of any material. The expansion of mercury, the electrical resistance of platinum, or the potential difference of a thermocouple do not allow the definition of an absolute scale, but only of a relative scale of temperature.

A physical quantity, for instance ΔQ, can be given an absolute value: $|\Delta Q|$, that is, without any sign. This is often because of intellectual laziness, justified when there is no ambiguity in the direction of the exchange, but is that always the case?

Absolute activity—The *chemical potential* of a component i in a mixture is given by: $\mu_i = \mu_i^\circ + RT \ln a_i$, with $a_i = f_i/f_i^\circ$, the *activity* of i. This activity, also expressed by:

$$a_i = \exp(-\Delta \mu_i / RT),$$

is a relative activity, since it is defined by reference to a *standard state*, which must be well specified.

In 1939, Fowler and Guggenheim introduced λ_i, the absolute activity of i, defined by the relation:

$$\lambda_i \equiv \exp(-\mu_i / RT).$$

The chemical potential being determined except for an additive constant, the absolute activity is *de facto* determined except for a multiplicative constant. However, it is independent of the choice of a standard state, which is the reason for its interest. The equilibrium of a constituent i between two phases α and β can be expressed by equating either chemical potentials or absolute activities, since $\mu_{i,\alpha} = \mu_{i,\beta} \Rightarrow \lambda_{i,\alpha} = \lambda_{i,\beta}$. On the other hand, the equality of activities implies the equilibrium of a constituent i between two phases only when the same standard state has been chosen for i in the two phases.

The *Gibbs energy of reaction* is simply expressed as a function of absolute activities. If one considers a mixture of compounds M_i and between them a *chemical reaction*: $\sum v_i M_i = 0$, the Gibbs energy of reaction is given by:

$$\Delta_r G = \Delta_r G^\circ + RT \ln \prod (a_i)^{v_i} = RT \ln \prod (\lambda_i)^{v_i}.$$

The equilibrium condition $\Delta_r G = 0$ is then merely expressed by:

$$\prod (\lambda_i)^{v_i} = 1.$$

Absolute temperature—A temperature scale is said to be absolute when it is not related to the physical properties of any particular material. Expansion of mercury, electrical resistance of platinum or electromotive force of a thermocouple, for instance, give only a relative temperature scale. The *thermodynamic temperature*, defined by the *second law*, gives an absolute temperature scale.

Absolute zero—The name given to the temperature of $0\,K$, defined as $-273.15°C$. Here, the word absolute is taken etymologically from the Latin *absolutus*, completed. Absolute zero corresponds to the lowest temperature conceivable. At absolute zero, there is no thermal energy: no molecular motion is available for transfer to other systems.

The notion that there is an ultimately coldest temperature was suggested by the behaviour of ideal gases. Using the Celsius temperature scale, the equation of state of an ideal gas gives:

$$PV = P_0 V_0 (1 + t/273.15).$$

Numerical application gives $PV = 0$ for $t = -273.15°C$. As it is difficult to conceive of a negative PV product, one concludes that it is impossible to reach temperatures lower than $-273.15°C$. The result is correct, but the argument is weak. The ideal-gas equation of state does not account for the condensation phenomenon which characterizes every real gas, so it is not applicable at low temperatures.

It is surely more straightforward to note that the *thermodynamic temperature* introduced from the *second law, corollary 7*, is physically a positive quantity, which is implicitly supposed in *corollary 9*. One of the statements of the *third law*, based on experimental measurements of *entropy* change, postulates the inaccessibility of absolute zero.

Acceleration—The second derivative of length with respect to time, expressed in $m\,s^{-2}$. The normal acceleration of free fall was defined in 1901 at the third Conférence Générale des Poids et Mesures and is conventionally fixed at $9.806\,65\,m\,s^{-2}$. It represents the mean value of acceleration of free fall at sea level at latitude 45°. It varies from about $9.78\,\mathrm{m\,s^{-2}}$ on the equator to about $9.83\,\mathrm{m\,s^{-2}}$ at the poles.

Acentric factor—For every fluid, the rule of *corresponding states* allows the use of the same *equation of state* in reduced coordinates, for example pressure divided by the critical pressure, etc. If this law were verified, the saturated vapour pressure in reduced coordinates would be the same for every liquid, a consequence unfortunately not confirmed by observation. The acentric factor has been introduced to preserve the rule of corresponding states by taking into account the effects of size, form, or polarity of various fluids.

The acentric factor ω is empirically defined by:

$$\omega = \log_{10}(P_c/P_s) - 1,$$

where P_c represents the critical pressure and P_s the saturated vapour pressure of the liquid at a reduced temperature of 0.7 (that is, $T = 0.7T_c$). This factor has been selected so that it takes the value zero for monatomic rare gases. It increases with the complexity of the molecule (O_2: 0.02; CO: 0.04; CO_2: 0.23; NH_3: 0.25). For saturated hydrocarbons with linear chains, it increases regularly with the chain length, from 0.013 for methane to 0.392 for *n*-octane, with a mean increment of 0.05 for each added carbon atom. The theory of the acentric factor allows a straightforward expression for the *compressibility factor*:

$$Z = PV/RT = Z_0 + \omega Z_1,$$

Z_0 and Z_1 depending only on the reduced coordinates T_r and P_r.

Empirical correlations exist between the critical compressibility factor and the acentric factor. For instance, for hydrocarbons:

$$Z_c = 0.291 - 0.07\omega.$$

Acidity constant—The name given to the dissociation *equilibrium constant* of an acid in aqueous solution:

$$AH + H_2O \rightarrow H_3O^+ + A^-$$

$$B^+ + 2H_2O \rightarrow H_3O^+ + BOH.$$

An acid in aqueous solution (for instance AH or B^+) is strong when its acidity constant is greater than unity: its *dissociation coefficient* is practically equal to unity. It is weak when its acidity constant lies between 10^{-3} and 10^{-11}: its dissociation coefficient is practically zero.

4 Action

Its acidity is negligible when its acidity constant is lower than 10^{-14}: it is then less acidic than water.

Action—An action is the product of work by time. For instance, the *Planck constant* represents the action quantum.

Activation energy—When a *reaction rate* for a *chemical reaction*:

$$\sum a_i A_i \rightarrow \text{Products}$$

is given by the expression:

$$v = \frac{\mathrm{d}\xi}{\mathrm{d}t} = k\varphi([A_i]),$$

the rate constant k varies exponentially with temperature following the Arrhenius law:

$$\frac{\mathrm{d}\ln k}{\mathrm{d}T} = +\frac{\Delta E}{RT^2}.$$

$\Delta E > 0$ is the activation energy of the reaction. Integration, with the hypothesis that ΔE is independent of T, gives:

$$k = A\exp(-\Delta E/RT).$$

The pre-exponential factor A can be related to an activation entropy:

$$A = \exp(\Delta S/R).$$

The higher the activation energy, the slower the reaction. Physically, an activation energy represents the energy required for bringing reactants to an excited state before obtaining reaction products. A catalyst increases or decreases a reaction rate by modifying its activation energy.

For an *elementary reaction*: $\sum a_i A_i \rightleftharpoons \sum b_i B_i$, equilibrium is obtained when the forward reaction rate (direction 1) equals the reverse reaction rate (direction -1):

$$A_1\exp\left(-\frac{\Delta E_1}{RT}\right)\prod_i a_{A_i}^{a_i} = A_{-1}\exp\left(-\frac{\Delta E_{-1}}{RT}\right)\prod_i a_{B_i}^{b_i},$$

$$\frac{\prod_i a_{B_i}^{b_i}}{\prod_i a_{A_i}^{a_i}} = \frac{A_1}{A_{-1}}\exp\left(\frac{\Delta E_{-1}-\Delta E_1}{RT}\right).$$

The difference $(\Delta E_1 - \Delta E_{-1})$ [pay attention to the signs!] between activation energies thus represents the reaction enthalpy.

Activity—A concept introduced by Lewis in 1913 which has become hard to circumvent, even for a novice in thermodynamics! The activity of a constituent i in a mixture is a dimensionless value defined by:

$$a_i = \frac{f_i}{f_i^\circ} = \frac{\text{fugacity of } i \text{ in the mixture}}{\text{fugacity of } i \text{ in its standard state}}.$$

Note: When a value of activity is quoted, the standard state chosen must be well specified. It may be that this specification seems so evident that authors refrain from giving it. Nevertheless, something that is clear in the author's mind may not always be clear in the reader's.

When i is gaseous, it is common practice to take $f_i^\circ = 1$ bar. The activity of i in a mixture is then given by: $a_i = f_i/\text{bar}$.

When i is in a condensed phase, it is often convenient to choose as standard state the pure compound i, under 1 bar pressure and in a precisely defined *aggregation state*; f_i° therefore represents the fugacity of i vapour in equilibrium with condensed phase i taken in its standard state.

In practice, when fugacities lie well below 1 bar, the fugacity and pressure can be assimilated: $a_i \approx p_i/p_i^\circ$. Thus, the activity of a condensed substance of low volatility can be expressed by:

$$a_i \equiv \frac{f_i}{f_i^\circ} \approx \frac{p_i}{p_i^\circ} = \frac{\text{vapour pressure of } i \text{ above the mixture}}{\text{vapour pressure of } i \text{ in its standard state}}.$$

Likewise, when i is a *solute* dissolved in a *solvent*, the activity is expressed by:

$$a_i \equiv \frac{f_i}{f_i^\circ} \approx \frac{c_i}{c_i^\circ} = \frac{\text{concentration of } i \text{ in the mixture}}{\text{concentration of } i \text{ in its standard state}}.$$

When the standard state selected is the solute i dissolved at 1 mole per litre of solution, the activity becomes $a_i \approx c_i/\text{mol L}^{-1}$. This expression implicitly supposes that the fugacity of i in the solution varies proportionally with concentration, a condition rarely verified except at the weakest concentrations, well below 1 mol L^{-1}.

Activity coefficient—This is defined by:

$$\gamma_i \equiv a_i \cdot c_i^\ominus/c_i,$$

the ratio of the *activity* of the substance i in a mixture to its concentration; $c_i^\ominus$ represents a *reference* concentration. As opposed to the activity, the activity coefficient depends on the expression of concentration:

- When concentrations are given in *mole fractions*:

$$\gamma_i \equiv a_i \cdot x_i^\ominus / x_i \equiv a_i / x_i \quad \text{(with } x_i^\ominus = 1\text{)}.$$

- When concentrations are given in *molalities*:

$$\gamma_i \equiv a_i \cdot m_i^\ominus / m_i \quad \text{(with } m_i^\ominus = 1 \, \text{mol kg}^{-1}\text{)}.$$

- When concentrations are given in *molarities*:

$$\gamma_i \equiv a_i \cdot c_i^\ominus / c_i \quad \text{(with } c_i^\ominus = 1 \, \text{mol L}^{-1}\text{)}.$$

Despite this complication, the advantage of dealing with activity coefficients rather than with activities arises from the fact that indeterminacies due to problems of limits are removed: when $c_i \to 0$, $a_i \to 0$, but $\gamma_i \to$ finite, non-zero limit. Thus, the *Gibbs–Duhem equation*, applied to activities: $\sum x_i \, \mathrm{d} \ln a_i = 0$ (T and P constant) is hardly integrable in the vicinity of a pure constituent i. On the other hand, when proceeding with $a_i = \gamma_i x_i$ and $\sum x_i = 1$, the term $\sum x_i \, \mathrm{d} \ln x_i \equiv 0$ vanishes and $\sum x_i \, \mathrm{d} \ln \gamma_i = 0$, an expression which is often easier to integrate.

Adiabatic—When a *system* undergoes a *process* in which only *work* is exchanged with the *surroundings*, such a process is called adiabatic. A boundary separating two systems is called adiabatic if it allows only exchange of work.

Remark 1: Although the Greek root of the word (α-$\delta\iota\alpha\beta\alpha\tau o\varsigma$: not allowing passage through) implicitly refers to *heat*, a rigorous presentation of thermodynamics begins with defining work, then considers the *first law*, and introduces heat after having ascertained the existence of exchanges which do not comply with the definition of work.

Remark 2: Adiabatic ($\mathrm{d}Q = 0$) is not synonymous with *isentropic*. Because of the relation $\mathrm{d}S = \mathrm{d}Q/T + \mathrm{d}\sigma$ (see *Second law, remark 2*), an adiabatic transformation is isentropic only when it is reversible ($\mathrm{d}\sigma = 0$).

Remark 3: It is quite possible to find an adiabatic process which makes a system at equilibrium evolve by supplying it with work ($\Delta W > 0$). On the other hand, it is impossible to find an adiabatic process which makes a system at equilibrium evolve by supplying work

to the surroundings ($\Delta W < 0$). This is a direct consequence of the
second law (*corollary 3*).

One can easily be convinced of this result by remembering *Joule*'s experiment: it is indeed possible to use the free fall of a mass to warm water (adiabatic transformation with $\Delta W > 0$), but it is impossible by an adiabatic transformation to use a fall in water temperature to raise a mass in a gravitational field.

Remark 4: For it to be possible to find an adiabatic process which transforms a system from a state 1 to a state 2, it is necessary that the transformation $1 \rightarrow 2$ be characterized by an increase in system *entropy* (*Second law, corollary 9*).

Adiabatic demagnetization—This technique for obtaining low temperatures was proposed independently by Debye (1926) and Giauque (1927), then put into application by the latter in 1935. The presence of a paramagnetic substance in a *magnetic field* has the effect of aligning the moment-carriers in the direction of the field, by which order is established and the *entropy* decreases.

The entropy variation under the influence of a temperature variation and of a variation in *magnetic excitation* $\mathcal{H}$ is (see *Magnetism*):

$$\mathrm{d}S = \frac{c_{\mathcal{H}}}{T}\,\mathrm{d}T + \mu_0\left(\frac{\partial M}{\partial T}\right)_{\mathcal{H}}\mathrm{d}\mathcal{H},$$

where μ_0 is the *permeability* of vacuum, M the *magnetic moment*, and $c_{\mathcal{H}}$ the *heat capacity* at constant magnetic excitation. For a paramagnetic material:

$$M = V\kappa\mathcal{H} - C\mathcal{H}/T,$$

where κ is the *magnetic susceptibility* and C the Curie constant. It follows that:

$$\left(\frac{\partial S}{\partial \mathcal{H}}\right)_T = \mu_0\left(\frac{\partial M}{\partial T}\right)_{\mathcal{H}} = -\frac{C\mu_0\mathcal{H}}{T^2}.$$

When $\mathrm{d}\mathcal{H} > 0$, $\mathrm{d}S < 0$: the presence of a magnetic field results in a decrease in the entropy of a magnetic material. The reversible ($\mathrm{d}S = 0$) adiabatic demagnetization induces cooling:

$$\left(\frac{\partial \mathcal{H}}{\partial T}\right)_S = -\frac{c_{\mathcal{H}}}{\mu_0 T}\left(\frac{\partial T}{\partial M}\right)_{\mathcal{H}} = -\frac{c_{\mathcal{H}}T}{C\mu_0\mathcal{H}}.$$

8 Affinity

If the moment-carriers are the electrons, it is possible to reach millikelvin temperatures. Nuclear adiabatic demagnetization gives access to the lowest temperatures ever obtained, about one micro-kelvin. However, higher magnetic fields are needed, the moment carriers then being the nuclei. Moments proceeding from nuclear spin alignment are 1000 times weaker than those proceeding from electron spin alignment.

Affinity—A concept introduced by T. De Donder in 1923. Consider a mixture of constituents M_i with the possibility of a *chemical reaction*: $\sum v_i M_i = 0$. During a time dt, the *extent* of the reaction varies by $d\xi$ and the *entropy* of the system changes by $dS = (dQ/T) + d\sigma$. Thus:

$$T \, d\sigma = T \, dS - dQ \geqslant 0,$$

$$T \frac{d\sigma}{dt} = T \frac{d\sigma}{d\xi} \times \frac{d\xi}{dt} = \mathscr{A} v \geqslant 0,$$

where $v = d\xi/dt$ is the *reaction rate*.

The affinity $\mathscr{A}$ of the reaction is defined by:

$$\mathscr{A} = T \frac{d\sigma}{d\xi}.$$

T, P and ξ being independent variables, $T \, d\sigma$ can be expressed as:

$$T \, d\sigma = \mathscr{A}_{T,\xi} \, dP + \mathscr{A}_{P,\xi} \, dT + \mathscr{A}_{T,P} \, d\xi.$$

For De Donder, the only cause of irreversibility arises from chemical effects, because it is always possible to change reversibly the pressure or temperature. It follows that $\mathscr{A}_{T,\xi} = \mathscr{A}_{P,\xi} = 0$. Different changes of variable give:

$$\mathscr{A}_{T,P} = \mathscr{A}_{T,V} = \cdots = \mathscr{A}_{x,y} = \mathscr{A}.$$

The affinity of a chemical reaction is expressed in $J \, mol^{-1}$.

Remark 1: The *second law* ($d\sigma \geqslant 0$) applied to a system in which a chemical reaction occurs is expressed by $\mathscr{A} v \geqslant 0$ or $\mathscr{A} \, d\xi \geqslant 0$. The sign of $\mathscr{A}$ implies that of $d\xi$. The affinity is positive when the reaction evolves spontaneously forward ($d\xi > 0$); it is zero at equilibrium and negative when the reaction is reversed ($d\xi < 0$).

Remark 2: When more than one reaction can occur inside one and the same system, the second law is expressed by $\sum \mathscr{A} \, d\xi \geqslant 0$. Such an inequality does not forbid, for one of the reactions, the possibility of observing, by coupling effects, $\mathscr{A}$ and $d\xi$ of opposite signs.

Remark 3: In a system with T and P constant:

$$T\,d\sigma = T\,dS - dH = -dG,$$

$$\mathscr{A} \equiv T\left(\frac{\partial\sigma}{\partial\xi}\right)_{T,P} = -\left(\frac{\partial G}{\partial\xi}\right)_{T,P} = -\sum v_i\mu_i = -\Delta_r G.$$

Likewise, in a system with T and V constant:

$$T\,d\sigma = T\,dS - dU = -dF,$$

$$\mathscr{A} \equiv T\left(\frac{\partial\sigma}{\partial\xi}\right)_{T,V} = -\left(\frac{\partial F}{\partial\xi}\right)_{T,V} = -\sum v_i\mu_i = -\Delta_r G.$$

Remark 4: In current usage, the concept of affinity seems to duplicate that of *Gibbs energy of reaction*. However, it is more general and infinitely more eloquent. For instance, my affinity for chocolate eclairs is positive before a meal, zero when I am satisfied, and negative if I have been too greedy!

Remark 5: The word 'affinity' is sometimes used incorrectly. The following statement may be inaccurate: 'Oxygen has a higher affinity for silicon than for manganese'. Actually, the author means that the *standard affinity* is higher for the reaction $Si + O_2 \rightarrow SiO_2$ than for the reaction $2Mn + O_2 \rightarrow 2MnO$. What is true for standard affinity may be wrong for affinity.

Aggregation state—A concept introduced by the International Union of Pure and Applied Chemistry (IUPAC). The aggregation state for a substance is defined by the *state of matter*, crystal structure (for a solid) and atomicity (for a gas).

Allotropy—A term introduced by Berzelius in 1840, from the Greek $\alpha\lambda\lambda o\varsigma$, different, and $\tau\rho o\pi o\varsigma$, direction. Allotropy is the property of an *element* of crystallizing with different structures. For instance, graphite and diamond are two allotropic varieties of carbon. For a *compound*, it is better to speak of polymorphism.

Allowed state—Said of every state in keeping with *constraints* imposed on the system. Gold is no longer, since the demise of the alchemist, an allowed state for lead. On the other hand, the air in a room all gathered in one corner is an allowed state, even if the probability of seeing such a state realized is very low!

10 Amagat diagram

Amagat diagram—A diagram of PV vs. V, from Emile Amagat (1841–1915), French physicist, author of important works on fluids under high pressures. An Amagat diagram gives, for a pure substance, the same information as a *Clapeyron diagram* (P vs. V), but allows better visualization of the behaviour of *real gases* at low pressures (the product $PV \to RT$ when $P \to 0$ with a positive or negative slope depending on whether T is higher or lower than the *Boyle temperature*) and at high pressures (the isotherms are nearly parallel, owing to the fact that when $P \to \infty$, the ratio $PV/P \to b$, the *covolume*, depending little on temperature).

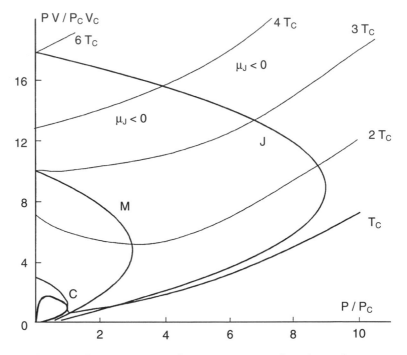

Amagat diagram: *General appearance in reduced coordinate.*

On an Amagat diagram, it is possible to draw three parabolic curves, characteristic of the behaviour of every real gas:

- The inner curve (C) is the locus of points representative of the liquid and vapour phases at equilibrium, meeting at the *critical* point with a vertical tangent. The critical isotherm at this point shows a vertical tangent with an inflection:

$$[\partial(PV)/\partial P]_T \to \infty \quad \text{and} \quad [\partial^2(PV)/\partial P^2]_T \to 0 \quad \text{when } P \to P_c.$$

- The intermediate curve (M) is the locus of the *PV* product minimum. The Boyle temperature is characterized by:

$$[\partial(PV)/\partial P]_T \to 0 \quad \text{when } P \to 0.$$

- The outer curve (J) is the locus of the *Joule–Thomson effect* inversion points, for which:

$$\mu_J = (\partial T/\partial P)_H = 0.$$

Amount of electricity—The integral of electric current over time. The amount of electricity in the SI system is expressed in coulombs:

$$1\,C = 1\,A\,s.$$

Amount of substance—Amount of substance was chosen as the seventh base unit in 1971 by the 14th Conférence Générale des Poids et Mesures. The unit of amount of substance in the SI system is the *mole* (symbol mol). By definition 1 mol of ^{12}C has a mass of 12 g.

Ampere—One of the seven SI base units, named from the French physicist André-Marie Ampère (1775–1836). The ampere (symbol A) was defined by the 9th Conférence Générale des Poids et Mesures in 1948 as the electric current which, if maintained in two straight parallel conductors of infinite length, of negligible circular cross-section, and placed one metre apart in vacuum, would produce between these conductors a force equal to 2×10^{-7} newtons per metre of length.

Angular momentum—Angular momentum, or moment of momentum, is the name given to the vector product $mv \times r$ of the *momentum* by the radius vector. The angular momentum of a mass m whose moment of inertia is I, rotating with angular velocity ω, is $I\omega$. In the SI system, angular momentum is expressed in $kg\,m^2\,s^{-1}$. For an isolated system, angular momentum is conserved.

Anion—A negatively charged ion, which, in an electrolytic cell, travels towards the *anode*.

Anode—A term introduced by Faraday in 1834, from the Greek ανοδος, rising (ανα: from low to high). The anode is the electrode at which oxidation occurs in a cell: the oxidation degree increases. The anode is linked not to the sign of an electrode but to the nature of the reaction that occurs; it represents the negative pole in a galvanic cell, but the positive pole in an electrolytic cell, towards which anions travel due to the electric potential.

Atmosphere—A unit of pressure outside any (1 atm = 101 325 Pa), which it is very hard to find a place for in the museum of antiquities! In fact, there exist numerous thermodynamic data tables and recent works in which the standard pressure is 1 atm instead of 1 bar. The difference is hardly perceptible in most practical applications, but sometimes such lack of rigour may be awkward. The *Gibbs energy* for an *ideal gas* increases by $RT \ln 1.01325$, or $32.6 \, \text{J} \, \text{mol}^{-1}$ at 298 K, when its pressure changes from 1 bar to 1 atm; its *entropy* decreases under the same conditions by $R \ln 1.01325$, or $0.11 \, \text{J} \, \text{mol}^{-1} \text{K}^{-1}$.

Autocatalysis—A chemical reaction is called autocatalytic when one species has the effect of increasing the reaction rate of the step in which it is produced, or decreasing the reaction rate of the step in which it is consumed. In order for a reaction to oscillate, it is necessary that there be an autocatalytic step. The Lotka system, investigated in 1920, represents a classical example, with an immediate biological interpretation, of a process in which the first two steps are autocatalytic:

$$A + X \rightarrow 2X \tag{1}$$

$$X + Y \rightarrow 2Y \tag{2}$$

$$Y + B \rightarrow B + E \tag{3}$$

These reactions model a predator–prey ecological system. A, X, Y, and B represent respectively available vegetation, herbivores, carnivores, and biomass supported by environment.

Denoting by k_1, k_2 and k_3 the respective rate constants of the three reactions, the process is governed by the system of differential equations:

$$\frac{d[X]}{dt} = k_1[A][X] - k_2[X][Y]$$

$$\frac{d[Y]}{dt} = k_2[X][Y] - k_3[B][Y].$$

The first equation expresses that the herbivore population increases proportionally to its density and to the available food, whereas it decreases proportionally to the probability of a predator–prey encounter. The second equation expresses that the carnivore population rises proportionally to this probability, but decreases proportionally to its density and to the biomass. The equations are solved by supposing A and B are constant. Initial conditions leading to a *stationary state* are easily obtained:

$$[X_0] = [B]k_3/k_2 \qquad [Y_0] = [A]k_1/k_2.$$

For initial conditions different from $[X_0]$ and $[Y_0]$, the system undergoes an oscillatory behaviour around $[X_0]$ and $[Y_0]$.

More elaborate models exist for describing the evolution of animal populations; however, Lotka's model gives an excellent account of the oscillations observed, with a 10 year period, in lynx and hare population in Canadian forests.

Available—An adjective often associated with 'energy' to distinguish between the energy which can be transformed to 'useful' work and the energy which is lost in 'useless' heat. However, this word must be used cautiously. Some authors do not include in available energy every kind of work that can be extracted from a given transformation. For instance, the word 'availability' sometimes designates the *exergy* concept; the expressions 'available work' or 'available energy' may be used to describe *Helmholtz energy*. Actually, only a change in Helmholtz energy ($\Delta F < 0$) can be used for obtaining work (see *Second law*):

$$\Delta F = \int_i^f dW_{rev} = F_f - F_i.$$

Avogadro, Amedeo (1776–1856)—Italian chemist, author of the hypothesis stating that equal volumes of two gases at the same pressure and temperature contain the same number of molecules.

Avogadro constant: $\mathcal{N} = 6.022\,136\,7 \times 10^{23}\,\text{mol}^{-1}$.

Axiom—From Greek root $\alpha\xi\iota o\mu\alpha$, that which is suitable. Name given to a statement that cannot be proved, but is sufficiently obvious to be accepted as a starting point for development of a theory. The main axioms used in thermodynamics are those of balance and conservation.

A balance axiom states that the variation of an *extensity* x is the sum of two terms: a *flux* term $d_e x$, expressing the exchange of x between the system and its surroundings, and a source term $d_i x$, expressing the production ($d_i x > 0$) or disappearance ($d_i x < 0$) of the quantity x in the system. When $d_i x = 0$, the extensity x is said to be conservative. Non-relativistic thermodynamics accepts without difficulty the conservation of mass and electricity. *Energy* conservation is not an axiom but a *principle* (see *First law*).

Azeotrope—A term introduced in 1911, constructed from the Greek prefix α-privative with the words ςεω, to boil, and τροπη, evolution; it may be translated by 'that which does not evolve on boiling'. An azeotrope is a liquid mixture whose composition is the same as that of the vapour in equilibrium with it. Azeotropic composition depends on temperature and pressure. The boiling temperature of an azeotrope is constant because the departure of the vapour has no influence on the remaining liquid composition. By extension, the word azeotrope is used for systems having an azeotropic composition. Azeotropy, encountered in 52% of the binaries and in 51% of the ternaries, so among known systems, is not an exceptional phenomenon.

GIBBS–KONOVALOV THEOREM: For the azeotropic composition, the *dew-point curve* and the *boiling* curve present an extremum.

Write successively the *Gibbs–Duhem equation*:

- for the liquid phase: $x_1 d\mu_1 + x_2 d\mu_2 = -s_1 dT + v_1 dP$;
- for the vapour phase: $y_1 d\mu_1 + y_2 d\mu_2 = -s_v dT + v_v dP$.

At equilibrium, *chemical potentials* μ_1 and μ_2 are identical in both phases. Let us eliminate $d\mu_2$ between the two equations:

$$(x_1 y_2 - x_2 y_1) d\mu_1 = -(y_2 s_1 - x_2 s_v) dT + (y_2 v_1 - x_2 v_v) dP.$$

The first member equals zero for the azeotropic composition. It follows that $dP = 0$ implies $dT = 0$: the dew-point curve and the boiling curve, on an isobaric diagram, present an extremum. The same is obviously true on an isothermal diagram, since $dT = 0$ implies $dP = 0$!

This result was established theoretically by Josiah Willard *Gibbs* (1878), then checked experimentally by the Russian chemist Dimitri Petrovich Konovalov (1856–1929), who in 1884 was the first to demonstrate the reality of azeotropy.

Remark 1: When T and P change simultaneously:

$$\left(\frac{dP}{dT}\right)_{az} = \frac{s_v - s_1}{v_v - v_1} = \frac{h_v - h_1}{T(v_v - v_1)}.$$

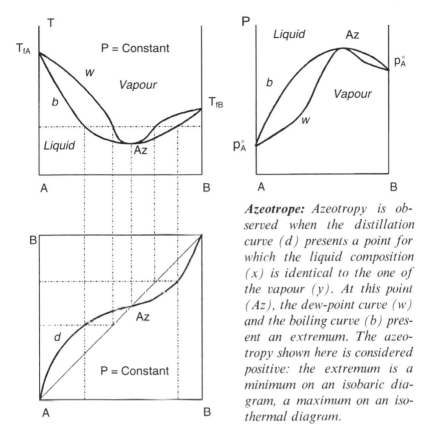

Azeotrope: *Azeotropy is observed when the distillation curve (d) presents a point for which the liquid composition (x) is identical to the one of the vapour (y). At this point (Az), the dew-point curve (w) and the boiling curve (b) present an extremum. The azeotropy shown here is considered positive: the extremum is a minimum on an isobaric diagram, a maximum on an isothermal diagram.*

This is the *Clapeyron equation*, which, by integration, gives the dependence of azeotropic temperature on pressure.

Remark 2: An azeotrope has the same behaviour as a pure substance towards boiling and distillation. The distillation of a mixture of two liquids with an azeotrope does not permit two pure components to be separated, but gives one pure component and the azeotropic mixture.

Remark 3: A mixture shows azeotropy when the *distillation* curve crosses the primary bisectrix (the liquid composition x equals that of the vapour y). The presence of an azeotrope inevitably implies a deviation from *ideality* for the liquid solution. This deviation may be very weak when the boiling points of the two liquids are very close.

Remark 4: The term azeotrope is confined to liquid–vapour equilibrium. When this phenomenon is encountered in condensed phases, one speaks of *congruence*.

B

Bar—A Unit of pressure, multiple of the *pascal* ($1\,\text{bar} = 10^5\,\text{Pa}$). Since 1983, the standard pressure, unless otherwise stated, is fixed at 1 bar instead of 1 atm (1.01325 bar). This convention, doubtless motivated by concern not to spread alarm among students and engineers, is nevertheless unfortunate. The choice of the pascal would be more judicious, despite the necessity for complete recasting of thermodynamic tables.

Barometric equation—By integrating the relation:

$$dG = Mg\,dz + v\,dp = 0,$$

one obtains the barometric equation (from Greek βαρος, gravity, a term introduced by Boyle in 1665):

$$p = p_\circ \exp(-Mgz/RT),$$

giving, for a gas of molar mass M, the atmospheric pressure p as a function of altitude z. This equation represents only a simplified approach to reality because it relies on three simplifying assumptions: the atmosphere is an ideal gas (satisfied as long as p is around 1 bar); the acceleration of free fall g is constant (satisfied as long as z is negligible with respect to the earth's radius); and the temperature does not depend on z, which is unrealistic. A better description can be obtained by assuming that a mass of gas changing in altitude undergoes an *isentropic* expansion ($dz > 0$) or compression ($dz < 0$).

Basicity constant—A name given to the *equilibrium constant* K_b for dissociation of a base in aqueous solution:

$$BOH \rightarrow B^+ + OH^-$$

$$A^- + H_2O \rightarrow AH + OH^-.$$

The basicity constant is becoming obsolete, because it duplicates the *acidity constant* K_a. In fact, the constant K_b of a base (for instance BOH or A^-) is easily deduced from the constant K_a of the conjugate acid (here B^+ or AH) by the relationship:

$$K_a K_b = K_e = 10^{-14} \text{ at } 24\,°C.$$

A strong acid corresponds to a negligible conjugate base:

$$K_a > 1 \Rightarrow K_b < 10^{-14}.$$

A weak acid corresponds to a weak conjugate base:

$$10^{-3} > K_a > 10^{-11} \Rightarrow 10^{-11} < K_b < 10^{-3}.$$

A negligible acid corresponds to a strong conjugate base:

$$K_a < 10^{-14} \Rightarrow K_b > 1.$$

Beau de Rochas cycle—See *Otto cycle*. Alphonse Beau de Rochas, French engineer (1815–1893), in 1862 put forward the theory of this cycle, a theory applied for the first time by Otto in 1876.

Belousov–Zhabotinskii (BZ) reaction—In 1958, B. P. Belousov showed that cerium ions catalysed the oxidation reaction of citric acid with bromate ions in water; however, surprisingly, the system oscillated, from yellow to colourless, with a period of about a minute. Belousov had some difficulty in publishing his observations on that 'chemical clock', because at that time such behaviour was considered to be thermodynamically impossible. The reaction was re-examined in 1960 by A. M. Zhabotinskii, who took it as the subject of his doctoral dissertation, with citric acid replaced by malonic acid.

The mechanism of the reaction, which is rather complex, comprises two processes: process [A], observed when the medium is enriched with bromide ions, followed by process [B], autocatalytic, established when the medium is depleted of bromide:

$$Br^- + BrO_3^- + 2H^+ \rightarrow HBrO_2 + HOBr \qquad [A1]$$

$$Br^- + HBrO_2 + H^+ \rightarrow 2HOBr \qquad [A2]$$

$$3Br^- + 3HOBr + 3H^+ \rightarrow 3Br_2 + 3H_2O \qquad [A3]$$

$$3Br_2 + 3CH_2(CO_2H)_2 \rightarrow 3BrCH(CO_2H)_2 + 3Br^- + 3H^+ \quad [A4]$$

$$2Br^- + BrO_3^- + 3H^+ + 3CH_2(CO_2H)_2 \rightarrow 3BrCH(CO_2H)_2 + 3H_2O \quad [A]$$

During process [A], the bromide–bromate reaction gives bromine, which reacts with malonic acid to form bromomalonic acid. Reaction [A1] governs the rate of the process. None of the reactants occurring in this process is able to oxidize Ce^{3+} to Ce^{4+}. When the bromide content becomes sufficiently low, process [B] is engaged. This step, which is non-linear, is the cause of the oscillatory behaviour of the system:

$$BrO_3^- + HBrO_2 + H^+ \rightarrow 2BrO_2^{\bullet} + H_2O \qquad [B1a]$$

$$2Ce^{3+} + 2BrO_2^{\bullet} + 2H^+ \rightarrow 2Ce^{4+} + 2HBrO_2 \qquad [B1b]$$

$$2Ce^{3+} + BrO_3^- + HBrO_2 + 3H^+ \rightarrow 2Ce^{4+} + 2HBrO_2 + H_2O \qquad [B1]$$

$$2HBrO_2 \rightarrow HOBr + BrO_3^- + H^+ \qquad [B2]$$

The electrode standard potential of the Ce^{3+}/Ce^{4+} couple ($+1.61$ V) being higher than that of the BrO_3^-/Br^- ($+1.48$ V) and Br_2/Br^- ($+1.09$ V) couples, $BrO_2^{\bullet}$ radicals form the most plausible species able to oxidize Ce^{3+} ions. Reaction [B1], the result of the two steps [B1a] and [B1b], autocatalytically generates bromous acid, $HBrO_2$. The autocatalytic process does not continue, because reaction [B2] in the steady state maintains only a low concentration of $HBrO_2$. Reactions [B1] and [B2] followed by [A3] and [A4] form the overall reaction of process [B]:

$$BrO_3^- + 4Ce^{3+} + CH_2(CO_2H)_2 + 5H^+ \rightarrow$$

$$4Ce^{4+} + BrCH(CO_2H)_2 + 3H_2O \qquad [B]$$

When the Ce^{4+} content is low, the step [B1a] governs the rate of the process. The mechanism presented here, one of the most likely among those proposed, is rather complicated, because at least twelve components or intermediate products are involved. It was described in 1974 with a model called the *oregonator* by its authors Field, Körös, and Noyes.

Berthelot, Marcelin (1827–1907)—French chemist, science historian, and statesman. His intense scientific activity (he signed some 1600 papers) was essentially devoted to organic chemistry and thermochemistry. His historical and philosophical work includes the translation of Greek and Arabic works on chemistry and alchemy. He introduced the terms *endothermic* and *exothermic* (1869).

Bifurcation—Let Y be a parameter describing the behaviour of a system and λ be a control parameter. Generally speaking, the evolution of the variable Y with time is given by:

$$dY/dt = f(Y, \lambda).$$

In the vicinity of *equilibrium*, the system rapidly becomes stationary and Y tends towards the thermodynamic branch, the limit depending on λ. The properties of the system are uniform in space and time; except for the particular case of phase transitions, perturbations are quickly damped.

Far from equilibrium, beyond a critical value λ_c, the equation $dY/dt = 0$ may have two solutions, corresponding to a first bifurcation on the curve giving the limiting value of Y as a function of λ: the system leaves the thermodynamic branch, fluctuations no longer regress and a new state appears, leading to a hysteresis phenomenon, a continuous variation of λ being expressed in a discontinuity of Y.

A straightforward example of behaviour showing a chain of bifurcations is provided by the rudimentary biological model of population evolution: if x represents the ratio between the present population and the maximum population that can be supported by the system, and with the hypothesis that the population at time $(t+1)$ is proportional to the population at time t, i.e. x, and to the available resources, i.e. $(1-x)$, one obtains the law of evolution:

$$x_{t+1} = \lambda x_t(1 - x_t).$$

For $\lambda < 1$, $x_t \to 0$ when $t \to \infty$.

For $1 < \lambda < 3$, x_t tends towards the thermodynamic branch when $t \to \infty$. This limit, the solution of the equation $[x - \lambda x(1-x)] = 0$, or $x = 1 - 1/\lambda$, is stable: by introducing $x = (1 - 1/\lambda) + \varepsilon$, the perturbation ε is absorbed, and x rapidly finds its equilibrium value $(1 - 1/\lambda)$.

For $\lambda = 3$, a first bifurcation appears. If $\lambda > 3$, the value $x = 1 - 1/\lambda$ is always a solution of the equation, but the solution is unstable: the introduction of a perturbation moves x away from the value $(1 - 1/\lambda)$.

For $3 < \lambda < 3.57$, there are two stable solutions; the critical value $\lambda = 3.57$ corresponds to the appearance of a second bifurcation beyond which the system presents four stable solutions. The third bifurcation is observed for $\lambda = 3.64$; the behaviour of the system then quickly becomes chaotic, the succession of bifurcations giving the successive values of x a random behaviour.

Binomial distribution—When the result of an experiment can only be A (probability p) or B (probability q) with $p + q = 1$, the probability of observing, in N trials, result A n_A times and result B n_B times (with $n_A + n_B = N$) is given by:

$$P_N(n_A) = \frac{N!}{n_A! \, n_B!} p^{n_A} q^{n_B}.$$

This *distribution* is the binomial distribution. This important relation can be applied equally well to the heads-or-tails game, in which case $p=q=0.5$, as to numerous problems in physics. It yields, for instance, the probability of observing, around a given site in a solid solution of composition $A_p B_q$ (with $p+q=1$), of coordination number N, the presence of n_A atoms of A and n_B atoms of B ($n_A + n_B = N$). The normalization condition is obeyed well:

$$\sum_{n_A=0}^{N} P_N(n_A) = (p+q)^N = 1.$$

The order-1 *moment* (or mean) of the result A is:

$$\langle n_A \rangle = \sum_{n_A=0}^{\infty} n_A P_N(n_A) = \sum_{n_A=0}^{\infty} \frac{N! n_A}{n_A!(N-n_A)!} p^{n_A} q^{N-n_A}$$

$$\langle n_A \rangle = p \frac{\partial}{\partial p}\left[\sum_{n_A=0}^{\infty} P_N(n_A) \right] = p \frac{\partial (p+q)^N}{\partial p} = Np.$$

The order-2 moment of the result A, the variance and the root mean square are respectively:

$$\langle n_A^2 \rangle = \sum_{n_A=0}^{\infty} n_A^2 P_N(n_A) = (Np)^2 + Npq$$

$$\sigma_N^2 = \langle n_A^2 \rangle - \langle n_A \rangle^2 = Npq; \quad \sigma_N = \sqrt{Npq}.$$

Bivariant—An *equilibrium* is said to be bivariant when it is impossible to modify independently more than two intensive parameters without modifying the phase number. An equilibrium of two constituents between two phases is bivariant, because there are two relations between the four intensive parameters: pressure, temperature, and two chemical potentials.

Boiling—A liquid boils when its saturation vapour pressure equals the atmospheric pressure imposed. Without an atmosphere, there is no boiling! By heating a liquid in a *closed* vessel under vacuum, the whole liquid–vapour equilibrium curve can be covered. By heating a liquid in an *open* vessel at atmospheric pressure, the liquid–vapour equilibrium curve is accessible only at a pressure lower than the atmospheric pressure or at temperatures lower than the boiling temperature.

The boiling point does not change with time when the vapour composition is identical to the liquid composition, which is the case for pure liquids or *azeotropic* mixtures. For *zeotropic* mixtures, the boiling point increases with time, because the remaining liquid is depleted of the more volatile compound. On an isobaric diagram the boiling curve gives the initial boiling point of a liquid mixture as a function of composition. On an isothermal diagram, more rarely used, it gives as a function of composition, the pressure under which boiling is observed. On both diagrams, the boiling curve is drawn with a *dew-point curve* giving the composition of the vapour in equilibrium with liquid.

If the liquid solution is ideal, the boiling curve has the equation:

$$P = p_1^\circ(1-x) + p_2^\circ x,$$

where x is the mole fraction of constituent 2 in the liquid, P is the imposed pressure, and p_1° and p_2° are the saturation vapour pressures of the pure constituents 1 and 2, as a function of temperature.

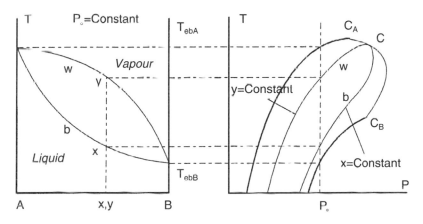

Boiling: *On the left, isobaric section of a binary mixture A–B showing dew-point (w) and bubble-point (b) curves. On a P–T diagram (right), dew-point and bubble-point curves for a given mixture meet at the critical point C with a common tangent. The envelope of the (w) and (b) curves, the locus of points C, joins the critical points of the pure A and B constituents. With an azeotropy, the (w) and (b) curves can be situated outside the region bounded by the saturation vapour pressure curves of the pure A and B constituents.*

For a non-ideal solution, the construction of the boiling curve requires a knowledge of the *activities* for constituents 1 and 2. Its equation is then given by:

$$P = p_1^\circ a_1 + p_2^\circ a_2.$$

The boiling curve might then present an extremum (azeotropic mixture).

Boltzmann, Ludwig Eduard (1844–1906)—Austrian physicist, who showed that the *second law* can be explained by applying the laws of mechanics and the theory of *probability*. He is considered as the father of statistical thermodynamics and, in 1875, proposed the well-known relationship linking *entropy* S and probability Ω, a relationship carved on his tombstone.

An *isolated* system evolves spontaneously towards its most probable configuration, which corresponds, according to the second law, to the state of maximum entropy. As entropy is an additive quantity, whereas probability is a multiplicative one, it is natural to imagine a logarithmic relation between these two quantities:

$$S = k \times \ln \Omega,$$

where k, the Boltzmann constant, is the ratio of the molar gas constant to the Avogadro constant and has the dimensions of entropy:

$$k = R/\mathcal{N} = 1.380658 \times 10^{-23} \, \text{J K}^{-1}.$$

Boltzmann distribution—The name given to the *distribution* of energies in an assembly of N particles following *Maxwell–Boltzmann statistics* and subject to *constraints* N (number of particles) and E (energy of the system) constant. If such a system is isolated, the distribution is obtained by seeking the maximum of the entropy function:

$$S = k \ln \Omega = k \ln \prod_i (g_i^N / N_i!).$$

On introducing the *Lagrange multipliers* α and β to take into account the constraints, we have (see *Microcanonical ensemble*):

$$N_i = g_i \exp(-\alpha) \cdot \exp(-\beta \varepsilon_i)$$

$$\frac{N_i}{N} = \frac{g_i \exp(-\varepsilon_i/kT)}{\sum g_i \exp(-\varepsilon_i/kT)},$$

where N_i is the population at the energy level ε_i, with a *degeneracy* g_i. $Z = \sum g_i \exp(-\varepsilon_i/kT)$ is the *partition function* of a particle:

$$N_i = \frac{N}{Z} g_i \exp\left(-\frac{\varepsilon_i}{kT}\right).$$

Bonding energy—The expression 'bonding energy' has for a long time referred to the dissociation enthalpy of an A–B bonding, which conferred on it a positive character. Present logic assimilates the bonding energy ε_{AB} to the formation enthalpy of the bonding: $A + B \rightarrow A - B$, which now confers on it a negative character. It is thus less ambiguous to refer to the formation enthalpy of the bonding. Many solution models, such as the quasi-chemical model, are based on bonding energy considerations. Let ε_{AA}, ε_{BB} and ε_{AB} be the formation enthalpies of the respective bondings:

- If $\varepsilon_{AB} = (\varepsilon_{AA} + \varepsilon_{BB})/2$, A–B interactions are said to be 'neutral'. From a thermodynamic point of view, this is expressed by ideality of the solution and by a statistical distribution of species A and B throughout the solution.
- If $\varepsilon_{AB} < (\varepsilon_{AA} + \varepsilon_{BB})/2$, A–B interactions are said to be 'attractive'. From a thermodynamic point of view, this is expressed by a negative deviation from ideality and by a distribution of A and B species throughout the solution favouring A–B bondings: *short-range order* is expected.
- If $\varepsilon_{AB} > (\varepsilon_{AA} + \varepsilon_{BB})/2$, A–B interactions are said to be 'repulsive'. From a thermodynamic point of view, this is expressed by a positive deviation from ideality and by a distribution of A and B species throughout the solution favouring A–A and B–B bondings: *clusters* are expected.

The terms 'repulsive', 'neutral' and 'attractive' are relative. Actually, a bond is always attractive within a condensed material as well as in a gaseous phase. Bonds can be weakly attractive, such as the van der Waals forces. At the limit, very 'attractive' bonds between A and B can lead to the formation of a new defined compound, whereas very 'repulsive' bonds between A and B lead to *unmixing*.

Bose–Einstein distribution—The name given to the *distribution* of energies in an assembly of N particles following *Bose–Einstein statistics* with imposed *constraints* N (number of particles) and E (energy of the system) constant. If such a system is isolated, the distribution is obtained by seeking the maximum of the entropy function:

$$S = k \ln \Omega = k \ln \prod_i \frac{(N_i + g_i - 1)!}{N_i!(g_i - 1)!}.$$

24 Bose–Einstein statistics

On introducing the *Lagrange multipliers* α and β to take into account the constraints, we have (see *Microcanonical ensemble*):

$$N_i = \frac{g_i}{\exp(\alpha) \cdot \exp(\beta \varepsilon_i) - 1}$$

$$N_i = \frac{g_i}{\exp(-G/NkT) \cdot \exp(+\varepsilon_i/kT) - 1},$$

where N_i is the population at energy level ε_i, with *degeneracy* g_i.

$Z = \sum g_i \exp(-\varepsilon_i/kT)$ is the *partition function* of a particle.

$G = -NkT \ln(Z/N)$ is the Gibbs energy of the system (see *Partition function*).

This distribution differs from the *Boltzmann distribution* only in the presence of -1 in the denominator. When $g_i \gg N_i$, which is often satisfied, except at low temperatures, the Bose–Einstein distribution merges with the Boltzmann distribution:

$$N_i = \frac{N}{Z} g_i \exp\left(-\frac{\varepsilon_i}{kT}\right).$$

Bose–Einstein statistics—The name given to the statistics followed by *bosons*, particles with integral or zero spin. Bosons are characterized by the fact that a *quantum state* can be occupied by several particles. If N_i indistinguishable particles occupy the ε_i energy level, with *degeneracy* g_i, the number of permutations is given by:

$$\Omega_i = \frac{(N_i + g_i - 1)!}{N_i!(g_i - 1)!}.$$

This relationship is obtained by calculating the number of permutations between N_i stars ($*$) representing particles and $(g_i - 1)$ dashes ($|$) representing separations between g_i quantum states, that is, the number of combinations between $(N_i + g_i - 1)$ objects taken N_i by N_i:

$$**|*||****||*|***|*****|*|**.$$

By taking the product at all energy levels ε_i, one obtains the basic relation giving the number of microscopic states in Bose–Einstein statistics:

$$\Omega_{BE} = \prod_i \frac{(N_i + g_i - 1)!}{N_i!(g_i - 1)!},$$

a relationship which can often be simplified by neglecting the 1 before g_i. By using the *Stirling formula* ($\ln N! \approx N \ln N - N$), the expression for entropy is obtained:

$$S = k \ln \Omega_{BE} = k \sum_i \left[N_i \ln\left(\frac{g_i}{N_i} + 1\right) + g_i \ln\left(1 + \frac{N_i}{g_i}\right) \right].$$

B

Boson—A particle having integral or zero spin angular momentum, named from Satyendranath Bose, Indian physicist (1894–1974). A boson does not obey the Pauli exclusion principle: any number of bosons may occupy each of the available discrete energy states. A boson follows *Bose–Einstein statistics*. Photons and phonons (spin $= 0$), π, τ and ζ mesons (spin $= 1$), helium-4 nuclei (spin $= 2$), nuclei of even nucleon number, etc. are bosons.

Boundary—A division, which may be material or not, between a *system* and its *surroundings*. Exchange of *work*, *heat* or matter between the system and its surroundings occurs through the boundaries. A boundary may be *adiabatic, isothermal, diathermanous, insulating, permeable, semipermeable*.

Boyle temperature—Named from Robert Boyle (1627–1691), English chemist and philosopher, known for his pioneer experiments on the properties of gases, the Boyle temperature, also called the 'metacritical temperature' of a real gas, is the temperature for which:

$$\lim_{P \to 0} \left(\frac{\partial(PV)}{\partial P}\right) = 0.$$

When using the virial expansion for a real-gas equation of state:

$$PV = RT + (B/V) + (C/V^2) + \cdots$$

or:

$$PV = RT + \frac{B}{RT}P + \frac{C - B^2}{(RT)^2}P^2 + \cdots,$$

the Boyle temperature corresponds to $B = 0$. Every real gas presents the same behaviour: $B < 0$ below the Boyle temperature, $B > 0$ above that temperature. B subsequently increases to a maximum and then tends towards zero when T tends towards infinity.

The ratio T_b/T_c (Boyle temperature/critical temperature) is about 2.7 for a monatomic gas and 2.6 for a diatomic gas. D. Berthelot in 1907 proposed an expression for the second virial coefficient:

$$B = \frac{9R}{128} \times \frac{T_c}{P_c} \times \left(1 - 6\frac{T_c^2}{T^2}\right),$$

which gives a ratio $T_b/T_c = 2.45$. A calculation from the van der Waals equation of state leads to $T_b/T_c = 3.37$.

Brayton cycle—The ideal cycle for the simple gas turbines and motors used in reaction propulsion, named from the American engineer George B. Brayton. Also called the Joule cycle, it is composed of two *isobaric* and two *adiabatic* paths. As opposed to the *Rankine cycle*, it uses only non-condensable gases. Designating by $Q_1(>0)$ the heat arising from the combustion and by $Q_2(<0)$ the heat given up to the surroundings by the combustion products and assuming ideal gases with constant heat capacity, the thermal efficiency of the Brayton cycle is found as follows:

$$\eta = \frac{Q_1 + Q_2}{Q_1} = 1 - \frac{mc_P(T_D - T_A)}{mc_P(T_C - T_B)} = 1 - \frac{T_A[(T_D/T_A) - 1]}{T_B[(T_C/T_B) - 1]}.$$

Using also the equations of the *isentropes* (with $\gamma = c_P/c_V$):

$$\frac{T_A}{T_B} = \left(\frac{P_A}{P_B}\right)^{(\gamma-1)/\gamma} = \left(\frac{P_D}{P_C}\right)^{(\gamma-1)/\gamma} = \frac{T_D}{T_C},$$

from which it follows that $T_D/T_A = T_C/T_B$. The expression of η is thus reduced to:

$$\eta = 1 - \frac{T_A}{T_B} = 1 - \left(\frac{P_A}{P_B}\right)^{(\gamma-1)/\gamma}.$$

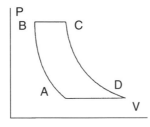
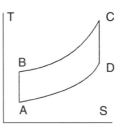

Brayton cycle: P–V (left) and T–S (right) diagrams.

The efficiency of a Brayton cycle increases with the compression ratio P_B/P_A. Actually, the efficiency shows a maximum, beyond which the work needed for compression increases more rapidly than the work supplied by gas expansion in the turbine.

Brillouin, Léon (1889–1969)—French physicist, known for the development of band theory of solids. In thermodynamics, he revealed, the link between *negentropy* and *information*.

Brusselator—The name given to one of the straightforward reaction schemes leading to the formation of *dissipative structures*, that is, a behaviour more interesting than a simple exponential evolution towards an *equilibrium* state. This model, developed by the Brussels school around Prigogine, presents the three necessary conditions for the appearance of *bifurcations*: that they be open, non-linear, and far from equilibrium:

$$A \rightarrow X \qquad [1]$$

$$B + X \rightarrow Y + D \qquad [2]$$

$$2X + Y \rightarrow 3X \qquad [3]$$

$$X \rightarrow E \qquad [4]$$

In practice, reactants A and B are present in excess, and final products D and E are discarded as soon as formed. By adjusting the flux of A, B, D, and E so that the system remains far from equilibrium, it is possible to obtain instabilities. The necessary condition for their formation is the presence in the reaction scheme of an *autocatalytic* step, here reaction [3].

Bulk expansion coefficient—Sometimes known as 'tension coefficient', it is defined as the ratio, reduced to unit pressure, of the pressure variation to the temperature variation. We should never forget to quote the conditions of the differential. So it is possible to define:

• the isochoric bulk expansion coefficient: $\beta_V \equiv \dfrac{1}{P}\left(\dfrac{\partial P}{\partial T}\right)_V$;

• the isentropic bulk expansion coefficient: $\beta_S \equiv \dfrac{1}{P}\left(\dfrac{\partial P}{\partial T}\right)_S$.

Bulk modulus—The reciprocal of isothermal *compressibility*:

$$\kappa = \frac{1}{\chi_T} = -V\left(\frac{\partial P}{\partial V}\right)_T.$$

For a solid, the compressibility, generally low, can be considered as pressure-independent. Bulk modulus is then easily calculated:

$$\kappa = -PV/\Delta V,$$

where ΔV represents the volume variation between the pressures 0 and P. The bulk modulus has dimensions of pressure. For steel, a mean value is 2×10^{11} Pa.

C

Caloric—The caloric theory, established in the eighteenth century, supplanted that of phlogiston for explaining many aspects of heat and combustion phenomena. Caloric is a weightless, conservative fluid, which builds up in hot materials. Its particles repel each other, but are attracted by matter. Around 1790, Benjamin Thompson, Count Rumford, raised doubts about the theory, by showing that caloric can be created. The theory was abandoned after 1824. By using caloric conservation, Sadi Carnot showed that every reversible engine working between two reservoirs at the same temperatures has the same efficiency. Actually, Carnot's demonstration can be disproved, but his report becomes clearer on substituting the term *entropy* for 'caloric'.

Calorie—The thermochemical calorie equals 4.1840 joules; the international calorie equals 4.1855 joules. Notwithstanding its drawbacks, scientists, young and not so young, feel much fondness for this unit, which was outlawed in 1978 and was recommended for withdrawal as long ago as 1948!

Calorimetric coefficient—The heat exchanged with the surroundings under conditions of *reversibility* can be expressed by differentials of the type:

$$\mathrm{d}Q = c_V\,\mathrm{d}T + l\,\mathrm{d}V; \quad \mathrm{d}Q = c_P\,\mathrm{d}T + h\,\mathrm{d}P; \quad \mathrm{d}Q = \lambda\,\mathrm{d}P + \mu\,\mathrm{d}V.$$

The coefficients c_V, c_P, l, h, λ, and μ are called calorimetric coefficients. The differentials $\mathrm{d}P$, $\mathrm{d}V$, and $\mathrm{d}T$ are related by the equation of state $f(P,V,T)=0$. Therefore it is possible to set up four relations between the six calorimetric coefficients:

$$l = (c_P - c_V)\left(\frac{\partial T}{\partial V}\right)_P, \qquad h = -(c_P - c_V)\left(\frac{\partial T}{\partial P}\right)_V,$$

$$\lambda = c_V\left(\frac{\partial T}{\partial P}\right)_V, \qquad \mu = c_P\left(\frac{\partial T}{\partial V}\right)_P.$$

Calorimetric coefficients are gimmicks liked by some students. However, it is possible to do without them, because they duplicate the differentials of the *entropy* $\mathrm{d}S$ which can be directly expressed by means of *Maxwell's equations*.

Candela—One of the seven SI base units, the candela (symbol cd) was defined in 1979 by the 16th Conférence Générale des Poids et Mesures as the luminous intensity of a source which emits, in a given direction, monochromatic radiation of frequency 540×10^{12} hertz and whose energetic intensity in that direction equals 1/683 watt per steradian.

C

Canonical ensemble—A name given to the Gibbs ensemble, which is an isolated assembly constituted by $\mathcal{N}$ identical systems, where each system is characterized by its number of particles N, its volume V, and its temperature T. The systems are in thermal contact with each other so that energy can pass from one system to the others. Under these conditions, the energy of each system varies, but their mean energy is known.

By considering this isolated assembly of $\mathcal{N}$ systems in which $\mathcal{N}_i$ systems shared between Ω_i macroscopic states have the same energy E_i and by seeking the most probable distribution, while taking into account the constraints:

$$\sum \mathcal{N}_i = \text{const.}, \qquad \sum \mathcal{N}_i E_i = \text{const.},$$

then the probability of finding a system in a state of energy E_i is given by:

$$P_i = \frac{\mathcal{N}_i}{\mathcal{N}} = \frac{\Omega_i \exp(-E_i/kT)}{\sum \Omega_i \exp(-E_i/kT)}.$$

$\mathscr{Z}(N,T,V) = \sum \Omega_i \exp(-E_i/kT)$ represents the canonical *partition function* of a system.

Remark: This *distribution* law is to be compared with the Boltzmann distribution. If one considers an isolated system with N particles which can occupy various energy levels ε_i with degeneracy g_i, a similar calculation leads to the relation (see *Microcanonical ensemble*):

$$P_i = \frac{N_i}{N} = \frac{g_i \exp(-\varepsilon_i/kT)}{\sum g_i \exp(-\varepsilon_i/kT)}.$$

$Z(T,V) = \sum g_i \exp(-\varepsilon_i/kT)$ represents the partition function of a particle.

Capacity—When an *extensity* moves from a source towards a sink under the influence of a potential gradient, it is possible to say, to a first approximation, that the *reduced extensity* gradient is proportional to the potential gradient. The proportionality coefficient is the reduced capacity of the medium at a given point. From a more general point of view, the term 'capacity' is used for the ratio of an extensity to an *intensity*:

$$\text{Capacity} = \frac{\text{Extensity}}{\text{Intensity}}.$$

A capacity is an extensive quantity. A reduced capacity is a capacity related to unit amount of substance (molar capacity), to unit mass (mass capacity), or to unit volume (volume capacity); therefore, a reduced capacity is a capacity which has swapped its extensive character for an intensive character.

ELECTRIC CAPACITY: The extensity is amount of electricity; the intensity is electric potential. In the SI system, electric capacity is expressed in farads: $1\,\text{F} = 1\,\text{C}\,\text{V}^{-1} = 1\,\text{m}^{-2}\,\text{kg}^{-1}\,\text{s}^4\,\text{A}^2$.

MOMENTUM CAPACITY: The extensity is momentum mv; the intensity is velocity v; the momentum capacity is therefore mass. The mass capacity equals unity, an obvious result coming from the fact that mass extensity identifies with intensity.

DIFFUSIONAL CAPACITY: The extensity is amount of substance; the intensity is concentration, or amount of substance divided by volume; the diffusional capacity is therefore volume. The volume diffusional capacity equals unity, an unsurprising result if one considers that intensity and volume extensity represent the same quantity.

HEAT CAPACITY: $c = \mathrm{d}Q/\mathrm{d}T$. See *Heat capacity*.

Caratheodory, Constantin (1873–1950)—Greek mathematician, author of a rigorous mathematical presentation of thermodynamic principles, which is noteworthy for the fact that it avoids the use of ill-defined concepts such as heat or temperature. Despite its logical and aesthetic qualities, the Caratheodory exposition has convinced neither teachers nor engineers, because of the rather abstract character of the concepts used.

Caratheodory proposed two ideas: a principle and a theorem, which must not be confused. The theorem gives the condition for $\mathrm{d}Q$ to be an *integrable differential*, i.e. near any given point P, one must find points inaccessible by a path along which $\int \mathrm{d}Q = 0$. The principle affirms the existence of points inaccessible by an adiabatic path, near a given point P.

CARATHEODORY'S PRINCIPLE:
Arbitrarily close to any given equilibrium state, there are states inaccessible by means of an adiabatic process.

This proposition is a possible statement of the second law; it is equivalent to the classical expression:

$$dS = \frac{dQ}{T} + d\sigma.$$

The only states accessible by an adiabatic process ($dQ = 0$, or $dS = d\sigma \geqslant 0$) are those which lead to an increase in the *entropy* of the system. When a process is *reversible*, the differential of the heat quantity dQ is integrable.

CARATHEODORY'S THEOREM: If a differential $dQ = \sum X_i dx_i$ possesses the property that in an arbitrarily close neighbourhood of a point P defined by its coordinates $(x_1, x_2, \ldots, x_n)$ there are inaccessible points, that is, points which cannot be connected to P along curves satisfying the equation $dQ = 0$, then dQ is integrable.

Proof: Cases $n = 1$ and $n = 2$ are trivial because a differential function of only one variable is necessarily total whereas a differential function of two variables is necessarily integrable (see *Integrable differential*). All points accessible to a given point P form around P a continuous domain. In an n-dimensional space ($n \geqslant 3$), this domain fills around P, a volume (n dimensions), or a surface $[(n-1)$ dimensions], or a curve $[\leqslant (n-2)$ dimensions]. The first possibility is excluded because it does not match the hypothesis: around P there are points which are inaccessible. The third possibility is also excluded because the expression $dQ = 0$ already defines a surface element containing only points accessible to P. Therefore, points close to P and accessible to P define only a surface. If we now consider a point P′ on that surface, it is impossible to go from P to P′ by a curve satisfying the condition $\int dQ = 0$ and not situated on this surface, otherwise every point situated within the immediate proximity of P would be accessible, which contradicts the hypothesis.

From a point P_1, it is possible to define a surface S_1, all points of which are accessible to P_1. Also, from a point P_2 not situated on S_1, it is possible to define a surface S_2. Surfaces S_1 and S_2 have no common point between them, otherwise it would be possible to go from P_1 to P_2

by a path such that $\int dQ = 0$. Therefore, there is a family of surfaces $\sigma(x_1, x_2, \ldots, x_n) = $ const. filling the space and having no common point among them. For this one-parameter family, $d\sigma = 0$ implies $dQ = 0$, from which, between dQ and $d\sigma$, there exists a relation of the type:

$$dQ = \tau(x_1, x_2, \ldots, x_n)\,d\sigma(x_1, x_2, \ldots, x_n),$$

where, because $dQ = \sum X_i\,dx_i$:

$$\tau = X_i \left(\frac{\partial x_i}{\partial \sigma}\right)_{x_{j \neq i}} \quad (i = 1, 2, \ldots, n).$$

Naturally, the family of surfaces $\sigma = $ const. may also be expressed by $S(\sigma) = $ const., where $S(\sigma)$ is an arbitrary function of σ:

$$dS = \frac{dS}{d\sigma}\,d\sigma = \frac{dS}{d\sigma} \times \frac{dQ}{T}.$$

Hence:

$$T = \tau \frac{d\sigma}{dS} = X_i \frac{\partial x_i}{\partial S}.$$

$(1/T)$ is the *integrating factor*. If a differential dQ has one integrating factor, it has an infinity, S being an arbitrary function of σ.

In summary: Caratheodory's theorem shows that if a differential dQ is integrable, the equation $dQ = 0$ characterizes in a space a family of surfaces sharing no common point. If one takes any point P on one of these surface, it is always possible to find, immediately near that point, points which do not belong to the surface and are therefore inaccessible by a curve solution of the equation $dQ = 0$. On the other hand, if dQ is not integrable, the equation $dQ = 0$ does not define any surface in the space and it is always possible to link any two points with a curve solution of the equation $dQ = 0$.

Carnot, Sadi (1796–1832)—French engineer, who in his pamphlet 'Considerations on the motive power of fire and on engines able to develop that power' (1824), provided the basis of the future *second law*. Scientists see him as the founder of thermodynamics and consider the year 1824 as the birthdate of this discipline.

Carnot cycle—A reversible cycle formed by two isotherms and two adiabats. On a *temperature–entropy* diagram, a Carnot cycle is represented by a rectangle described, in the case of a heat engine, in the direction ABCD:

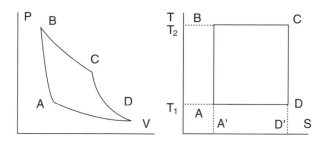

Carnot cycle: P–V (left) and T–S (right) diagrams.

AB: adiabatic compression
BC: isothermal expansion during which the fluid exchanges a quantity of heat $Q_2 > 0$ with the hot source at temperature T_2
CD: adiabatic expansion
DA: isothermal compression during which the fluid exchanges a quantity of heat $Q_1 < 0$ with the cold source at temperature T_1.

With a cycle described in the reverse direction, DCBA, the system works not as a heat engine but as a refrigerator or heat pump.

The efficiency of a Carnot cycle (see *Carnot's theorem* and *Thermodynamic temperature*) is given by: $\eta = (T_2 - T_1)/T_2$, or, on a temperature–entropy diagram: $\eta = \text{area(ABCD)}/\text{area(A'BCD')}$.

Carnot's theorem—Reversible heat engines describing a cycle between two constant temperatures have the same *efficiency*.

A heat engine exchanges Q_1 with a cold source, Q_2 with a hot source and W with the surroundings. If $Q_1 < 0$ and $Q_2 > 0$, then $W < 0$: the engine works as a motor. Over a cycle, the *first law* imposes: $W + Q_1 + Q_2 = 0$. The thermal efficiency is given by:

$$\eta = |W|/|Q_2| = (Q_1 + Q_2)/Q_2.$$

Suppose now that two engines, A and B, working between the same two heat reservoirs at temperatures T_1 and T_2 do not have the same efficiency. Let them work so that they abstract during one cycle the same heat quantity Q_2 at the hot reservoir, and let A be the engine with the higher efficiency:

$$(Q_{1A} + Q_2)/Q_2 > (Q_{1B} + Q_2)/Q_2,$$

with $Q_2 > 0$, which implies: $Q_{1A} > Q_{1B}$, or $|Q_{1A}| < |Q_{1B}|$, since heat quantities Q_1 are negative.

Since the engines are reversible, it is possible to couple them so that A works as a motor: $Q_{1A} < 0$, $Q_{2A} = Q_2 > 0$; and B as a refrigerator: $Q_{1B} > 0$, $Q_{2B} = -Q_2 < 0$. The quantities of work exchanged are:

$$W_A = -Q_{1A} - Q_{2A} = -Q_{1A} - Q_2; \quad W_B = -Q_{1B} - Q_{2B} = -Q_{1B} + Q_2.$$

The application of the first law to the assembly (A + B) gives:

$$Q_{1A} + Q_{1B} + Q_{2A} + Q_{2B} + W_A + W_B = 0$$

$$Q_{1A} + Q_{1B} = -W_A - W_B.$$

The assembly (A + B) behaves as an engine which takes from only one reservoir at a temperature T_1 a heat quantity $(Q_{1A} + Q_{1B} > 0)$ to obtain work $(W_A + W_B < 0)$, which contradicts the *second law*. Two reversible engines working between the two same heat reservoirs therefore have the same efficiency.

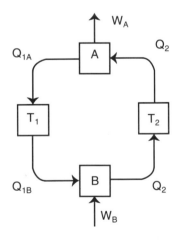

Carnot's theorem: *The engine A works as a heat engine; the engine B works as a refrigerator or as a heat pump. By coupling both engines in order to short-circuit the heat reservoir at temperature T_2, one gets an assembly $(A + B)$ working with only one heat reservoir and which, according to the second law, cannot provide work to the surroundings.*

Corollary: The efficiency of an irreversible thermal engine working between two temperatures T_1 and T_2 is lower than that of a reversible engine working between the same temperatures. It is sufficient again to take the demonstration of Carnot's theorem by coupling an irreversible engine working as a heat engine with a reversible engine working as a refrigerator. It is easily shown that the second law does not forbid the irreversible engine to have a lower efficiency, but forbids the irreversible engine from having an efficiency higher than that of the reversible engine.

Remark: Carnot's theorem has been demonstrated here from the second law. Historically, this result was given by *Carnot* himself in 1824 in his pamphlet on the motive power of fire from considerations of *caloric* conservation. This paper is now hard to understand owing to the vocabulary used by Carnot, but that does not detract at all from the perspicacity of his intuitions.

Catalysis—A word created by Berzelius in 1836, from the Greek καταλυσις, destruction, as opposed to analysis. A catalyst is a substance whose presence in a medium modifies the reaction rate, but which does not participate directly in the reaction and is present intact when the reaction is completed. A catalyst does not modify the *equilibrium constant*, but acts on the *activation energy* of the reaction.

Catatectic—A word proposed by J. B. Wagner in 1974, in place of the old 'metatectic', for describing the transformation on heating:

$$\langle \text{Solid-}\alpha \rangle + (\text{Liquid}) \rightarrow \langle \text{Solid-}\beta \rangle.$$

This judicious term (from the Greek κατα, going down, and τηκτος, melting) well describes the observed phenomenon: the solid compound β melts when its temperature is decreased. It is possible to speak of *incongruent* and *retrograde* fusion.

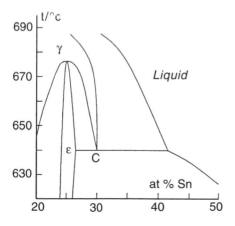

Catatectic: In the binary diagram Cu–Sn, the intermetallic compound ε-Cu$_3$Sn undergoes a transition ε → γ at 676°C. The γ solid solution containing 30 at. % Sn, on cooling at 640°C, undergoes catatectic melting: γ → ε + liquid.

Examples of catatectic transformations are:

$$\text{Al}_3\text{Sm (25 at.\% Sm)} + \text{L(4 at.\% Sm)} \xrightarrow{1066°C} \text{Al}_{11}\text{Sm}_3 \text{ (21 at.\% Sm)}$$

$$\varepsilon\text{-Cu}_3\text{Sn (25 at.\% Sn)} + \text{L(43 at.\% Sn)} \xrightarrow{640°C} \gamma\text{-CuSn (30 at.\% Sn).}$$

Cathode—A term introduced by Faraday in 1834, from the Greek $\kappa\alpha\theta o\delta o\varsigma$, descent ($\kappa\alpha\tau\alpha$, going down). The cathode corresponds to the electrode where reduction occurs: the oxidation degree of the reduced species decreases. The cathode is linked not to the sign of an electrode but to the nature of the reaction. It is the positive pole in a galvanic cell, and the negative pole in an electrolysis cell.

Cation—Name given to positively charged ions, because they are attracted towards the *cathode* of an electrolysis cell.

Cell—See Galvanic cell.

Celsius temperature scale—See *Degree Celsius*.

Chemical potential—The *first law* implies there is a state function U, the internal energy, defined by its differential:

$$dU = \sum(dQ + dW) = \sum X_i \, dx_i.$$

X_i is a variable of *tension* and x_i a variable of *extensity*. The chemical potential μ_i of the substance i is the tension associated with the extensity amount of substance n_i. It is thus expressed in J mol^{-1}.

Since the chemical potential is a variable of tension, the equilibrium of a substance i between two phases α and β is expressed by $\mu_{i,\alpha} = \mu_{i,\beta}$, a relation which has the merit of simplicity. However, this relation can be useful only if we are able to express the chemical potential as a function of physical parameters which characterize the system.

CHEMICAL POTENTIAL $\equiv$ PARTIAL MOLAR GIBBS ENERGY: Write, assuming reversibility, the differentials:

$$dU = T\,dS - P\,dV + \sum \mu_i \, dn_i + \cdots$$

$$dH = T\,dS + V\,dP + \sum \mu_i \, dn_i + \cdots$$

$$dF = -S\,dT - P\,dV + \sum \mu_i \, dn_i + \cdots$$

$$dG = -S\,dT + V\,dP + \sum \mu_i \, dn_i + \cdots .$$

It is thus possible to get various expressions for μ_i:

$$\mu_i=\left(\frac{\partial U}{\partial n_i}\right)_{S,V,n_{j\neq i}}=\left(\frac{\partial H}{\partial n_i}\right)_{S,P,n_{j\neq i}}=\left(\frac{\partial F}{\partial n_i}\right)_{T,V,n_{j\neq i}}=\left(\frac{\partial G}{\partial n_i}\right)_{T,P,n_{j\neq i}}.$$

C

Of these expressions, the last is by far the most interesting, because it allows the chemical potential to be identified with the partial molar Gibbs energy of i in the system (see *Partial molar quantity*).

If, for a given system at equilibrium characterized by its *Gibbs energy G*, we add δn_i mol of i at constant T and P, then its Gibbs energy becomes:

$$G+\left(\frac{\partial G}{\partial n_i}\right)_{T,P,n_{j\neq i}}\delta n_i=G+\mu_i\,\delta n_i.$$

$\mu_i\,\delta n_i$ is the contribution of δn_i mol of i to the Gibbs energy of the system.

μ_i, the partial molar Gibbs energy of i, is the contribution of one mole of i.

It is important to note that the chemical potential identifies with the partial molar Gibbs energy, because the differentiation of G is performed by keeping the intensive parameters P and T constant. Chemical potential thus satisfies the *Euler identity*: $G=\sum n_i\mu_i$. Clearly, it appears that $\mu_i=(\partial U/\partial n_i)_{S,V,n_{j\neq i}}$ cannot be identified with a partial molar internal energy, because the differentiation is performed by keeping the extensive parameters S and V constant. The partial molar internal energy is defined by $u_i-(\partial U/\partial n_i)_{T,P,n_{j\neq i}}$ and satisfies the Euler identity: $U=\sum n_i u_i$. The remark is also valid for $\mu_i=(\partial H/\partial n_i)_{S,P,n_{j\neq i}}$ and for $\mu_i=(\partial F/\partial n_i)_{T,V,n_{j\neq i}}$, which, for the same reason, represent neither a partial molar enthalpy nor a partial molar Helmholtz energy.

EXPRESSIONS OF THE CHEMICAL POTENTIAL: Because the chemical potential identifies with partial molar Gibbs energy, it is possible to state:

$$dG=-S\,dT+V\,dP$$

$$d\mu_i=-s_i\,dT+v_i\,dP,$$

where s_i and v_i represent respectively the partial molar entropy and the partial molar volume of i in the system. This last relation can easily be integrated in the following two ways:

- at constant pressure: $\mu_i(T,p_i)=\mu_i(T_o,p_i)+\int_{T_0}^{T}-s_i\,dT$;
- at constant temperature: $\mu_i(T,p_i)=\mu_i(T,p_i^\circ)+\int_{p_i^\circ}^{p_i}v_i\,dP$.

38 Chemical potential

The latter expression is by far the most useful, because the integral can easily be dropped. The attentive reader will notice the different symbols used for the integration limits: $T_\circ$ ('$\circ$' as subscript) and p_i° ('$\circ$' as superscript). This is not a whim. Superscript '$\circ$' means '*standard*', but the well-understood interests of thermodynamicists commands them to define only a standard pressure p_i° and not a *standard temperature*! $\mu_i(T, p_i^\circ)$, the value taken by μ_i when $p_i = p_i^\circ$, is the standard chemical potential. One writes, for simplicity, $\mu_i(T, p_i^\circ) = \mu_i^\circ(T)$, where the standard pressure p_i° is implicitly included in $\mu_i^\circ(T)$. The integral expression of the chemical potential is thus:

$$\mu_i(T, p_i) = \mu_i^\circ(T) + \int_{p_i^\circ}^{p_i} v_i \, dP.$$

To continue the integration, three possibilities have to be examined:

- If i is an ideal gas: $v_i = RT/p_i$, whence:

$$\mu_i(T, p_i) = \mu_i^\circ(T) + RT \ln(p_i/p_i^\circ).$$

- If i is a real gas, one puts, by analogy:

$$\mu_i(T, p_i) = \mu_i^\circ(T) + RT \ln(f_i/f_i^\circ),$$

where f_i is the *fugacity* of the real gas and f_i° its standard fugacity.
- If i is a condensed phase, its chemical potential is that of its vapour in equilibrium with the condensed phase.

By definition, the *activity* of i is given by:

$$a_i \equiv \frac{f_i}{f_i^\circ} \equiv \frac{\text{fugacity of } i \text{ in the system}}{\text{standard fugacity of } i},$$

so the general expression for the chemical potential of i is given by:

$$\mu_i(T, p_i) = \mu_i^\circ(T) + RT \ln a_i.$$

Remark 1: μ_i°, a function of T and p_i°, is explicitly present in the expression of μ_i. However, the value of μ_i cannot be a function of p_i°, a standard pressure arbitrarily selected, otherwise, it would be possible to modify μ_i in a phase by changing the standard pressure, which contradicts the expression of the equilibrium of i between two phases by the equality of the chemical potentials. Actually, a change of standard state modifies μ_i°, and also modifies the value of $a_i \equiv f_i/f_i^\circ$. These two modifications cancel one another and the value of μ_i remains unchanged.

Remark 2: When i is gaseous, it is judicious to take as the standard state pure gaseous i at a fugacity $f_i^\circ = 1$ bar. For the condensed phase, it is often better to take as standard state the condensed substance i in a well-defined aggregation state; f_i° then represents the fugacity of pure gaseous i in equilibrium with i condensed in its standard state.

C

Remark 3: The calculation of the saturation vapour pressure of iron near room temperature gives 10^{-50} bar, a value far too low to be determined experimentally by direct measurement. It may seem utopian, in such conditions, to use the definition $a_i \equiv f_i/f_i^\circ$. Nevertheless, simple experimental methods, generally based on equilibrium measurements, give direct access to activities, without the need for fugacity determinations.

Remark 4: When a substance i is present in several phases, it is possible to select different standard states for i in each phase, for instance, pure liquid for i in the liquid phase and vapour under $f_i^\circ - 1$ bar for i in the vapour phase. Application to the equilibrium of a pure liquid substance with its vapour gives: $\mu_l^\circ - \mu_v^\circ + RT \ln(f_l/f_v^\circ)$, which is no other than the *Clapeyron equation*, because $\mu_v^\circ - \mu_l^\circ = \Delta_{vap}G^\circ$.

Remark 5: By selecting the same standard state for i, whatever the phase considered, the equality of chemical potentials, the equilibrium condition for i between several phases, would be expressed by the equality of the activities of i in the different phases. Such a situation often leads to an impasse! It is clear that experimenters are in a situation similar to that of the 'enlightened despots' of the eighteenth century: arbitrariness is often limited by reason, which dictates a good choice.

Chemical reaction—A chemical reaction is characterized by a balance equation:

$$a_1A_1 + a_2A_2 + \cdots \rightarrow b_1B_1 + b_2B_2 + \cdots$$

Species A_i are the reactants 'starting materials'; species B_i are the products ('final products'); a_i and b_i are the stoichiometric numbers (the expression 'stoichiometric coefficient' is inappropriate).

It may be convenient to write a chemical reaction in a more condensed symbolic form: $\sum v_i M_i = 0$. In this case, constituents M_i may represent reactants or products; numbers v_i are positive for the products (right-hand side constituents), negative for the reactants (left-hand side constituents).

40 Clapeyron, Emile

When speaking about a chemical reaction, it is necessary to specify the balance equation and the *standard states* used. However, it is clear that an equilibrium state resulting from a physicochemical process cannot be dependent on an arbitrary selection of standard states.

Clapeyron, Emile (1799–1864)—French engineer, considered along with Clausius, Joule, and Kelvin as one of the founders of thermodynamics. He made known *Sadi Carnot*'s ideas. He showed in 1834 how these ideas can be mathematically expressed. He introduced the notion of *reversibility* and, in 1858, put forward the equation bearing his name.

Clapeyron diagram—The name generally given to a pressure–volume diagram. For a pure substance, isotherms corresponding to a temperature much higher than the critical temperature are hyperbolic. When the temperature is reduced, two inflection points appear which meet at the critical temperature. The critical isotherm has a horizontal tangent:

$$[\partial P / \partial V]_T \to 0 \quad \text{and} \quad [\partial^2 P / \partial V^2]_T \to 0 \text{ when } P \to P_{\mathrm{c}}.$$

A horizontal line LV corresponding to the liquid–vapour equilibrium appears below the critical temperature. The points L and V meet at the critical point with a horizontal tangent.

From *Maxwell's equations*, it is easily shown that, on a Clapeyron diagram, the slope of the *isentropes* is always greater than that of the *isotherms*:

$$\frac{(\partial P / \partial V)_S}{(\partial P / \partial V)_T} = \frac{c_P}{c_V} = \gamma > 1.$$

Clapeyron equation—A differential equation giving, in its most common form, the slope $\mathrm{d}P/\mathrm{d}T$ of the equilibrium curve between the two phases α and β of a pure substance.

The equilibrium condition is given by equating the *chemical potentials*: $\mu_\alpha = \mu_\beta$. To maintain equilibrium after changing one constraint acting on the system, equality of chemical potentials must be conserved: $\mathrm{d}\mu_\alpha = \mathrm{d}\mu_\beta$.

When the constraints are temperature and pressure, then:

$$-s_\alpha \, \mathrm{d}T + v_\alpha \, \mathrm{d}P = -s_\beta \, \mathrm{d}T + v_\beta \, \mathrm{d}P$$

$$\left(\frac{\mathrm{d}T}{\mathrm{d}P}\right)_{\mathrm{eq}} = \frac{\Delta_{\mathrm{tr}}S}{\Delta_{\mathrm{tr}}V} = \frac{\Delta_{\mathrm{tr}}H}{T\Delta_{\mathrm{tr}}V}.$$

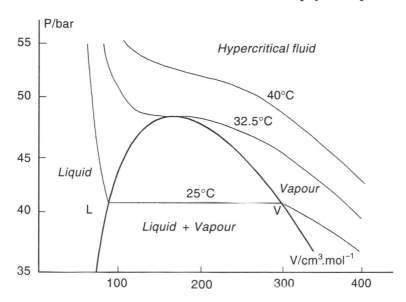

Clapeyron diagram: *Ethane close to the critical point.*

$\Delta_{tr}H$ and $\Delta_{tr}V$ represent respectively the *enthalpy* and the volume changes during the transition $\alpha \to \beta$. Indeed, it is possible to write $\Delta_{tr}S = \Delta_{tr}H/T$ when the transition $\alpha \to \beta$ occurs at equilibrium. Since the Clapeyron equation is a differential equation, a knowledge of one point is necessary to obtain the equilibrium curve by integration.

Application to liquid–vapour equilibrium of a pure substance, where $\Delta_{vap}H$, the enthalpy of vaporization, and T_b, the boiling point under atmospheric pressure, are known, leads to the well-known expression for the saturation vapour pressure:

$$\ln\left(\frac{P}{\text{atm}}\right) = \frac{\Delta_{vap}H}{R}\left(\frac{1}{T_b} - \frac{1}{T}\right).$$

This relation implies three simplifying assumptions which must be kept in mind:

1. $\Delta_{vap}H$ is independent of T, which implies: $\Delta c_P = c_{P,vap} - c_{P,liq} \approx 0$;
2. $v_{liq} \ll v_{vap}$, which is realized only far from the critical state;
3. $v_{vap} = RT/P$, acceptable for pressures lower than ≈ 1 bar.

The Clapeyron relation can be generalized. For instance, in a system subjected to elastic constraints for which the expression of the elementary work is $dW = \mathscr{F}dl$, the *thermodynamic potential* at constant P, T and $\mathscr{F}$ is the 'Gibbs elastic energy': $G^{\#} = G - \mathscr{F}l$. For a variation of the constraints P, T and $\mathscr{F}$:

$$dG^{\#} = -S\,dT + V\,dP - l\,d\mathscr{F},$$

and, for one mole: $d\mu^{\#} = -s\,dT + v\,dP - l_m\,d\mathscr{F}$.

Under a hydrostatic pressure, $P = \text{const.}$, the equilibrium between two phases is expressed by $\mu_\alpha^{\#} = \mu_\beta^{\#}$, or $d\mu_\alpha^{\#} = d\mu_\beta^{\#}$. By specifying:

$$-s_\alpha\,dT - l_{m,\alpha}\,d\mathscr{F} = -s_\beta\,dT - l_{m,\beta}\,d\mathscr{F},$$

we obtain:

$$\left(\frac{dF}{dT}\right)_{eq} = -\frac{\Delta_{tr}S}{\Delta_{tr}l} = -\frac{\Delta_{tr}H}{T\Delta_{tr}l}.$$

Clausius inequality—The name given to the inequality $\oint dQ/T \leqslant 0$, deduced from the classical expression of the *second law*:

$$dS = \frac{dQ}{T} + d\sigma, \quad \text{with } d\sigma \geqslant 0.$$

The Clausius inequality may also be expressed by:

$$dS = (dQ/T)_{rev} > (dQ/T)_{irrev}.$$

Clausius, Rudolf (1822–1888)—German physicist, who developed the *kinetic theory of gases*, stated explicitly in 1850 the two first laws of thermodynamics, and introduced in 1868 the state functions internal energy and entropy.

Closed—A closed system is a system which does not exchange matter with its *surroundings*. A closed system may exchange *energy* in the form of *work* or *heat* with its surroundings.

Cluster—In an A–B solid solution, clusters are observed when the number of A–A bondings is higher than the number calculated statistically. Indeed, the same must be true for the number of B–B bondings, whereas the probability of finding an A–B bonding must be lower than that calculated statistically. The presence in the solution of clusters richer in A or in B than the overall solution implies the existence of repulsive interactions between A and B species:

$$\varepsilon_{AB} > (\varepsilon_{AA} + \varepsilon_{BB})/2.$$

The *enthalpies of formation* of the bondings ε_{AA}, ε_{BB}, and ε_{AB} are naturally negative, otherwise the solution could not exist!

In the presence of repulsive interactions, the number of A–B bondings is reduced. However, this reduction does not lead to total suppression; there exists an *equilibrium* value, which can be calculated by looking for the minimum of the function $G = H - TS$. This minimum exists because the two terms H and S have opposite influences. A decrease in the number of A–B bondings has the effect of decreasing H and S. The detailed calculation of the number of A–B bondings at equilibrium is given under *Short-range order*. When interactions between A and B become too repulsive, the solid solution is no longer stable: there is *unmixing*.

Coefficient—When two physical quantities A and B are proportional and do not have the same dimension, i.e. $A = kB$, the quantity k is called a coefficient or sometimes 'modulus'. If A and B have the same dimension, k is thus a dimensionless number and is called a 'factor' or sometimes 'index'.

Colligative—From Latin *colligere*, to bring together. Colligative identifies a property of a group of substances which is independent of the nature of each substance. For instance, the *osmotic pressure* of a solution, the *vapour pressure* of a solvent (*tonometry*), the depression of its melting temperature (*cryometry*) or the elevation of its boiling temperature (*ebulliometry*) depend only on the *activity* of the solvent and not on the nature of the solute present in the solution.

Combustion—Tables of thermodynamic data generally give the standard *enthalpies of formation* of chemical compounds. Sometimes, mainly for organic compounds, the tables provide the standard enthalpies of combustion, which are more easily accessible experimentally. By convention, the products of the combustion of a compound composed of C, H, O, N, S, and Cl are CO_2, H_2O, N_2, SO_2, and Cl_2. Their enthalpy of combustion is therefore zero. The *standard enthalpy of reaction*: $\sum v_i M_i = 0$ is given, from the standard enthalpies of combustion $\Delta_c H^\circ(M_i)$, by:

$$\Delta_r H^\circ = -\sum v_i \Delta_c H^\circ(M_i).$$

Notice the minus sign. Its presence arises from the fact that the product formed is on the right-hand side of the formation reaction whereas the product burnt is on the left-hand side of the combustion reaction.

44 Complexion

Complexion—In an assembly of N discernible particles, each one is characterized by its quantum state. A complexion, in the Boltzmann sense, is defined by a knowledge of the quantum state of each particle. The exchange of two particles between two quantum states creates a new complexion.

The probability of a given macroscopic state increases with Ω, the number of complexions allowing the realization of this state. The *Boltzmann* relation $S = k \ln \Omega$, proposed in 1875, links the *entropy* to the number of complexions.

The statistics based on this relation gives acceptable results if particles can effectively be considered as discernible. This is the case for solids where each atom occupies a position in a crystal characterized by its spatial coordinates. Consider for instance the transformation:

$$N_A \langle A \rangle + N_B \langle B \rangle \rightarrow \langle \text{Solid solution} \rangle$$

$$\Delta S = S_{SS} - S_A - S_B = k \ln \frac{\Omega_{SS}}{\Omega_A \Omega_B}.$$

There are $N!$ ways to dispose N discernible particles among N sites. As a consequence: $\Omega_A = N_A!$, $\Omega_B = N_B!$, and $\Omega_{SS} = N!$ (with $N = N_A + N_B$), and:

$$\Delta S = k \ln \frac{N!}{N_A! \, N_B!}.$$

By using the *Stirling formula* $(N! \approx N \ln N - N)$:

$$\Delta S = k(N \ln N - N_A \ln N_A - N_B \ln N_B) = -k(N_A \ln x_A + N_B \ln x_B),$$

with $x_i = N_i/N$. Putting $k = R/\mathcal{N}$, we find the well-known expression for the *entropy of mixing* for an *ideal solution*:

$$\Delta S = -R(n_A \ln x_A + n_B \ln x_B).$$

Here, k is the Boltzmann constant, R the ideal-gas constant, $\mathcal{N}$ the Avogadro constant, n the number of moles, and N the number of particles, with $\mathcal{N} = R/k = N/n$.

The assumption of indiscernible particles would lead to the same expression for the entropy of mixing, because:

$$\Omega_A = 1, \quad \Omega_B = 1, \quad \text{but } \Omega_{SS} = N!/N_A! \, N_B!.$$

Generally, the discernibility hypothesis leads to difficulties. The entropy loses its additive character owing to the introduction of a term 'entropy of mixing' even for identical particles (See *Maxwell–Boltzmann statistics*). It is thus necessary to abandon the division of a macroscopic state into complexions.

The solution, proposed by Bose, consists of dividing the macroscopic state into microscopic states, where a microscopic state is defined by the number of particles occupying a quantum state. The probability Ω of a macroscopic state is thus given by the number of microscopic states which realize it.

Composition—The composition of a mixture is defined by the nature of the phases present and by the quantity of each constituent in each phase. A composition may be expressed in terms of *titres* or *concentrations*.

Compound—A substance formed of several elements, whose properties are more or less different from those of its constituents. An extreme example is that of common salt (NaCl), which resembles neither the metal (Na) nor the gas (Cl_2) which permit its synthesis! An intermetallic compound such as Cu_5Zn_8 differs from its constituents in its physical and crystallographical properties, but retains the same metallic character as copper and zinc. Compound is not synonymous with *species*: the compound NaCl, crystallized or dissolved in water, is constituted of species Na^+ and Cl^-.

Compressibility (or compression coefficient)—The ratio, in terms of unit volume, between the volume variation and the pressure variation. It is necessary to specify the conditions of the transformation. Hence it is possible to define:

- the *isothermal* compressibility: $\chi_T \equiv -\dfrac{1}{V}\left(\dfrac{\partial V}{\partial P}\right)_T$;

- the *isentropic* compressibility: $\chi_S \equiv -\dfrac{1}{V}\left(\dfrac{\partial V}{\partial P}\right)_S$.

It is easy to show, from *Maxwell's equations*, that the two compressibilities are linked by the relation: $\chi_T/\chi_S = c_P/c_V = \gamma$.

Compression factor—$Z = PV/RT$. It may be useful to present the equation of state for a real gas as an expansion of the compression factor Z:

$$Z = 1 + (B/V) + (C/V^2) + \cdots$$

46 Concave

An important parameter characterizing the behaviour of a real gas is the critical compression factor: $Z_c = P_c V_c / RT_c$. For hydrogen, $Z_c = 0.304$; however, in most cases (monatomic gases, N_2, O_2, CO, CO_2, etc.), $Z_c \approx 0.29$. For NH_3, $Z_c = 0.242$, whereas for water vapour, $Z_c = 0.230$. With the *van der Waals equation*, one calculates $Z_c = 3/8 = 0.375$, a value appreciably removed from experimental reality.

Concave—A curve $y = f(x)$ is said to be concave when $d^2 f/dx^2 < 0$. If y represents a *thermodynamic potential* of the system and x represents a state variable, the concavity of the function $y = f(x)$ means an instability of the system in relation to a small fluctuation of the variable x (see *Stability*).

Concentration—A concentration is generally expressed in *moles* per litre of solution (*molarity*). It can be expressed in moles per kg of solvent (*molality*) or in kg per m^3 of solution (mass concentration). Incorrectly, the word concentration is used to denote composition, even if the compositions are expressed in mass *fractions* or in mole fractions. It is better to use *titre*.

Condensation—Transition from the vapour state to the condensed state (solid or liquid). Condensation is an exothermic phenomenon.

Conductivity—When an *extensity* flows from a source towards a sink under the influence of a potential gradient, it is possible to state, to a first approximation, that the *flux* of extensity is proportional to the potential gradient. The proportionality coefficient is the conductivity of the medium at the point considered. From a more general point of view, conductivity is an intensive quantity; it is the ratio between a flux of extensity and an intensity gradient:

$$\text{Conductivity} = \frac{\text{Flux of extensity}}{\text{Intensity gradient}}.$$

ELECTRIC CONDUCTIVITY: The extensity being quantity of electricity, the flux of extensity is expressed in $C m^{-2} s^{-1}$ or in $A m^{-2}$. The intensity is electric potential, electric potential gradient, or electric field and is expressed in $V m^{-1}$ or in $m kg s^{-3} A^{-1}$. Electric conductivity is thus expressed in $S m^{-1}$; the *siemens* is the unit of electric conductance:

$$1 S m^{-1} = 1 m^{-3} kg^{-1} s^3 A^2.$$

CONDUCTIVITY OF MOMENTUM: This quantity is best known as *dynamic viscosity*. The extensity is momentum mv; the flux of extensity is expressed in $(kg\,m\,s^{-1})(m^{-2}\,s^{-1})$, or $kg\,m^{-1}\,s^{-2}$; the intensity is velocity, and the intensity gradient is reciprocal time. The conductivity of momentum, or dynamic viscosity, is thus expressed in $kg\,m^{-1}\,s^{-1}$ or in Pa s. The name poiseuille (symbol Pl) is sometimes given to this unit:

$$1\,Pl = 1\,Pa\,s = 1\,kg\,m^{-1}\,s^{-1}.$$

C

THERMAL CONDUCTIVITY: The extensity is heat quantity, and the flux of extensity is expressed in $J\,m^{-2}\,s^{-1}$, or in $W\,m^{-2}$; the intensity is temperature, and the intensity gradient is expressed in $K\,m^{-1}$. The thermal conductivity λ is thus expressed in watts per metre and per kelvin:

$$1\,W\,m^{-1}\,K^{-1} = 1\,kg\,m\,s^{-3}\,K^{-1}.$$

Fourier's law of heat conduction states that the heat flux is proportional to the temperature gradient: $J_Q = -\lambda\,\mathrm{grad}\,T$.

There is a straightforward relation between the thermal conductivity and the *phenomenological coefficient* L_{QQ}. Indeed, the equation related to heat transfer under the only influence of a temperature gradient is:

$$J_Q = L_{QQ}\,\mathrm{grad}\,(1/T).$$

By comparison, we have: $L_{QQ} = \lambda T^2$.

DIFFUSIONAL CONDUCTIVITY: The extensity is amount of substance, and the flux of extensity is expressed in $mol\,m^{-2}\,s^{-1}$. The intensity is concentration, and the intensity gradient is expressed in $mol\,m^{-4}$. The diffusional conductivity, or diffusion coefficient D_i, is thus expressed in $m^2\,s^{-1}$.

Fick's first law of diffusion states that the flux of matter is proportional to the concentration gradient:

$$J_i = -D_i\,\mathrm{grad}\,c_i.$$

It is not surprising that diffusional conductivity identifies with diffusivity, because intensity identifies with volumetric extensity.

It is possible to link the diffusion coefficient diffusion D_i to the phenomenological coefficient L_{ii}, but the relations obtained are often complex. Indeed, the phenomenological equation corresponding to diffusion under the influence of only the *chemical potential* gradient is:

$$J_i = -L_{ii}\cdot\frac{1}{T}\cdot\mathrm{grad}\,\mu_i.$$

48 Configurational entropy

Configurational entropy—The configuration of a *solid solution* is determined by the position of atoms or groups of atoms in the lattice. The configurational entropy is thus the contribution of the atomic positions to the *entropy of mixing*.

In the straightforward case of a solid solution AB with only one kind of site, at which A and B are distributed at random (A–B interactions are thus neutral), the configurational entropy is given by:

$$\Delta_{mix}S^{conf} = \Delta_{mix}S^{id} = -R\sum x_i \ln x_i.$$

With two kinds of sites, a moment's thought obviously yields:

$$x\langle GaAs\rangle + (1-x)\langle InSb\rangle \rightarrow \langle Solution\rangle$$

$$\Delta_{mix}S^{conf} = 2\Delta_{mix}S^{id}.$$

However:

$$x\langle Mg_2SiO_4\rangle + (1-x)\langle Fe_2SiO_4\rangle \rightarrow \langle Solution\rangle.$$

$\Delta_{mix}S^{conf} = 2\Delta_{mix}S^{id}$, or $\Delta_{mix}S^{conf} = \Delta_{mix}S^{id}$: which is true? If we consider the mixture of clusters Mg_2SiO_4 and Fe_2SiO_4, then $\Delta_{mix}S^{conf} = \Delta_{mix}S^{id}$. Whereas if we consider the mixture of ionic species Mg^{2+} and Fe^{2+} at cation sites, then $\Delta_{mix}S^{conf} = 2\Delta_{mix}S^{id}$. Experiment supports the second hypothesis.

A configurational entropy is not always so easily determined (see for example *Short-range order*), but it can often be evaluated with precision. The configurational entropy is not the only contribution to the entropy of mixing. Other contributions (vibration, electronic, magnetic, etc.) are more laboriously evaluated.

Congruence—From Latin *congruus*, concordant. A transition $\alpha \rightarrow \beta$ is said to be congruent when the two phases α and β have the same composition at equilibrium. If the α phase represents a liquid solution and the β phase is its vapour in equilibrium, then the congruence is called *azeotropy*.

Conservative—A quantity is said to be conservative if it can be neither created nor destroyed. If a quantity x_i is exchanged between two systems A and B, and the assembly $(A+B)$ is *isolated*, then the quantity x_i is conservative if:

$$dx_{iA} + dx_{iB} = 0.$$

Remark 1: We show (see *Extensity*) that at equilibrium, extensive quantities are conservative. This result is generally true, even for transformations out of equilibrium (e.g. mass, volume, electricity). *Entropy* is a notable exception: an irreversible transformation manifests itself in entropy creation.

Remark 2: A force *field* is said to be conservative if:

$$\int_i^f \mathscr{F} \mathrm{d}l = V(f) - V(i)$$

independently of the path followed. Force $\mathscr{F}$ derives from a *potential*.

Remark 3: A *flux* is conservative if it equals zero when emanating from a closed surface:

$$\Phi = \iint_S \mathscr{F} \cdot \mathrm{d}\mathscr{S} = \iiint_V \mathrm{div}\, \mathscr{F} \mathrm{d}V = 0,$$

where $\mathrm{d}\mathscr{S}$ is a vector normal to the surface $\mathscr{S}$. A flux is therefore conservative if:

$$\mathrm{div}\, \mathscr{F} \quad (\partial A/\partial x)_{y,z} + (\partial B/\partial y)_{x,z} + (\partial C/\partial z)_{x,y} - 0,$$

where A, B, C are the components of the vector $\mathscr{F}$. For instance, in the absence of magnetic material, the flux of the *magnetic field* vector is conservative:

$$\Phi = \iint_S \mathscr{B} \cdot \mathrm{d}\mathscr{S} = \iiint_V \mathrm{div}\, \mathscr{B} \mathrm{d}V = 0.$$

Constrained system—A system is said to be constrained if it is subject to specified conditions, called *constraints*, which are not altered during the evolution of the system. Examples of constrained systems are *adiabatic, isobaric, isochoric, isothermal*, and *isolated* systems.

Constraint—A restriction that prevents a system from reaching *states* which otherwise would be allowed. Constraints may be external, for instance T and P constant (as imposed by the experimenter), or they may be internal, as imposed by the laws of physics. For example, a system containing initially solid ZnO susceptible to decomposition into gaseous Zn and O_2 is subject to the implicit constraint: $P_{Zn} = 2P_{O_2}$. Electrical neutrality is a constraint imposed on all electrolytic solutions.

Convex—A curve $y = f(x)$ is convex when $\mathrm{d}^2 f/\mathrm{d}x^2 > 0$. If y represents a *thermodynamic potential* of the system and x a variable of state, then concavity of the function $y = f(x)$ means stability of the system in relation to a small fluctuation of the variable x (see *Stability*).

Corresponding states—The law of corresponding states was stated for the first time in 1873 by van der Waals in a rather terse form:

'Real gases have the same equation of state in reduced coordinates'.

This statement means that the *van der Waals equation*:

$$[P + (a/V^2)](V - b) = RT,$$

which represents, on a pressure–volume diagram, a critical isotherm characterized by an inflexion point with a horizontal tangent, can be expressed in reduced coordinates ($T_R = T/T_c$, $P_R = P/P_c$, $V_R = V/V_c$) by:

$$[P_R + (3/V_R^2)](3V_R - 1) = 8T_R.$$

Actually, every equation of state with three parameters, or even with two parameters if one considers the gas constant R as a third parameter, can undergo such a transformation.

The theoretical justification of the law of corresponding states rests on an interaction model for molecules in the fluid phase which introduces an attractive term and a repulsive term. Both terms increase when the distance between molecules decreases, but the repulsive term increases more steeply. The cohesive energy ε can be expressed, as a function of the mean distance r separating the molecules, by the relation:

$$\varepsilon = \varepsilon_o \varphi(r/r_o),$$

where r_o represents the distance corresponding to the minimum ε_o of the cohesive energy. Since ε_o and r_o are linked to the nature of the substance, it is possible to correlate them with the critical coordinates or with any group of characteristic physical quantities. The function φ is a universal function; the expression given by Lennard-Jones in 1926 is one of the best known:

$$\varphi(r/r_o) = (r_o/r)^{12} - 2(r_o/r)^6.$$

Coulomb—The unit of amount of electricity in the SI system (symbol C), named after the French physicist Charles de Coulomb (1736–1806). The coulomb represents the charge under 1 *volt* of a condenser whose capacity is 1 *farad*:

$$1\,C = 1\,A\,s = 1\,F\,V.$$

Coupling—Experiment shows that a *generalized force*, caused by a *potential* gradient, gives rise to an *extensity* current. For instance, a temperature gradient creates a heat current, an electric potential gradient creates an electric current, a *chemical potential* gradient creates a current of matter, etc. They are direct effects.

C

However, these effects are not independent. Thus, a chemical potential gradient may create a pressure gradient (*osmosis*), an electric potential gradient (*galvanic cell*), or a temperature gradient (*Dufour effect*); conversely, a temperature gradient may cause a chemical potential gradient (*Soret effect*), an electric potential gradient (*Seebeck effect*), or a pressure gradient (*thermomolecular effect*). These effects are coupling effects: each flux is linked to the ensemble of generalized forces which are able to act independently on the system.

Coupling effects—See *Coupling*. The main coupling effects are *thermodiffusion* (*Dufour* and *Soret effects*), *thermoelectric effects* (*Peltier*, *Seebeck*, and *Thomson effects*), *thermomechanical effects*, and *thermomolecular effects* (*Fountain* and *Knudsen effects*).

Covolume—The theoretical limit for the volume of molecules in a gas when the pressure tends towards infinity. In the *van der Waals equation* of state: $(P + a/V^2)(V - b) = RT$, the parameter b represents the covolume of the gas, which is the limit of the molar volume when $P \to \infty$. This limit cannot be attained, because, if the pressure tends towards infinity, a black hole is created!

Critical—A critical point is the name given to a second-*order* phase transition. Critical points can be isolated or localized along a curve. The best known are the critical points of liquid–vapour equilibrium and the critical points of mixtures.

The liquid–vapour transition for a pure substance is a first-order transition, characterized by a non-zero enthalpy of vaporization. On increasing the temperature, the molar volume of the liquid in equilibrium with its vapour decreases owing to the expansion, whereas the molar volume of the vapour in equilibrium with the liquid decreases owing to the exponential increase in the saturated vapour pressure. At the critical temperature, the properties of liquid and vapour become identical: the liquid–vapour equilibrium curve stops (molar volumes are equal), the vaporization enthalpy becomes zero, and the transition becomes of the second order.

An equation of state $f(P, V, T) = 0$ takes into account the critical phenomenon if there exists a critical point (P_c, V_c, T_c) satisfying the relations:

$$f(P,V,T)=0, \quad \left(\frac{\partial P}{\partial V}\right)_T=0, \quad \left(\frac{\partial^2 P}{\partial V^2}\right)_T=0.$$

On a *Clapeyron diagram*, a critical point lies on an *isotherm* called the critical isotherm, characterized by an inflexion point with a horizontal tangent. The critical point presents noteworthy properties.

The *heat capacities* c_P of liquid and vapour in equilibrium tend towards infinity when $T \to T_c$. The molar volumes of the liquid and the vapour in equilibrium (index x) tend towards V_c when $T \to T_c$ with $(\partial V/\partial T)_x \to +\infty$ for the liquid and $-\infty$ for the vapour. Thus, curves $V=f(T)$ for the liquid and the vapour in equilibrium meet at the critical point with a vertical tangent.

The enthalpy of vaporization $\Delta_{vap}H \to 0$ with $[\partial(\Delta_{vap}H)/\partial T]_x \to -\infty$ when $T \to T_c$. Verification is easy by differentiating with respect to temperature the expression for $\Delta_{vap}H$ extracted from the *Clapeyron equation*, then by allowing $T \to T_c$:

$$\Delta_{vap}H = T\left(\frac{\partial P}{\partial T}\right)_x (v_v - v_l)$$

$$\left(\frac{\partial \Delta_{vap}H}{\partial T}\right)_x = \left(\frac{\partial P}{\partial T}\right)_x (v_v - v_l) + T\left(\frac{\partial^2 P}{\partial T^2}\right)_x (v_v - v_l)$$

$$+ T\left(\frac{\partial P}{\partial T}\right)_x \left(\frac{\partial v_v}{\partial T} - \frac{\partial v_l}{\partial T}\right).$$

The first two terms on the right-hand side tend towards 0 when $T \to T_c$, whereas the third term tends towards $-\infty$.

The concept of the critical point can also be applied to liquid or solid solutions undergoing an *unmixing*. The critical point of a binary mixture, characterized by its critical coordinates of temperature and composition, is the point where the properties of the two phases at equilibrium become identical. For instance, liquid water–phenol solutions are single-phase above 67°C, and two-phase below. Likewise, NaCl and KCl are in a solid solution at all possible proportions above 480°C, and show unmixing below 480°C.

Remark: If two phases in equilibrium at a critical point are identical, then it is necessary that they possess the same symmetries. A solid–liquid equilibrium cannot show a critical point; an equilibrium between two *enantiomorphic* forms does not generally show a critical point. Following the Clapeyron relation, when two enantiomorphic phases have the same molar volume, the equilibrium curve on a pressure–temperature diagram has a vertical tangent.

Critical opalescence—Opal is a natural colloid composed of silica and water; the beauty of its lustre makes it appreciated as a semi-precious stone. When a system consisting of two liquids or one liquid in equilibrium with its vapour approaches its critical point, it loses its transparency and takes on an opalescent appearance. This phenomenon, called critical opalescence, arises from density fluctuations close to the critical point. It disappears when the critical point is passed.

Cryometry—Cryometry measures the decrease in the crystallization temperature of a *solvent* (index 1) caused by the presence of a *solute*. If the solute is insoluble in the pure solid solvent, the equilibrium between the pure solid solvent (index s) and the solvent in solution (index l) at the temperature T is given by equality between the *chemical potentials* $\mu_{1s} = \mu_{1l} \Rightarrow$

$$\mu_{1s}^{\circ} = \mu_{1l}^{\circ} + RT \ln a_{1l}$$

$$a_{1l} = \exp[(\mu_{1s}^{\circ} - \mu_{1l}^{\circ})/RT] = \exp[-\Delta_{fus} G_1^{\circ}/RT],$$

where $\Delta_{fus} G_1^{\circ} = \mu_{1l}^{\circ} - \mu_{1s}^{\circ}$, the *standard Gibbs energy* of fusion of the pure solvent. Putting $\Delta_{fus} G_1^{\circ} = \Delta_{fus} H_1^{\circ}(1 - T/T_{fus1})$, neglecting the difference between heat capacities of solid and liquid, and with the assumption of a solution dilute enough to obey *Henry's law*, it follows that:

$$x_{1l} = \exp\left[\frac{\Delta_{fus} H_1^{\circ}}{R}\left(\frac{1}{T_{fus1}} - \frac{1}{T}\right)\right].$$

Crystal—From the Greek κρυσταλλος, ice. The crystallized state is the stable state for a solid material, as opposed to the amorphous or vitreous states. A crystal is an ordered, solid phase, which is built up from an elementary lattice, and reproduced identically, in the three directions of space. When a crystal melts, disorder appears in the orientation and in the position of the molecules, but the simultaneous disappearance of both kinds of order is not a general rule. A 'plastic crystal' is characterized, before melting, by the disappearance of order in the orientation of molecules. A 'liquid crystal' retains order in orientation after melting. This oxymoron, proposed by Lehmann in

1900, is now well-known to the public since the invasion of liquid–crystal display screens in our everyday life. It is nevertheless better to speak of *mesomorphous* phases, or mesophases.

Curie, Pierre (1859–1906)—French physicist, who discovered piezoelectricity in 1881 with his brother Paul (1855–1941), stated the *symmetry principle* in 1894, and in 1903 shared the Nobel prize for physics for his work on radioactivity.

Cycle—From the Greek $κυκλος$, circle; 'closed cycle' is a pleonasm! A system undergoes a cycle of *transformations* when the final state is identical with the initial state. A heat engine and a refrigerator work over a cycle. From the *first law*, when a system undergoes a cycle of transformations: $\oint dU = \oint (dQ + dW) = 0$.

D

Debye, Peter Joseph William (1884–1966)—Dutch–US physical chemist, author of important works on dipolar moments, X-ray scattering, and ionic interactions in solutions. Nobel prize in chemistry in 1936.

Debye–Hückel theory—The interaction energy between two neutral molecules, governed by London forces, follows a d^{-6} law, whereas the interaction energy between two ions, governed by Coulomb forces, follows a d^{-1} law. As a consequence, ionic solutions show important deviations from *Henry's law* which are perceptible even at very weak concentrations, say $10^{-4} \, mol \, L^{-1}$. The Debye–Hückel theory, proposed in 1923, provides an assessment of the *activity coefficient* of an ion i in solution. The *chemical potential* of the ion i is:

$$\mu_i = \mu_i^\circ + RT \, \ln(c_i/c_i^\ominus) + RT \, \ln \gamma_i.$$

$RT \ln \gamma_i$, representing the deviation from Henry's law, expresses the electrostatic interaction between ions of opposite charge. As a first approximation, Debye and Hückel, neglecting interactions between ions having the same charge and interactions between ions and solvent, showed that charge density around ion i decreases following a

Boltzmann distribution. They calculated the potential of ion i alone and that of ion i surrounded by its atmosphere, and arrived at the *work* needed to bring the ion alone to its new potential:

$$RT \ln \gamma_i = - \frac{z_i^2 e^2 \mathcal{N} b}{8 \pi \varepsilon (1 + ab)},$$

D

where $z_i e$, $\mathcal{N}$ and ε represent respectively the charge of the ion i, the *Avogadro constant* and the *permittivity* of the medium.

The quantity a represents the ionic radius of species i in solution, i.e. the minimum distance of approach of an ion of opposite charge.

The quantity b satisfies the relation $b^2 = e^2 \mathcal{N}^2 \rho \sum n_i z_i^2 / \varepsilon RT$ (ρ is the volumetric mass of the solvent, $I = \frac{1}{2} \sum n_i z_i^2$ is the ionic strength of the medium. In the SI system, n_i is the quantity of substance (expressed in *moles*) of ions i per cubic metre of solution.); b has the dimensions of reciprocal length. The sum $l = a + 1/b$ is the *Debye length*; $1/b$ represents the thickness of the ionic atmosphere.

Numerical application to water at 298 K, with $I = \frac{1}{2} \sum c_i z_i^2$ (c_i in moles per litre), gives: $\log_{10} \gamma_i = -0.5092 z_i^2 \sqrt{I}/(1 + \sqrt{I})$.

This relation implies $a \cdot b = \sqrt{I}$, or $a = 3.1 \times 10^{-10}$ m, following Guggenheim's suggestion. Debye and Hückel's theory is best satisfied by experiment when $\sqrt{I} \ll 1$.

Remark: Direct experimental access to the *activity* or to the activity coefficient of an ion is impossible. Only the mean activity coefficient $\gamma_\pm$ of a salt $A_x B_y$ can be measured·

$$(x + y) \ln \gamma_\pm = y \ln \gamma_+ + x \ln \gamma_-.$$

The model applied to the mean activity coefficient gives:

$$\log_{10} \gamma_\pm = \frac{0.5092 z_+ z_- \sqrt{I}}{1 + \sqrt{I}}.$$

As a rule, $\gamma_\pm < 1$, because the charges of the ions are of opposite sign.

Debye length—The distance beyond which an ion is no longer influenced by the electric field of another ion. The Debye length decreases when the ionic strength of the medium increases, because of a screening effect. It appears in the calculation of the *activity* of an ion in solution by means of the *Debye–Hückel theory*, and in the case of a decimolar solution, varies between 1 and 20 nm. The mean ionic radius in solution is about 0.3 nm.

56 Debye temperature

Debye temperature—At low temperatures, the *internal energy* of a crystal derives mainly from vibrational modes of the crystal. The heat capacity of the crystal is given by:

$$c_V = K(T/\Theta)^3,$$

where Θ is a temperature characterizing the crystal, called the Debye temperature, whose order of magnitude is a few hundred kelvins (95 K for lead, 2230 K for graphite). This relationship, well satisfied at low temperature $(T \lesssim \Theta/100)$ may nevertheless be applied up to $T \approx 0.1\Theta$. With metals, it is necessary to introduce a contribution from conduction electrons adding to the crystal vibration. The electron heat content is proportional to the temperature.

The contribution of the vibration at low temperature is obtained from statistical thermodynamics. The *partition function* for an oscillator of frequency v is given by:

$$Z = \sum_{n=0}^{\infty} \exp\left[-\left(n + \frac{1}{2}\right)\frac{hv}{kT}\right] = \frac{\exp(-hv/2kT)}{1 - \exp(-hv/kT)}.$$

The internal energy associated with $3N$ oscillators of frequency v (N is the number of atoms and $3N$ the number of degrees of freedom) is:

$$U = 3NkT^2\left(\frac{\partial \ln Z}{\partial T}\right)_V = 3N\left[\frac{hv}{2} + \frac{hv}{\exp(-hv/kT) - 1}\right],$$

from which the heat capacity at constant volume is:

$$c_V = \left(\frac{\partial U}{\partial T}\right)_V = 3Nk\left(\frac{hv}{kT}\right)^2 \frac{e^{hv/kT}}{(e^{hv/kT} - 1)^2} = \frac{3Nk(hv/kT)^2}{2\,\text{ch}(hv/kT) - 1}.$$

This is Einstein's solid model. The quantity hv/k, whose dimensions are that of temperature, is the Einstein characteristic temperature. The model explains well the empirical law of Dulong and Petit ($c_V \to 3R$ when $T \to \infty$), but does not explain experimental observations when $T \to 0$.

Debye, using a model of elastic deformation propagation in a crystal lattice, replaces the single frequency by a distribution; c_V is then given by:

$$c_V = \int 3Nk\left(\frac{hv}{kT}\right)^2 \frac{e^{hv/kT}}{(e^{hv/kT} - 1)^2} g(v)\, dv,$$

with $g(v) = 3v^2/v_r^3$, v_r being the frequency of breakage of the bond. By introducing $\Theta = hv_r/k$ and $x = hv/kT$, we obtain ($Nk = R$):

$$c_V = 9R \left(\frac{T}{\Theta} \right)^3 \int_0^{\Theta/T} \frac{x^4 e^x}{(e^x - 1)^2} \, dx.$$

Pure mathematicians will notice at once that $c_V \to 3R$ when $T \to \infty$; after integrating by parts, they will show, when $T \to 0$, that:

$$c_V \approx 3R \frac{4\pi^4}{5} \left(\frac{T}{\Theta} \right)^3.$$

In spite of improvements to Einstein's model and the satisfactory results obtained, Debye's theory is only an approximation which accounts poorly for the observed variations of c_V at intermediate temperatures. Max Born improved the model with an expression of $g(v)$ taking into account the crystal anisotropy.

De Donder, Théophile (1872–1957)—Belgian thermodynamicist, who introduced the notion of *affinity* in 1923. He is also considered as the father of the thermodynamics of irreversible process, very well developed subsequently by the Belgian thermodynamic school around Prigogine.

Degeneracy—Degeneracy, sometimes called the statistical weight of an energy level, represents the number of quantum states having this energy.

When energy levels are close enough to be considered continuous, the degeneracy g_i of the level having an energy ε_i is defined by $g_i = \mathrm{d}n_i/\mathrm{d}\varepsilon_i$, where $\mathrm{d}n_i$ represents the number of levels with an energy between ε_i and $\varepsilon_i + \mathrm{d}\varepsilon_i$. Degeneracy is calculated by quantum-mechanical considerations which cannot be developed here (every one to his trade!). We give only the main results:

TRANSLATIONAL ENERGY: Let v_i, v_y, v_z be the components of the velocity vector for a particle having a mass m. Its energy is:

$$\varepsilon_{v_x, v_y, v_z} = \frac{1}{2} m (v_x^2 + v_y^2 + v_z^2).$$

The number of particles in a volume V, whose velocity components are situated in the interval $(v_x, v_x + \mathrm{d}v_x), (v_y, v_y + \mathrm{d}v_y), (v_z, v_z + \mathrm{d}v_z)$, is:

$$\mathrm{d}N_{v_x, v_y, v_z} = \left(\frac{m^3 V}{h^3} \right) \mathrm{d}v_x \mathrm{d}v_y \mathrm{d}v_z.$$

58 Degree Celsius

ROTATIONAL ENERGY: When a particle rotates around a centre, its rotational energy is:

$$\varepsilon_{rot} = J(J+1)h^2/8\pi^2 I,$$

where I is the moment of inertia of the particle ($I = mr^2$) and J the rotation quantum number.

The degeneracy of a rotational level is: $g_{rot} = 2J + 1$.

When the particle is constrained to move in a plane:

$$\varepsilon_{rot} = m^2 h^2/8\pi^2 I,$$

m being an integer. The degeneracy is then: $g_{rot} = 2$.

VIBRATIONAL ENERGY: This is given by: $\varepsilon_{vib} = (v + 1/2)hv$ where v is then vibrational frequency and v the vibrational quantum number. Vibrational energy levels are not degenerate: $g_{vib} = 1$.

ELECTRONIC ENERGY: For a given electronic level: $\varepsilon_{el} = hv$, where v is the frequency of the transition between the level considered and the fundamental level. The degeneracy of each electronic level is $g_{el} = 2j_i + 1$, where j_i is the quantum number, a whole number or half-integer.

Degree Celsius—The Celsius temperature, expressed in degrees Celsius (not degrees 'centigrade') is defined by:

$$(t/°C) \equiv (T/K) - 273.1500.$$

The unit degree Celsius equals the unit *kelvin*. A temperature difference may be expressed indifferently in °C or in K.

Historically, the Celsius temperature scale was proposed in 1742 by Anders Celsius (1701–1744), Swedish astronomer. The 0°C and 100°C were defined respectively as the melting point of ice and the boiling point of water under atmospheric normal pressure. However, these temperatures are hardly reproducible with great precision because of the solubility of air in liquid water. This is the reason why the more easily reproduced triple point of water, for which by definition $T = 273.1600$ K, or $t = +0.01$°C, has been selected as a fixed point of the thermodynamic scale of temperature.

Remark: With $T = 273.1600$ K for the triple point of water, the calculation, using the *Clapeyron equation*, of the normal melting temperature of water gives $T = 273.1525$ K. The difference of 0.0025 K arises from the air dissolved in liquid water. By convention, the normal

Diagram 59

melting point of water is equal to 273.1500 K exactly. Actually, this temperature is the crystallization temperature of water saturated with air under normal atmospheric pressure, i.e. water having dissolved about 1.3 millimoles of air per litre.

Dew point—The name given to the temperature below which vapour condensation in a wet gas is observed. A knowledge of the dew point allows the vapour content of a gas to be determined.

D

Dew-point curve—On an *isobaric* diagram for a binary mixture, the dew-point curve gives, as a function of the vapour composition, the temperature at which the first liquid drop appears. On an *isothermal* diagram, less often used, the dew-point curve gives, as a function of composition, the pressure at which the first liquid drop appears. On both diagrams, the *boiling* curve gives the composition of the liquid in equilibrium with the vapour.

Let us consider, for a binary solution, p_1° and p_2°, the vapour pressures of the pure constituents 1 and 2, x the composition of the liquid $(x-x_2, 1-x=x_1)$, and y that of the vapour $(y=y_2, 1-y=y_1)$. If the liquid solution is *ideal*, the vapour pressure above the liquid is given by $P=p_1^\circ(1-x)+p_2^\circ x$ and the vapour composition by $y=p_2^\circ x/P$.

The dew-point curve is easy to obtain by eliminating x between both equations. Since p_1° and p_2° are dependent on temperature, a relationship $f(P,T,y)=0$ of the first degree in P and y remains, from which it is possible to draw dew-point curves on an isothermal or on an isobaric diagram. If the liquid solution is not ideal, it is necessary to know the *activity*–composition relations. The two equations allowing the dew-point curve to be constructed become:

$$P=p_1^\circ a_1+p_2^\circ a_2, \qquad y=p_2^\circ a_2/P.$$

In such conditions, the dew-point curve may, with the boiling curve, show an extremum. Both extrema coincide and the mixture is called *azeotropic*.

Diabatic—From the Greek διαβατος, that which can be crossed. Diabatic characterizes a *process* implying heat exchange with the surroundings. Although rarely used, this adjective is more suitable than the two negations: 'not *adiabatic*'!

Diagram—From the Greek διαγραμμα, drawing. A clear diagram is always more pleasant to contemplate than a nice differential equation and in practice provides services more directly accessible to engineers. Mathematicians will comfort themselves by thinking that diagrams are generally constructed from differential equations!

A pure substance may be characterized by variables such as P, V, T, G, H, S, etc. Of these, only two are independent, owing to the equation of state $f(P, V, T) = 0$ and the fact that the thermodynamic quantities G, H, S, etc. are state functions. It is thus possible to represent the behaviour of a pure substance by projections on a two-dimensional diagram.

The description of a binary mixture needs one extra parameter: the composition. Its behaviour is thus represented on a two-dimensional diagram, drawn by assigning to the system two *constraints* which have to be kept in mind.

Diathermanous—A term introduced in 1833 from the Greek $\delta\iota\alpha$, across, and $\theta\varepsilon\rho\mu\alpha\iota\nu\omega$, to heat. A diathermanous boundary is permeable to heat.

Dielectric constant—The name sometimes given to the relative *permittivity*, the ratio of the permittivity of a material to the permittivity of vacuum:

$$D = \varepsilon_r = \varepsilon/\varepsilon_o.$$

Diesel cycle—Rudolf Diesel, German engineer (1858–1913), in 1897 introduced the motor whose theory he developed. It is an internal combustion, compression-ignition engine, with a high compression ratio and working with heavy oils (or gas oil). The diesel cycle is a four-stroke cycle formed of one *isobaric* stroke, one *isochoric* stroke, and two *adiabatic* strokes:

AB: air intake at low pressure
BC: adiabatic compression of air
CD: combustion of gas oil
DE: adiabatic expansion of the combustion products
EA: exhaust of burnt gases.

The diesel cycle can be compared to the *Otto cycle* because it is made up of the same phases, but there the likeness stops. It differs in the fact that only air is introduced in the AB intake stroke, hence allowing a higher compression ratio without fear of detonation. Gas oil, injected at C, burns spontaneously because of the higher temperature of the air, without the need to induce combustion with a spark. Finally, fuel is injected at a controlled rate so that the maximal pressure inside the cylinder is never higher than the maximal pressure of the air at the end of compression. Stroke CD is thus isobaric instead of being isochoric as in the Otto cycle.

D

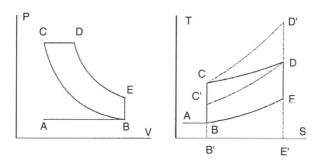

Diesel cycle: P–V (left) and T–S (right) diagrams.

Designating by $Q_1 (>0)$ the heat evolved by the combustion and by $Q_2 (<0)$ the heat given to the surroundings by the exhaust gases, and assuming ideal gases with constant heat capacity, the thermal efficiency of this cycle is found as follows:

$$\eta = \frac{Q_1 + Q_2}{Q_1} = 1 - \frac{mc_V(T_E - T_B)}{mc_P(T_D - T_C)} = 1 - \frac{1}{\gamma}\frac{T_B[(T_E/T_B) - 1]}{T_C[(T_D/T_C) - 1]}$$

It is easy to use the expressions for isentropics to express η as a function of the compression ratio V_B/V_C. The expressions obtained are a little more complex than in the case of the Otto cycle. On a temperature–entropy diagram, the thermal efficiency is simply expressed by:

$$\eta = \text{area (BCDE)/area (B'CDE')}.$$

If we compare, for the same compression ratio, the efficiency of a spark-ignition engine (Otto cycle: BCD'E) with that of a compression-ignition engine (diesel cycle: BCDE), the advantage lies with the Otto cycle, because the area of the cycle is larger. It must not be forgotten however that the two motors do not have the same compression ratio. If we compare the efficiency for the same maximal temperature in the cylinder (temperature at point D), we must look at the areas of the cycles BC′DE for the gasoline motor and BCDE for the diesel motor, which restores the higher efficiency.

Difference—By convention, ΔY represents the difference between the value of Y in a final state and the value of Y in an initial state:

$$\Delta Y = \int_i^f dY = Y_f - Y_i.$$

This definition is consistent with the *sign convention*: if $\Delta Y > 0$, the quantity Y characterizing the system increases. Recall that if dY is an *exact differential*, the function Y exists, it is a state function for the system, and ΔY does not depend on the *path* followed between initial and final states; otherwise function Y does not exist and $\int_i^f dY$ depends on the path followed.

Differential—The differential dF of a function $F(x, y, z)$ is given by:

$$dF = \left(\frac{\partial F}{\partial x}\right)_{y,z} dx + \left(\frac{\partial F}{\partial y}\right)_{x,z} dy + \left(\frac{\partial F}{\partial z}\right)_{x,y} dz.$$

It is always possible to calculate the differential of a given function. The use of differentials allows easy calculation of partial derivatives with implicit functions. When an equation of state is given in the form $F(P, V, T) = 0$, it is possible, by letting $dF = 0$, to obtain all partial derivatives, for instance:

$$\left(\frac{\partial P}{\partial T}\right)_V = -\left(\frac{\partial F}{\partial T}\right)_{P,V} \Big/ \left(\frac{\partial F}{\partial P}\right)_{T,V}.$$

There are six partial derivatives and four relations between them. Three of these relations are of the type: $(\partial P/\partial T)_V = 1/(\partial T/\partial P)_V$. The fourth is the chain rule:

$$\left(\frac{\partial P}{\partial T}\right)_V \cdot \left(\frac{\partial T}{\partial V}\right)_P \cdot \left(\frac{\partial V}{\partial P}\right)_T = -1.$$

Hence a relationship is obtained between the isobaric *expansivity* α_P, the isothermal *compressibility* χ_T and the isochoric *bulk expansion coefficient* β_V:

$$\alpha_P = \beta_V \chi_T P.$$

Diffusion—Transport of matter under the influence of a *chemical potential* gradient. From a phenomenological point of view, diffusion is governed by Fick's two laws:

FICK'S FIRST LAW is similar to Ohm's law for electrical conduction, or to Fourier's law for thermal conduction. It states that the *flux* of matter is proportional to the concentration gradient:

$$J = -D \operatorname{grad} c.$$

It is rigorous only in the case of autodiffusion, because strictly speaking the motive force for diffusion is the chemical potential gradient, not the concentration gradient. In systems with three constituents or more, diffusion can even be observed against the concentration gradient.

FICK'S SECOND LAW is merely the expression of the conservation of matter. It states that the variation of concentration c (in $\operatorname{mol} m^{-3}$) with time is related to the flux of matter J (in $\operatorname{mol} m^{-2} s^{-1}$) by:

$$\frac{dc}{dt} = -D \operatorname{grad} J.$$

D is the diffusion coefficient, expressed in $m^2 s^{-1}$. If D is constant and if the matter diffuses in a single direction, it is possible to write:

$$\frac{\partial c}{\partial t} = D \frac{\partial^2 c}{\partial \ell^2}.$$

THE DIFFUSION COEFFICIENT is related to the thermodynamic properties of the system. Let us consider a pair of species A and B which diffuse countercurrently, normal to a surface Σ, A diffusing from 1 towards 2. At constant volume, in an isolated system:

$$dU = T dS + \sum \mu_i dn_i - 0.$$

The conservation of matter through Σ is expressed by:

$$dn_{i2} = -dn_{i1} = dn_i \quad (i = A, B),$$

whence:

$$T \frac{dS}{dt} = -\Delta\mu_A \frac{dn_A}{dt} - \Delta\mu_B \frac{dn_B}{dt} \quad \text{with } \Delta\mu_i = \mu_{i2} - \mu_{i1}.$$

Let us fix the position of Σ such that $J_A = -J_B$, then let us refer the flux to unit volume by defining: $J_i = d(n_i/V)/dt$. We obtain the *entropy production* per unit volume: $\dot{\sigma} = -J_B \operatorname{grad}(\mu_B - \mu_A)/T$: $\dot{\sigma} > 0$. From the *second law*, $\dot{\sigma}$ may be considered as the product of a flux J_B by a

64 Diffusion coefficient

generalized force X_B: $\dot{\sigma} = J_B X_B > 0$ with:

$$X_B = -\text{grad}\frac{\mu_B - \mu_A}{T} = -\frac{\partial[(\mu_B - \mu_A)/T]}{\partial x}.$$

Using the linear approximation of the thermodynamics of irreversible processes, which postulates proportionality between fluxes and forces, $J_B = L_B X_B$, the entropy creation per unit volume is given by:

$$\dot{\sigma} = L_B X_B^2 = \frac{J_B^2}{L_B} > 0.$$

Here L_B is the mobility coefficient. On the other hand:

$$\mu_B - \mu_A = \left(\frac{\partial G}{\partial x_B}\right),$$

G being the integral molar Gibbs energy and x_B the mole fraction of constituent B. The flux of B can thus be expressed by:

$$J_B = -L_B \cdot \frac{1}{T} \cdot \frac{\partial^2 G}{\partial x_B \partial \ell} = -L_B \cdot \frac{1}{T} \cdot \frac{\partial^2 G}{\partial x_B^2} \cdot \frac{\partial x_B}{\partial \ell}$$

$$J_B = -L_B \cdot \frac{1}{T} \cdot \left(1 + \frac{\partial \ln \gamma_B}{\partial \ln x_B}\right) \cdot \frac{\partial x_B}{\partial c_B} \cdot \frac{\partial c_B}{\partial \ell},$$

which gives the diffusion coefficient as:

$$D_B = -L_B \cdot \frac{1}{T} \cdot \left(1 + \frac{\partial \ln \gamma_B}{\partial \ln x_B}\right) \cdot \frac{\partial x_B}{\partial c_B},$$

where γ_B is the *activity coefficient* of constituent B. The influence of the concentration on the diffusion coefficient is basically due to the thermodynamic factor: $[1 + (\partial \ln \gamma_B / \partial \ln x_B)]$.

Diffusion coefficient—The name given to the *diffusivity* for the *diffusion* of a substance i under the influence of a *chemical potential* gradient. Fick's first law, accurate only to a first approximation, states that the flux of the substance i is proportional to the concentration gradient:

$$J_i = -D_i \,\text{grad}\, c_i.$$

D_i, the diffusion coefficient, is expressed, like every diffusivity, in $m^2 s^{-1}$.

J_i, the flux of i, is expressed in $\text{mol}\,m^{-2}\,s^{-1}$ and c_i, the concentration, in $\text{mol}\,m^{-3}$.

Diffusivity—Fick's first law related to *diffusion* expresses a proportionality, satisfied to a first approximation, between the flux of matter and the concentration gradient: $J_i = -D_i \operatorname{grad} c_i$. The diffusion coefficient D_i is expressed in $m^2 s^{-1}$, provided that concentrations are in $mol\, m^{-3}$ and flux is in $mol\, m^{-2} s^{-1}$. More generally, diffusivity $\mathscr{D}$ is the ratio of a flux of *extensity* to a volume extensity gradient:

$$\text{Diffusivity} = \frac{\text{Flux of extensity}}{\text{Volume extensity gradient}}.$$

D

If $[x]$ represents the unit of extensity, $[x] m^{-2} s^{-1}$ is the flux of extensity and $[x] m^{-4}$ the volume extensity gradient. Thus diffusivity has the same dimensions as diffusion coefficient, that of surface divided by time.

Remark 1: Diffusivity, the ratio of two *intensive* quantities, is an intensive quantity. It may be necessary to recall that a volume extensity, that is, an extensity per unit volume, is a *reduced extensity*; as a consequence, its properties are not those of an extensive quantity. An 'extensity gradient' would be a monstrosity without any physical meaning!

Remark 2: It amounts to the same thing to define the diffusivity by the ratio:

$$\text{Diffusivity} = \frac{\text{Conductivity}}{\text{Volume capacity}}.$$

Indeed, dimensional analysis gives:

$$\frac{\text{Flux of extensity}}{\text{Volume extensity gradient}} = \frac{\text{Conductivity} \times \text{Potential gradient}}{\text{Volume extensity gradient}}$$

$$\text{Diffusivity} = \frac{\text{Conductivity} \times \text{Intensity}}{\text{Volume extensity}} = \frac{\text{Conductivity}}{\text{Volume capacity}}.$$

Remark 3: A flux of extensity J_x is easily linked to the potential (U) gradient and to the volume extensity (Γ) gradient by means of conductivity (λ) and diffusivity $(\mathscr{D})$:

$$J_x = -\lambda \operatorname{grad} U = -\mathscr{D} \operatorname{grad} \Gamma.$$

By introducing the volume capacity $C = \Gamma/U$, the ratio of volume extensity to potential, we again find the expression for diffusivity:

$$\mathscr{D} = \lambda/C.$$

66 Dilution

The most common diffusivities are as follows:

DIFFUSION COEFFICIENT D_i defined by $J_i = -D_i \operatorname{grad} c_i$. In this case, diffusivity identifies with conductivity, because the tension c_i is also the volume extensity. The volume capacity thus equals unity.

THERMAL DIFFUSIVITY α, defined from the flux of heat J_Q, is directly linked to the thermal conductivity λ:

$$J_Q = -\lambda \operatorname{grad} T = -\alpha \operatorname{grad} (c_P \rho T);$$

$c_P \cdot \rho$ is the heat capacity of unit volume. We again find the general relation between diffusivity and conductivity: $\lambda = \alpha c_P \rho$.

DIFFUSIVITY OF MOMENTUM v identifies with *kinematic viscosity*. Defined from the flux of momentum J_p, it is directly related to the conductivity of momentum η, better known under the name of *dynamic viscosity*:

$$J_p = -\eta \operatorname{grad} v = -v \operatorname{grad} (\rho v),$$

where v is the velocity and ρ the mass per unit volume; ρv thus represents the momentum per unit volume. The relation between the dynamic viscosity η and the kinematic viscosity v is thus straightforward: $\eta = v\rho$.

ELECTRICAL DIFFUSIVITY $\mathscr{D}$ is derived from the same relations:

$$J_e = -\lambda E = -\mathscr{D} \operatorname{grad} (n_e/V)$$

J_e: flux of electric charges, in $\mathrm{A\,m^{-2}}$
λ: electrical conductivity, in $\mathrm{A^2\,m^{-3}\,kg^{-1}\,s^3}$
E: potential gradient, or *electric field*, in $\mathrm{kg\,m\,s^{-3}\,A^{-1}}$
n_e/V: charge per unit volume, in $\mathrm{A\,s\,m^{-3}}$
$\mathscr{D}$: electrical diffusivity, always in $\mathrm{m^2\,s^{-1}}$
C: volume capacity, in $\mathrm{F\,m^{-3}}$, or in $\mathrm{A^2\,m^{-5}\,kg^{-1}\,s^4}$
We again find the relation: $\mathscr{D} = \lambda/C$.

Dilution—Addition of *solvent*. Ostwald's dilution law, a direct consequence of *Le Chatelier's law*, states that the *dissociation coefficient* for an electrolyte increases with dilution:

$$A_m B_p \rightarrow m A^{p-} + p B^{m+} \quad \text{with} \quad K(T) = [a(A^{p-})]^m \cdot [a(B^{m+})]^p / a(A_m B_p).$$

With the introduction of *activity coefficients* $\gamma_i = a_i/c_i$, which can be supposed constant if the solutions remain dilute, the product of concentrations $[c(A^{p-})]^m \cdot [c(B^{m+})]^p / c(A_m B_p)$ is constant. It can be expressed as a function of c_o, the initial concentration of $A_m B_p$, and of the dissociation coefficient α:

$$(\alpha m)^m (\alpha p)^p \cdot c_o^{m+p-1}/(1-\alpha) = \text{const.}$$

The constant being independent of the initial concentration c_o, a decrease in c_o, that is, an increase in dilution, results in an increase in α, thus displacing the equilibrium towards dissociation of the electrolyte.

D

Dirac delta function—The name given to the function $\delta(x)$ defined by:

$$\delta(x) = 0 \text{ for } x \neq 0 \quad \text{and} \quad \int_{-\infty}^{+\infty} \delta(x)\,dx = 1.$$

It satisfies the fundamental relationship:

$$\int_{-\infty}^{+\infty} \varphi(x)\delta(x-x_o)\,dx = \varphi(x_o).$$

Its advantage arises from the fact that it allows discontinuous functions to be considered as continuous ones. The step function ($y=0$ for $x<0$ and $y=1$ for $x>0$) for instance is discontinuous, and hence undifferentiable at $x=0$. However, $y'(x)=\delta(x)$. Thus:

$$y(x) = \int_{-\infty}^{x} \delta(x)\,dx = \begin{cases} 1 \text{ for } x>0 \\ 0 \text{ for } x<0. \end{cases}$$

The Dirac δ function must not be confused with the *Kronecker delta* δ_{ij} ($\delta_{ij}=1$ for $i=j$ and $\delta_{ij}=0$ for $i\neq j$).

Dirac, Paul Adrien Maurice (1902–1984)—English physicist, who made a major contribution in physics by devising a new version of quantum mechanics. He developed, with Enrico Fermi, statistics modelling the behaviour of *fermions*. Nobel prize for physics in 1933.

Disorder—The disorder of a system is characterized by the quantity of information needed to describe it. A *crystal* is a well-ordered system, a knowledge of the positions of the atoms in an elemental lattice allowing the whole crystal to be reconstituted; a *liquid* is more disordered because its complete description would need a knowledge at a given time of the position and velocity of every molecule. The highest disorder is represented by *gases*, which, in contrast to liquids, occupy the whole volume available. The word 'gas' is derived from the Latin word *chaos* ($\chi\alpha o\varsigma$ in Greek).

68 Dissipative structure

Transitions *solid* → liquid and liquid → *vapour* are characterized by an increase in *entropy*, because:

$$\Delta_{\text{fus}}S = \Delta_{\text{fus}}H/T_{\text{fus}} > 0 \quad \text{and} \quad \Delta_{\text{vap}}S = \Delta_{\text{vap}}H/T_{\text{vap}} > 0.$$

This result allows the statement: 'entropy is a measure of disorder'. This is not untrue of course, but we must be careful not to believe that we understand thermodynamics after having made such a discovery!

In a solid solution with only one kind of site, disorder will be complete if, in the neighbourhood of an atom A, the probability of finding another atom i ($i = A, B$) equals x_i, the mole fraction of i. If there is no interaction between A and B, the solid solution will be *ideal* and its molar entropy given by:

$$S = x_A S_A + x_B S_B - R(x_A \ln x_A + x_B \ln x_B).$$

Dissipative structure—A term proposed in 1969 by Prigogine to describe the spontaneous appearance of ordered structures in the non-linear domain, far from equilibrium. Irreversibility may then be the source of order evidenced by the organization of space or time on a macroscopic scale, of the order of centimetres or minutes. Organization may also be space–time, with the appearance of chemical waves. The *Belousov–Zhabotinskii reaction* provides a classical example of a dissipative structure.

Dissociation coefficient—This concept applies to a substance, more often an electrolyte, capable of being dissociated in a solvent. By definition, the dissociation coefficient α is the ratio:

$$\alpha \equiv \frac{\text{quantity of dissociated substance}}{\text{quantity of initial substance}}.$$

The dissociation coefficient is an experimental quantity, directly related to the dissociation *equilibrium constant*. However, it is not a characteristic of the behaviour of the substance dissolved, because, as opposed to the equilibrium constant, it depends on the *dilution*.

Distillation—A separation technique for a mixture of two or more liquids based on their difference in volatility. For a binary mixture, the distillation curve is the isobar giving the liquid composition as a function of the vapour composition. Each point of the curve corresponds to a distillation temperature for which the saturated vapour pressure of the mixture equals the distillation pressure.

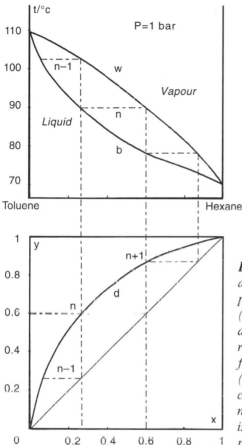

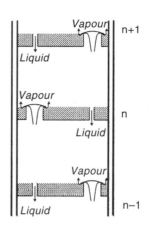

Distillation: *Construction of the distillation curve (d) from dew-point (w) and bubble-point (b) curves. In a theoretical distillation column, equilibrium is assumed to be realized for each plate. Liquid on the (n+1)th plate has the same composition as vapour on the nth plate. A plate is characterized by its temperature and a point on the distillation curve.*

Let p_1° and p_2° be the vapour pressures of pure substances 1 and 2, x being the liquid composition ($x=x_2; 1-x=x_1$) and y that of the vapour ($y=y_2; 1-y=y_1$). With the assumption of an *ideal solution*, the vapour pressures p_1 and p_2 above the liquid are given by:

$$p_1 = p_1^{\circ}(1-x), \qquad p_2 = p_2^{\circ}x.$$

If P is the imposed distillation pressure, the distillation temperature for the mixture of composition x is obtained by solving the equation:

$$P = p_1 + p_2 = p_1^{\circ}(1-x) + p_2^{\circ}x.$$

Hence the composition of the vapour in equilibrium with the liquid is:

$$y = \frac{p_2^\circ x}{p_1^\circ (1-x) + p_2^\circ x},$$

representing the equation of the distillation curve. If the ratio p_1°/p_2° is independent of the temperature, the distillation curve is a hyperbola. Generally the ratio is a function of T and it is necessary to use activities.

To each point of the distillation curve corresponds a boiling temperature calculated from the equation: $P = p_1^\circ a_1 + p_2^\circ a_2$, P being the imposed distillation pressure. The composition y of the vapour in equilibrium with the liquid is then given by: $y = p_2^\circ a_2/P$. In such calculations, the vapour is considered as a mixture of ideal gases.

The distillation curve of an ideal solution always lies above the bisectrix $y = x$ if constituent 2 is more volatile than no. 1 or below the bisectrix if no. 2 is the less volatile. In a real solution, the distillation curve may cross the bisectrix. The mixture is then said to be *azeotropic*.

Distribution—If a stochastic variable X takes different values $\{x_1, x_2, \ldots, x_i, \ldots\}$, $\{x_i\}$ having a probability $f(x_i) > 0$ [with $\sum f(x_i) = 1$], the ensemble of the values $f(x_i)$ forms the probability distribution.

With continuous variables, it is possible to define the probability of observing, for the variable X, a value situated between a and b:

$$P(a \leqslant x \leqslant b) = \int_a^b f_X(x)\,dx \text{ with } f_X(x) \geqslant 0 \quad \text{and} \quad \int_0^1 f_X(x)\,dx = 1.$$

Generally, the probability $f(x_i)$ is unknown for a discontinuous variable X and the *probability density* $f_X(x)$ is unknown for a continuous variable X. However, the *moments* of X are more easily available.

Droplet—A drop of rain has a diameter of about 1 mm, but its thermodynamic properties are no different from those of water; droplets whose diameter is of the order of nanometres are characterized by a *saturated vapour pressure* which depends on the radius of curvature.

Consider a drop with radius r and volume V, subjected to its internal pressure p. The surface tension at the liquid–vapour interface is Γ and the surface area of the drop is $\mathscr{S}$. If the volume of the drop increases by dV, work is supplied which is used to increase the surface by $d\mathscr{S}$:

$$p\,dV = \Gamma\,d\mathscr{S}$$

$$p\,d(4\pi r^3/3) = \Gamma\,d(4\pi r^2).$$

Hence the Laplace relationship: $p = 2\Gamma/r$.

The *Gibbs energy* for a drop consisting of n moles is given by:

$$G = F + PV = nf + 4\pi r^2 \Gamma + p_r nv.$$

f: molar *Helmholtz energy*, neglecting surface tension effects
Γ: surface tension (Helmholtz energy per unit surface)
v: molar volume of the liquid ($nv = 4\pi r^3/3$)
p_r: vapour pressure of a drop whose radius of curvature is r.

The *chemical potentials* of the liquid and vapour are respectively:

$$\mu_1 = \left(\frac{\partial G}{\partial n}\right)_{T,P} = f + \frac{2v\Gamma}{r} + vp_r$$

$$\mu_v = \mu_v^{\circ} + RT \ln p_r .$$

The liquid–vapour equilibrium is expressed by the equality $\mu_1 = \mu_v$:

$$\mu_v^{\circ} + RT \ln p_r = f + (2v\Gamma/r) + vp_r$$

This relationship is also valid for a planar surface ($r = \infty$):

$$\mu_v^{\circ} + RT \ln p_{\infty} = f + vp_{\infty}.$$

By subtracting both sides:

$$RT \ln(p_r/p_{\infty}) = v[(2\Gamma/r) + p_r - p_{\infty}],$$

an equation which can be solved if $(p_r - p_{\infty}) \ll p = 2\Gamma/r$:

$$p_r/p_{\infty} = \exp(2\Gamma v/RTr).$$

By introducing M, the molar mass of the liquid, and ρ, the density:

$$p_r/p_{\infty} = \exp(2\Gamma M/RT\rho r).$$

The vapour pressure of a droplet increases when its radius of curvature shrinks, but this effect is noticeable only when $r \lesssim 1\,\mu m$. The reasoning may also be applied to the solubility s of a *solute* in a *solvent*: it depends on the radius of curvature of the solid solute, according to the same relation:

$$s_r/s_\infty = \exp(2\Gamma M/RT\rho r),$$

where M and ρ are molar mass and volumetric mass of the solid solute.

CHARGED DROPLETS: The same reasoning applies when droplets have a surface electric charge q; then an electrostatic term $-q^2/8\pi\varepsilon_\circ r$ must be added to the energies F and G. If $(p_r - p_\infty) \ll p = 2\Gamma/r$:

$$RT \ln \frac{p_r}{p_\infty} = v\left(\frac{2\Gamma}{r} - \frac{q^2}{8\pi\varepsilon_\circ r^4}\right).$$

When the radius of curvature shrinks, the vapour pressure goes through a maximum, returns to its former value p_∞ when $\Gamma = q^2/16\pi\varepsilon_\circ r$, and tends towards 0 when r does so. Ionization of a supersaturated vapour promotes condensation. This is the working principle of a Wilson chamber: the passage of an ionizing particle through supersaturated vapour induces the formation of a string of drops which may be photographed.

Dufour effect—Discovered in 1872, the Dufour effect is an effect of *thermodiffusion* characterized by the appearance of a temperature gradient under the influence of a concentration gradient. In practice, this effect is noticeable only with gases.

Duhem, Pierre (1861–1916)—French philosopher and thermodynamicist. His works are in the spirit of Josiah Willard Gibbs. For his treatise on chemical mechanics, he is considered as the founder of energetics. In epistemology, he expressed his ideas in *Physical theory, its object, and structure* (1906).

E

Ebulliometry—Measurement of the initial boiling point of a mixture of *solvent* (index 1) plus *solutes*. If the solutes are not volatile, the equilibrium between the gaseous solvent (index 1v) and the solvent in solution (index 1l) at a temperature T is expressed by equal *chemical potentials*: $\mu_{1v} = \mu_{1l}$, whence:

$$\mu_{1v}^\circ + RT \ln(f_1/f_1^\circ) = \mu_{1l}^\circ + RT \ln a_{1l}.$$

If boiling occurs under atmospheric pressure, it is then possible to select this pressure (*fugacity f_1*) as the *standard pressure* (fugacity f_1°)

and therefore $f_1 = f_1^\circ$, whence:

$$a_{1l} = \exp[(\mu_{1v}^\circ - \mu_{1l}^\circ)/RT] = \exp(+\Delta_{vap}G_1^\circ/RT),$$

with $\Delta_{vap}G_1^\circ = (\mu_{1v}^\circ - \mu_{1l}^\circ)$, the Gibbs energy of vaporization for pure solvent. Putting:

$$\Delta_{vap}G_1^\circ \approx \Delta_{vap}H_1^\circ(1 - T/T_{b1}),$$

implicitly neglecting the difference between the heat capacities of the liquid and the vapour, and assuming that the solution is dilute enough for the solvent to obey *Raoult's law* ($a_{1l} = x_{1l}$), we have:

$$x_{1l} = \exp\left[\frac{\Delta_{vap}H_1^\circ}{R}\left(\frac{1}{T} - \frac{1}{T_{b1}}\right)\right].$$

This relationship gives the initial boiling temperature T of a solvent in which a non-volatile solute is dissolved; x_{1l} is the mole fraction of the solvent after taking into account the possible dissociation of the dissolved solutes, and $\Delta_{vap}H_1^\circ$ is the *enthalpy* of vaporization of pure solvent at its boiling temperature T_{b1} under atmospheric pressure $p_1 = p_1^\circ$.

Efficiency—Generally, efficiency is measured by the ratio of the useful energy delivered to the energy supplied.

A thermal motor which takes a heat quantity ($Q_2 > 0$) from a hot source for returning ($Q_1 < 0$) to a cold sink and converting the difference into useful work ($W < 0$) has an efficiency defined by:

$$\eta = -W/Q_2 < 1.$$

A refrigerator which absorbs work ($W > 0$) by taking a heat quantity ($Q_1 > 0$) from a cold source and supplying a heat quantity ($Q_2 < 0$) to a hot sink has an 'efficiency' defined by:

$$\eta = +Q_1/W > 1.$$

A heat pump working as a refrigerator, but with the aim of supplying heat to a hot sink, has an 'efficiency' defined by:

$$\eta = -Q_2/W.$$

For a refrigerator and a heat pump these 'efficiencies' are often termed 'work ratios'.

Ehrenfest equations—When a compound undergoes a polymorphic transition, the equilibrium curve $f(P,T) = 0$ is obtained by integration of *Clapeyron equation*: $(\mathrm{d}P/\mathrm{d}T)_{eq} = \Delta_{tr} H/T \Delta_{tr} V$, but the technique is not applicable to a second-*order* transition, because $\Delta_{tr} H$ and $\Delta_{tr} V$ are zero for such transitions. Paul Ehrenfest, Austrian physicist (1880–1933), nevertheless obtained relations by using the fact that the second derivatives of the *Gibbs energy*, that is, the first derivatives of the V and H functions, are discontinuous:

- From the volume V: $V_1 = V_2 \Rightarrow \mathrm{d}V_1 = \mathrm{d}V_2$;

$$\left(\frac{\partial V_1}{\partial T}\right)_P \mathrm{d}T + \left(\frac{\partial V_1}{\partial P}\right)_T \mathrm{d}P = \left(\frac{\partial V_2}{\partial T}\right)_P \mathrm{d}T + \left(\frac{\partial V_2}{\partial P}\right)_T \mathrm{d}P.$$

By using the isobaric *expansivity* α_P and the isothermal *compressibility* χ_T, one obtains the first Ehrenfest equation:

$$\left(\frac{\partial P}{\partial T}\right)_{eq} = \frac{\alpha_{P2} - \alpha_{P1}}{\chi_{T2} - \chi_{T1}}.$$

- From the *entropy* S: $S_1 = S_2 \Rightarrow \mathrm{d}S_1 = \mathrm{d}S_2$:

$$\left(\frac{\partial S_1}{\partial T}\right)_P \mathrm{d}T + \left(\frac{\partial S_1}{\partial P}\right)_T \mathrm{d}P = \left(\frac{\partial S_2}{\partial T}\right)_P \mathrm{d}T + \left(\frac{\partial S_2}{\partial P}\right)_T \mathrm{d}P.$$

Now:

$$\left(\frac{\partial S}{\partial T}\right)_P = \frac{c_P}{T} \quad \text{and} \quad \left(\frac{\partial S}{\partial P}\right)_T = -\frac{\partial^2 G}{\partial T \, \partial P} = -\alpha_P V,$$

whence the second Ehrenfest equation:

$$\left(\frac{\partial P}{\partial T}\right)_{eq} = \frac{1}{TV} \cdot \frac{c_{P2} - c_{P1}}{\alpha_{P2} - \alpha_{P1}}.$$

Einstein, Albert (1879–1955)—American physicist, born in Germany, surely one of the most creative intellects in human history. He is not only the father of *relativity*, which in 1905 revealed to the world the equivalence of mass and energy; he also presented the theory of Brownian movement, an explanation of the photoelectric effect for which he won the Nobel prize for physics in 1921, a model for the heat capacity of solids, and gases, and the basis, with Bose, of statistics allowing the description of the behaviour of *bosons*.

Electric energy—Elementary *work* exchanged with surroundings when the potential of an electric charge $\mathrm{d}\mathcal{Q}$ varies by $\Delta\mathcal{E}$ is given by:

$$\mathrm{d}W = \Delta\mathcal{E} \, \mathrm{d}\mathcal{Q}.$$

A more general relation exists:

$$\Delta W = \iiint_V \int_D E \, dV \, dD,$$

where E is the *electric field* and dD the increment of the *electric excitation* vector in the volume dV. This expression simplifies if the electric field does not depend on the coordinates:

$$dW = VE \, dD.$$

Excitation can also be expressed as a function of the electric field:

$$D = \varepsilon_o E + P = \varepsilon E.$$

In vacuum, $(D = \varepsilon_o E)$, ε_o is a scalar and the work exchanged with surroundings when the electric field varies from 0 to E is given by:

$$\frac{dW}{dV} = \int_0^E \varepsilon_o E \, dE = \frac{1}{2} \varepsilon_o E^2;$$

$\varepsilon_o E^2/2$, in J m^{-3}, is the energy density of vacuum in the electric field E.

In the presence of a homogeneous electric field, ε is a scalar; it is then possible to express the energy density by $dW/dV = \varepsilon E^2/2$.

When a dielectric is introduced into a region initially devoid of matter, but with a pre-existing electric field, the work exchanged by the dielectric is expressed by:

$$dW = VE \, dD - V\varepsilon_o E \, dE = VE \, dP.$$

In summary, electric energy is made up of two contributions:

• an external energy which is the electrostatic energy of vacuum and which is expressed by:

$$\Delta W = \iiint_V \int_E \varepsilon_o E \, dV \, dE;$$

• an internal energy which is the electrostatic energy given to a dielectric introduced into an electric field and which is expressed by:

$$\Delta W = \iiint_V \int_P E \, dV \, dP.$$

Electric excitation—The *electric field* inside a plane capacitor with a charge $\mathcal{Q}$ and a surface $\mathcal{S}$ is given by: $E = \mathcal{Q}/\varepsilon_o \mathcal{S}$ (ε_o is the *permittivity* of vacuum). By introducing the volume V of the capacitor and the distance d between the electrodes, $\varepsilon_o E = 2d/V$, which represents an *electric moment* per unit volume. In the presence of a dielectric, $(\varepsilon_o \times E)$ is the resultant of two contributions: that of the charge on the electrodes and that of the *polarization P* of the material:

$$\varepsilon_o E = (2d/V) - P;$$

$(\varepsilon_o E + P)$ is related to the capacitor charge in the presence of a dielectric as $(\varepsilon_o \times E)$ was related to the capacitor charge in vacuum. Generally, E and P are vectors, not collinear. By definition:

$$D \equiv \varepsilon_o E + P.$$

The vector D is called 'electric excitation'. It is still encountered under its historical appellation 'electric displacement'. In the SI system, electric excitation is expressed in $A\,s\,m^{-2}$.

Electric field—More precisely the 'electric field strength', following Sommerfeld's suggestion, this is defined as the ratio between the force applied to an electric charge and the value of this charge. In the SI system, an electric field is expressed in $V\,m^{-1}$; it is a vector quantity.

The force exerted by an electric charge q_a on a charge q_b at a distance d is given by:

$$F = \frac{1}{4\pi\varepsilon_o} \frac{q_a q_b}{d^2} r_u,$$

where $\varepsilon_o = 1/\mu_0 c^2 = 1/4\pi \times 10^{-7} c^2 \approx 8.854 \times 10^{-12}\,F\,m^{-1}$ is the *permittivity* of vacuum, and r_u is the unit vector directed from the source point (a) towards point (b) where F is calculated.

The electric field created by the charge q_a is then:

$$E = \frac{q_a}{4\pi\varepsilon_o d^2} r_u.$$

The work needed to move a charge in a field is:

$$\Delta W = - \int_A^B q E \cdot dx.$$

Over a cycle $\oint dW = 0$, the electric field generated by a distribution of charges is conservative: $E = - \operatorname{grad} \Phi$.

The potential Φ is a scalar function of the position; the minus sign takes into account the fact that field is directed towards decreasing potentials.

The electric field generated by a distribution of charges is thus irrotational: rot $E = 0$; however, in the presence of a variation of the *magnetic field*:

$$\text{rot } E = -d\mathscr{B}/dt.$$

Electric moment—Two electric charges $\mathcal{Q}$, equal and with opposite signs, separated by a distance d, form a dipole whose electric moment (or 'dipolar electric moment') equals the product $\mathcal{Q}d$. It is a vector, directed from the negative charge towards the positive charge. The unit of electric moment in the SI system is the coulomb–metre ($1\,C\,m = 1\,A\,s\,m$). The Debye, a unit discouraged but sometimes encountered for expressing the electric moment of a molecule, equals $3.336 \times 10^{-30}\,C\,m$.

E

Electric potential—A quantity of *tension*, conjugate with the *extensity* quantity of electricity. The unit of electric potential in the system SI is the volt:

$$1\,V = 1\,m^2\,kg\,s^{-3}\,A^{-1}.$$

Electrochemical affinity—The electrochemical affinity of a reaction performed in a galvanic cell, a concept introduced by Prigogine, is defined by:

$$\tilde{\mathscr{A}} = \mathscr{A} + n\mathscr{F}(\mathscr{E}_1 - \mathscr{E}_2) = -\sum v_i \tilde{\mu}_i.$$

$\mathscr{A}$: *affinity* of the *chemical reaction* $\sum v_i M_i = 0$
$\mathscr{E}$: *electrode potential*
n: number of electrons involved in writing down the reaction
$\tilde{\mu}_i$: *electrochemical potential*: $\tilde{\mu}_i = \mu_i + z_i \mathscr{F}\mathscr{E}$.

z_i, $\mathscr{F}$ and $\mathscr{E}$ represent respectively the charge of the ion i, the Faraday constant and the *Galvani potential* (or internal potential).

Note: Electrochemical affinity is merely the new name given to the affinity of a chemical reaction performed in a galvanic cell. The new 'affinity' of the reaction:

$$\mathscr{A} = \tilde{\mathscr{A}} + n\mathscr{F}(\mathscr{E}_2 - \mathscr{E}_1)$$

corresponds in fact to the affinity defined by De Donder minus an electrochemical term. A non-specialist must pay attention to this lack of rigour in terminology.

Electrochemical potential—When a *system* exchanges electrical work with the *surroundings*, it is traditional among electrochemists to modify the symbol for the thermodynamic functions associated with the system by applying a tilde, $\sim$, to them. For instance:

$$\mathrm{d}\tilde{G} = -S\,\mathrm{d}T + V\,\mathrm{d}P + \varphi\,\mathrm{d}\mathcal{Q} + \sum \mu_i\,\mathrm{d}n_i + \cdots,$$

with $\mathrm{d}W_{el} = \varphi\,\mathrm{d}\mathcal{Q} = \varphi\sum z_i\mathcal{F}\,\mathrm{d}n_i$, z_i being the electric charge of the ion i and $\mathcal{F}$ the Faraday constant. φ is the potential of a point inside the system (inner potential or *Galvani potential*), which must not be confused with the outer potential (or *Volta potential*), the potential of a point on the surface.

This change of notation is not harmless, because the quantity so modified is labelled 'electrochemical' although nothing in its nature has changed. The electrochemical Gibbs energy is divided into two components:

$$\mathrm{d}\tilde{G}_{\text{electrochemical}} = \mathrm{d}G_{\text{chemical}} + \mathrm{d}W_{\text{electrical}},$$

which may be simplified by writing: $\mathrm{d}\tilde{G} = \mathrm{d}G + \mathrm{d}W_{el}$ or, worse, $\tilde{G} = G + W_{el}$ (!!). Such a presentation, usual in electrochemistry, is open to criticism, because neither G, nor W_{el}, is a state function! The electrochemical potential $\tilde{\mu}_i$ is defined as the electrochemical partial molar Gibbs energy:

$$\tilde{\mu}_i = \left(\frac{\partial \tilde{G}}{\partial n_i}\right)_{T,P,n_{j\neq i}} = \left(\frac{\partial G}{\partial n_i}\right)_{T,P,n_{j\neq i}} + \left(\frac{\partial W_{el}}{\partial n_i}\right)_{T,P,n_{j\neq i}},$$

whence $\tilde{\mu}_i = \mu_i + z_i\mathcal{F}\varphi$. Actually, the electrochemical potential is the new name for the *chemical potential* divided into two components: its chemical component μ_i and its electrical component $z_i\mathcal{F}\varphi$. One must be aware of the implications of this new jargon: only the electrochemical potential (former chemical potential) is a quantity of *tension*, because the equilibrium of a constituent i between two phases is given by the equality of electrochemical potentials, and not by the equality of its chemical component μ_i.

Electrode potential—Let us consider the *galvanic cell*:

(1) Pt, $H_2(f=1\,\text{bar})/H^+(a=1)$//solution/electrode (2).

The potential of electrode (2) is, by convention, equal to the electromotive force of the cell. The potential of electrode (1), called the 'standard hydrogen electrode', by convention equals zero. Indeed, although it is possible to measure a potential difference, it is not possible to have direct experimental access to the potential of an electrode. To let $\mathscr{E}°=0$ for the standard hydrogen electrode amounts to putting $\Delta G°=0$ for the reaction:

$$2H^+ + 2e^- \to H_2.$$

In 1983, the standard pressure was changed from 101 325 Pa (1 atm) to 100 000 Pa (1 bar). Rigorously, this results at 298 K in a difference of 0.17 mV in the value of an electrode potential, a difference smaller than experimental inaccuracies, and which need not bother the user of electrochemical data.

E

If, in a galvanic cell, the overall reaction:

$$Ox_2 + Red_1 \to Ox_1 + Red_2 \qquad (r)$$

occurs, the electromotive force of the cell $\Delta\mathscr{E} = \mathscr{E}_2 - \mathscr{E}_1$ is linked to the *Gibbs energy of reaction* by:

$$\Delta_r G = -n\mathscr{F}\Delta\mathscr{E} = \Delta_r G° + RT \ln \frac{[Ox_1][Red_2]}{[Ox_2][Red_1]}.$$

If each constituent of the reaction is in its *standard state*: $\Delta_r G = \Delta_r G°$ and $\Delta\mathscr{E} = \Delta\mathscr{E}°$, the standard electromotive force of the cell, whence:

$$n\mathscr{F}\Delta\mathscr{E} = n\mathscr{F}\Delta\mathscr{E}° + RT \ln \frac{[Ox_2][Red_1]}{[Ox_1][Red_2]}.$$

By separating the terms relating to the two electrodes:

$$n\mathscr{F}(\mathscr{E}_2 - \mathscr{E}_1) = n\mathscr{F}(\mathscr{E}_2° - \mathscr{E}_1°) + RT \ln \frac{[Ox_2]}{[Red_2]} - RT \ln \frac{[Ox_1]}{[Red_1]},$$

we obtain the *Nernst relation* for the electrode potential:

$$\mathscr{E} = \mathscr{E}° + \frac{RT}{n\mathscr{F}} \ln \frac{[Ox]}{[Red]}.$$

$\mathscr{E}°$ is the *electrode standard potential*, n the number of electrons exchanged at the electrode in writing down the reaction, and $\mathscr{F}$ the Faraday constant. [Ox] and [Red] symbolize the products of the activities for the species reacting at the electrode. For instance, for the half-reaction:

$$\sum \alpha_i A_i + ne^- \to \sum \beta_i B_i$$

$$\mathscr{E} = \mathscr{E}° + \frac{RT}{n\mathscr{F}} \ln \frac{\prod [a_{A_i}^{\alpha_i}]}{\prod [a_{B_i}^{\beta_i}]}.$$

80 Electrode standard potential

Remark: An electrode potential is obviously independent of the balanced equation for the reaction at the electrode. To balance an equation with only one electron results in removing n from the prelogarithmic term and raising the argument of the logarithm to the power $1/n$.

Electrode standard potential—The name given to the *electrode potential* when every constituent taking part in the reaction at the electrode is in its *standard state*. Electrode standard potentials, generally given in tables, allow equilibrium constants to be calculated for redox reactions. Electrode standard potential is related to the *standard Gibbs energy* of the half-reaction $Ox + ne^- \rightarrow Red$, by: $\mathscr{E}^\circ = -\Delta_r G^\circ / n\mathscr{F}$. By convention, $\Delta_r G^\circ = 0$, thus $\mathscr{E}^\circ = 0$ for the half-reaction: $2H^+ + 2e^- \rightarrow H_2$.

For a *galvanic cell*: Electrode 1/ Solution 1// Solution 2/ Electrode 2, which operates according to the two half-reactions:

$$Ox_1 + ne^- \rightarrow Red_1 \quad (r1), \qquad Ox_2 + ne^- \rightarrow Red_2 \quad (r2),$$

and thus by the overall reaction:

$$Ox_2 + Red_1 \rightarrow Ox_1 + Red_2 \quad (r),$$

the standard electromotive force $\Delta\mathscr{E}^\circ = \mathscr{E}_2^\circ - \mathscr{E}_1^\circ$ is related to the standard Gibbs energy of the reaction $\Delta_r G^\circ = \Delta_{r2} G^\circ - \Delta_{r1} G^\circ$:

$$\Delta_r G^\circ = -n\mathscr{F}\Delta\mathscr{E}^\circ = -RT \ln K$$

$$K = \exp\left[\frac{n\mathscr{F}}{RT}(\mathscr{E}_2^\circ - \mathscr{E}_1^\circ) \right].$$

Note: It is very easy to introduce a mistake in the sign of the exponent when using this equation. It is wise to note that $\mathscr{E}_2^\circ > \mathscr{E}_1^\circ \Rightarrow K > 1$: the Ox_2 form is then more oxidizing than the Ox_1 form, whence a reaction favoured in the direction in which the Ox_2 form is reduced.

Element—A general term, often used instead of *simple substance* if one does not take into account the state of aggregation. Graphite and diamond represent two different simple substances, but also two allotropic forms of the same element carbon. There exist 81 elements having at least one stable isotope; in 1997, the Mendeleev periodic table identified 112 elements.

Elementary charge—$e = 1.602\,177\,33 \times 10^{-19}$ C.

Elementary reaction—A *chemical reaction*:

$$a_1 A_1 + a_2 A_2 + \cdots \rightarrow b_1 B_1 + b_2 B_2 + \cdots$$

is called elementary when the *reaction rate* obeys the relation:

$$v = \frac{d\xi}{dt} = k_1 \prod a_{A_i}^{a_i} - k_{-1} \prod a_{B_i}^{b_i},$$

where k_1 and k_{-1} are the 'rate constants', dependent on T and P but independent of concentrations. When a reaction is elementary, the sums of the stoichiometric numbers $\sum a_i$ and $\sum b_i$ generally equal 1 or 2, rarely 3. An overall reaction is seldom elementary; however, it is often possible to find a mechanism which breaks it down into several steps, each of which constitutes an elementary reaction.

E

Ellingham diagram—A diagram of $\mu_X = RT \ln P_X$ versus T. Most often, X represents dioxygen, in which case such a diagram allows the relative stabilities of metals and oxides to be visualized. However, X may represent any gas. To quote by way of example the commonest ones, there are Ellingham diagrams for sulfides, nitrides, chlorides, and carbonates.

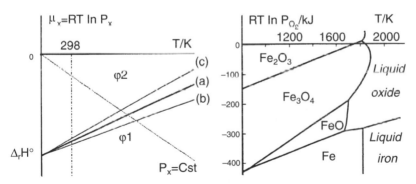

Ellingham diagram: On the left, the monovariant equilibrium curves $\langle\varphi_1\rangle + X \rightarrow \langle\varphi_2\rangle$ *obey the equation:* $\mu_X = \Delta G° + RT \ln [a(\varphi_2)/a(\varphi_1)]$; *they separate the monophase fields* φ_1 *and* φ_2, *the curves (a), (b), and (c) corresponding to ratios* $a(\varphi_2)/a(\varphi_1)$ *respectively equal to, lower than, and higher than unity. On the right, the monophase fields of iron and its oxides in the liquid and solid states.*

82 Ellingham diagram

A pure component φ with a single phase is stable into a domain in which T and P_X can vary independently: the equilibrium is bivariant. Two phases $\langle\varphi1\rangle$ and $\langle\varphi2\rangle$ coexist at equilibrium along a border: the equilibrium is then monovariant: $\langle\varphi1\rangle + X \rightarrow \langle\varphi2\rangle$.

If the phases $\langle\varphi1\rangle$ and $\langle\varphi2\rangle$ at equilibrium are in their *standard state*, the equation of the border separating the two fields is given straightforwardly by:

$$\mu_X = \Delta_r G^\circ = -RT \ln K = +RT \ln P_X.$$

By neglecting heat capacities ($\Delta c_P \approx 0$), $\Delta_r G^\circ$ is a linear function of T, whose ordinate at the origin equals $\Delta_r H^\circ$ and whose slope is $(-\Delta_r S^\circ)$, with $\Delta_r S^\circ = S^\circ(\langle\varphi2\rangle) - S^\circ(\langle\varphi1\rangle) - S^\circ(X)$.

In addition, if the phases $\langle\varphi1\rangle$ and $\langle\varphi2\rangle$ are condensed, their entropy is generally much lower than that of a gas; the slope of the curve $\Delta_r G^\circ = f(T)$ is then nearly equal to the entropy of a gas. This explains the roughly parallel appearance on the same Ellingham diagram of all equilibrium curves between condensed phases taken in their standard state.

When taking into account heat contents, the curves $\Delta_r G = f(T)$ deviate little from a straight line. The tangent to a point of the equilibrium curve ($\langle\varphi1\rangle$, $\langle\varphi2\rangle$) always has $\Delta_r H^\circ$ as its ordinate at the origin and $(-\Delta_r S^\circ)$ as its slope, but $\Delta_r H^\circ$ and $\Delta_r S^\circ$ then depend on T.

When condensed phases are not in their standard state:

$$\Delta_r G^\circ = -RT \ln K = RT \ln(P_X \cdot a_{\varphi1}/a_{\varphi2})$$

$$\mu_X = RT \ln P_X = \Delta_r G^\circ + RT \ln(a_{\varphi2}/a_{\varphi1}).$$

The monovariant equilibrium curve between two phases $\langle\varphi1\rangle$ and $\langle\varphi2\rangle$ may differ notably from a straight line if the *activities* of the phases $\langle\varphi1\rangle$ and $\langle\varphi2\rangle$ at equilibrium depend strongly on temperature. This is what is observed on plotting the monovariant equilibrium curve between a phase of quasi-constant composition and a phase of variable composition.

The reduction of the activity of a phase, for instance by dissolution in a solvent or by chemical combination without changing the activity of the other phase at equilibrium, has the effect of increasing its range of stability. Thus wüstite, $Fe_{1-x}O$, has a more limited existence domain than fayalite, Fe_2SiO_4, or the solid solution magnesiowüstite, $[yFe_{1-x}O, (1-y)MgO]$.

When the existence domains of two phases $\langle\varphi1\rangle$ and $\langle\varphi2\rangle$ have no common points on an Ellingham diagram, they cannot coexist at equilibrium. For instance, Al is unstable together with Cr_2O_3 and undergoes the aluminothermic reaction: $2Al + Cr_2O_3 \rightarrow 2Cr + Al_2O_3$.

Enantiomorphic—From the Greek εναντιος, opposite, and μορφος, form. It is used of two crystal forms able to coexist at equilibrium.

Endothermic—From the Greek ενδον, inside and θερμοω, to heat. It is used of a transformation in which a system receives heat from the surroundings: $Q > 0$. When the transformation occurs at constant pressure: $\Delta H > 0$; at constant volume: $\Delta U > 0$. If the surroundings do not supply heat, an endothermic transformation leads to a drop in the temperature of the system.

Energy—From the Greek ενεργεια, force in action (εργον, work). Energy, a fundamental concept of physics, characterizes the ability of a system to modify the state of its surroundings. Such a definition, very popular and nicely turned, satisfies an artist or an economist without refuting our daily experience. Energy transfer across a boundary may be made evident in the form of heat, electricity, mechanical work, etc., and be accompanied by a change in the state of the system and its surroundings. A thermodynamicist cannot be satisfied with this statement: it lacks rigour, defining a very vague concept. In a word, it may satisfy only those who already believe they know what is covered by this notion!

In thermodynamics, the first concept introduced is that of *work*. The concept of energy results directly from the statement of the *first law*:

Energy is a state function whose differential equals the work exchanged with the surroundings during an adiabatic process.

Remark 1: Energy is a state function defined except for a constant. It is possible to measure only the energy change of a system which undergoes a transformation from a state i to a state f:

$$\Delta E = E_f - E_i = \int_i^f dW_{ad}.$$

Remark 2: It follows from the *second law*, *corollary 9*, that during an *adiabatic process*, the *entropy* of the system cannot decrease. If the entropy of a system decreases during a transformation i → f, there exists no adiabatic process allowing this transformation. The energy change of the system is then calculated by:

$$\Delta E = E_f - E_i = \int_f^i -dW_{ad}.$$

Remark 3: It is often convenient to separate the energy of a system into two precisely defined contributions: *internal energy* and *external energy*, the latter being itself divided into *kinetic energy* and *potential energy*:

$$E = U + E_k + E_p.$$

Remark 4: Whereas it is quite possible to define precisely internal, kinetic or potential energies which are state functions of a system, it is on the other hand impossible to define chemical, elastic, electric magnetic, mechanical, surface, thermal, etc., energies. Let us imagine a transformation from the same state i (for instance: $H_2 + 0.5 O_2$) towards the same state f (for instance: H_2O) by various processes: combustion, explosion, electrochemistry, etc. It is obvious that $\int_i^f dW_{el}$, $\int_i^f dW_{mech}$, and $\int_i^f dQ$ depend on the process selected. Neither heat, nor electrical work, nor mechanical work is a state function.

Enthalpy—Enthalpy, H, from the Greek $\theta\alpha\lambda\pi\omega$, to warm, a function introduced by Kammerlingh Onnes, is defined by:

$$H \equiv E + PV.$$

E represents the *energy* of the system. In the absence of an external field, the enthalpy may be defined by:

$$H = U + PV.$$

ENTHALPY, THERMODYNAMIC POTENTIAL: By differentiating H, we obtain:

$$dH = dE + P\,dV + V\,dP$$

$$dH = T(dS - d\sigma) + V\,dP + \sum \mu_i\,dn_i + \cdots.$$

When a system is bound by constraints of constant entropy, pressure and amount of substance (a closed system without chemical reactions), and in the absence of other energy exchange (magnetic, electric, etc.), then $dH = -T\,d\sigma \leqslant 0$. In such conditions, the function H may only decrease and tends towards a minimum which corresponds to the equilibrium state of the system. Enthalpy is thus the *thermodynamic potential* of a system bound by the constraints (S, P, n_i) constant.

Such constraints are unusual in practice and the function H used only as thermodynamic potential would not be of great interest. On the other hand, with the constraints of constant pressure and amount of substance, we have $dH = dQ_P$. The enthalpy change of a system evolving in such conditions equals the amount of heat exchanged with the surroundings.

E

Enthalpy–entropy diagram—Or Mollier diagram. On such a diagram are shown the *isobaric, isothermal,* and *isochoric* curves for a pure substance. Its characteristics are deduced from the differentials dH and dS, expressed, thanks to *Maxwell's equations,* as a function of dT and dP or of dT and dV:

$$dS = c_P \frac{dT}{T} - \left(\frac{\partial V}{\partial T}\right)_P dP; \quad dH = c_P dT + \left[V - T\left(\frac{\partial V}{\partial T}\right)_P\right] dP.$$

Whence:

- for $V = $ const.: $\Rightarrow \left(\frac{\partial H}{\partial S}\right)_V = T\left[1 + \frac{V}{c_V}\left(\frac{\partial P}{\partial T}\right)_V\right] = T\left(1 + \frac{\beta_V PV}{c_V}\right);$

- for $P = $ const.: $\Rightarrow \left(\frac{\partial H}{\partial S}\right)_P = T;$

- for $T = $ const.: $\Rightarrow \left(\frac{\partial H}{\partial S}\right)_T = T - V\left(\frac{\partial V}{\partial T}\right)_P = T - \frac{1}{\alpha_P}.$

α_P: isobaric *expansivity;* β_V: isochoric *bulk expansion coefficient.*

The slopes of isochoric, isobaric, and isothermal curves decrease in the order:

$$\left(\frac{\partial H}{\partial S}\right)_V > \left[\left(\frac{\partial H}{\partial S}\right)_P = T\right] > \left(\frac{\partial H}{\partial S}\right)_T.$$

The slope of the isotherms may be positive or negative, that is, higher or lower than that of the isenthalps, which is zero by construction. The isotherm slope has the same sign as that of the Joule–Thomson coefficient $\mu_J = (\partial T/\partial P)_H$ (see *Joule–Thomson effect*). By plotting on an enthalpy–entropy diagram the locus of the extrema on the isotherms, a curve is obtained separating the plane into two

86 Enthalpy–entropy diagram

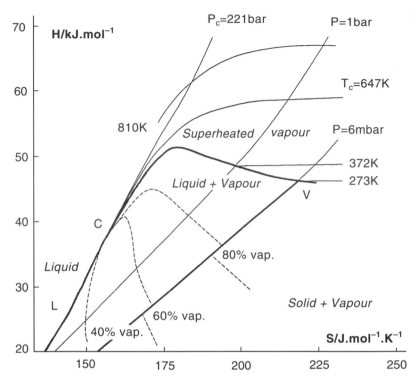

$H/kJ.mol^{-1}$

$P_c=221bar$

$P=1bar$

$T_c=647K$

810K

Superheated / vapour

$P=6mbar$

372K

273K

Liquid + Vapour

V

C

80% vap.

Liquid

Solid + Vapour

L

60% vap.

40% vap.

$S/J.mol^{-1}.K^{-1}$

Enthalpy–entropy diagram: In this diagram concerning water, the enthalpy origin (H=0) has been arbitrarily taken as the point representing liquid water in equilibrium with its vapour at 273 K under 6 mbar.

domains. Inside the curve, the isotherm slope is positive: an isenthalpic expansion of the gas induces cooling. Outside the curve, the isotherm slope is negative: an isenthalpic expansion of the gas induces heating.

In a two-phase domain (solid + liquid, solid + vapour, or liquid + vapour), the isothermal and isobaric curves are the same and their slope equals the temperature. In crossing the border between a two-phase and a single-phase domain, there is no break in the slope of the isobars, but there is a break in the slopes of the isotherms and of the isochores.

In the case of an *ideal gas*, it is found that the isotherm slope is zero; the isochore slope equals $T(1 + R/c_V)$.

On a P–T diagram, the equilibrium curve between two phases is unique and its slope is given by the *Clapeyron equation*, whereas on an enthalpy–entropy diagram, it consists of two branches. At liquid–vapour equilibrium, the branch for the liquid and that for the vapour meet at the critical point with a slope equal to the critical temperature.

The differentials dH and dS give at once the slopes of the two branches at the liquid–vapour equilibrium:

$$\left(\frac{\partial H}{\partial S}\right)_x = \frac{c_P + [V - T(\partial V/\partial T)_P](\partial P/\partial T)_x}{(c_P/T) - (\partial V/\partial T)_P(\partial P/\partial T)_x} = T\left[1 + \frac{V}{c_x}\left(\frac{\partial P}{\partial T}\right)_x\right],$$

where c_x represents the *heat capacity* along the saturated vapour curve. The above expression may be applied equally well to the liquid as to the vapour. For the liquid, far from the critical point, the second term in the last bracket is generally small compared with unity. The slope of the curve is thus slightly higher than T. For the saturated vapour, c_x is generally negative; the slope is then lower than T. When two inversion points exist for which $c_x = 0$, the curve corresponding to the saturated vapour presents two vertical tangents. When $(\partial H/\partial S)_x > 0$, which corresponds to $c_x > 0$, the point corresponding to the saturated vapour is then situated between the two inversion points: an *adiabatic* compression of the saturated vapour has the effect of making the system enter the two-phase domain (liquid + vapour). Otherwise, when $c_x < 0$, adiabatic compression of the saturated vapour has the effect of making the system enter the domain of the superheated vapour.

Enthalpy of formation—The enthalpy change, denoted $\Delta_f H$, observed when a compound is synthesized from its elements. For instance:

$$S + H_2 + 2O_2 \rightarrow H_2SO_4, \quad \Delta_f H°(298.15\,K) = -814.1\,kJ\,mol^{-1}.$$

Reactions of formation are most often exothermic; however, there are some exceptions. Nitrogen monoxide, NO, and acetylene, C_2H_2, are the best known examples of endothermic compounds ($\Delta_f H° > 0$).

Enthalpies of formation given in tables of thermodynamic data are generally standard enthalpies of formation, that is, enthalpies of formation of compounds taken in their *standard state* from elements taken in their standard state. The standard state must be clearly defined! For instance:

$$\Delta_f H°(H_2O,\ liq,\ 298.15\,K) = -285.9\,kJ\,mol^{-1}$$

$$\Delta_f H°(H_2O,\ vap,\ 298.15\,K) = -242.5\,kJ\,mol^{-1}.$$

88 Enthalpy of mixing

When elements are in their standard reference state (*standard element reference*), the standard state is the stable state at the temperature considered.

The enthalpy of formation gives access to the *enthalpy of reaction*:

$$\sum v_i \langle M_i \rangle = 0 \quad \Rightarrow \quad \Delta_r H^\circ = \sum v_i \Delta_f H^\circ (\langle M_i \rangle),$$

with $\Delta_f H^\circ(\langle M_i \rangle) = 0$, if M_i is an element taken in its standard state.

Enthalpy of mixing—The integral enthalpy of mixing $\Delta_{mix} H$ is the enthalpy change related to the transformation:

$$\sum n_i \langle A_i \rangle \rightarrow \langle \text{Mixture} \rangle.$$

The integral molar enthalpy of mixing is the integral enthalpy of mixing reduced to one mole:

$$\sum x_i \langle A_i \rangle \rightarrow \langle \text{Mixture} \rangle \quad \left(\text{with} \sum x_i = 1 \right).$$

The partial molar enthalpy of mixing of i, or the partial molar enthalpy of dissolution of i, is Δh_i, the difference between h_i, the partial molar enthalpy of i in the mixture, and h_i°, the molar enthalpy of i outside the mixture. δH, the enthalpy change corresponding to the transformation:

$$\langle \text{Mixture} \rangle + \delta n_i \langle A_i \rangle \rightarrow \langle \text{Mixture} \rangle$$

is given by: $\qquad \delta H = (h_i - h_i^\circ) \delta n_i = \Delta h_i \, \delta n_i.$

Between the partial molar enthalpies of dissolution and the integral enthalpy of mixing there exist:

- the definition relation: $\Delta h_i = (\partial \Delta_{mix} H / \partial n_i)_{T, P, n_{j \neq i}}$;
- the *Euler identity*: $\Delta_{mix} H = \sum n_i \Delta h_i$;
- the *Gibbs–Duhem equation*: $\sum n_i \, d\Delta h_i = 0$ (T and P constant).

Enthalpies of mixing derive from *Gibbs energies of mixing*:

$$\Delta h_i = -T[\partial(\Delta \mu_i / T) / \partial T]_P = -RT^2(\partial \ln a_i / \partial T)_P$$

$$\Delta_{mix} H = -T[\partial(\Delta_{mix} G / T) / \partial T]_P = -RT^2 \sum n_i (\partial \ln a_i / \partial T)_P.$$

If a solution is *ideal* ($a_i = x_i$), the activities are independent of temperature, therefore the enthalpy of mixing is zero.

Enthalpy of reaction—The enthalpy of a reaction $\sum v_i \langle M_i \rangle = 0$ is the heat quantity exchanged with the surroundings when the reaction takes place at constant pressure. An enthalpy of reaction is also well defined if the pressure varies, but it no longer equals the heat quantity exchanged with the surroundings.

Tables of thermodynamic data give access to the *standard enthalpy of reaction* $\Delta_r H°(T)$. If the constituents of the reaction are not in their *standard state*, it is necessary to apply a correction. For instance, if the constituents M_i are under a pressure p_i, the standard pressure being $p_i°$:

$$\Delta_r H(T, p_i) = \Delta_r H°(T) + \sum_i v_i \int_{p_i°}^{p_i} \left[v_i - T \left(\frac{\partial v_i}{\partial T} \right)_P \right] dP.$$

In the same way, if constituents M_i are *solutes* in concentrations c_i, the standard concentration being $c_i°$:

$$\Delta_r H(T, c_i) = \Delta_r H°(T) + \sum_i v_i (h_i - h_i°),$$

with $\Delta h_i = h_i - h_i°$, the partial molar enthalpy of dissolution of the solute i taken from its standard state (concentration $c_i°$) to the solution (concentration c_i).

The correction, zero when M_i is an *ideal gas* or when the solutions are *ideal*, is most often negligible; it is thus possible to use, without too much compunction, $\Delta_r H(T) \approx \Delta_r H°(T)$! One must be conscious that such an approximation, although legitimate for enthalpies, is not allowed for *entropies of reaction*, nor, taking into account the relation $\Delta_r G = \Delta_r H - T \Delta_r S$, for *Gibbs energies of reaction!*

Enthalpy–pressure diagram   This diagram, well known to engineers, is useful in dealing with heat exchangers, heat pumps, or refrigerators. Curves generally drawn on an enthalpy–pressure diagram are *isotherms*, *isentropes* and *isochores*. From the expressions of dS obtained by *Maxwell's equations*:

$$dS = c_P \frac{dT}{T} - \left(\frac{\partial V}{\partial T} \right)_P dP = \frac{c_V}{T} \left(\frac{\partial T}{\partial P} \right)_V dP + \frac{c_P}{T} \left(\frac{\partial T}{\partial V} \right)_P dV$$

$$dH = T\,dS + V\,dP,$$

it follows that:

- at $T = $ const.: $\Rightarrow \left(\dfrac{\partial P}{\partial H} \right)_T = \dfrac{1}{V(1 - \alpha_P T)}$;

- at $S = $ const.: $\Rightarrow \left(\dfrac{\partial P}{\partial H} \right)_S = \dfrac{1}{V}$;

90 Enthalpy–pressure diagram

- at $V=$ const.: $\Rightarrow \left(\dfrac{\partial P}{\partial H}\right)_V = \dfrac{1}{V+c_V/(\beta_V P)}$.

α_P: isobaric *expansivity*; β_V: isochoric *bulk expansion coefficient*.

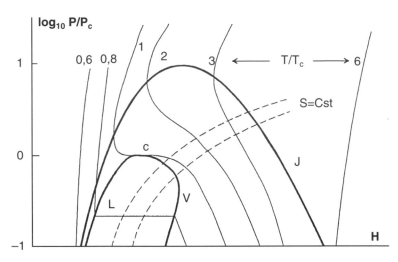

Enthalpy–pressure diagram: *Generally presented in the form* $\log_{10} P$
versus H. *In bold are the liquid–vapour equilibrium curve* (*LCV*) *and
the Joule–Thomson effect inversion curve* (*J*).

The slopes of the isothermal, isentropic, and isochoric curves
decrease in the order:

$$\left(\frac{\partial P}{\partial H}\right)_T > \left[\left(\frac{\partial P}{\partial H}\right)_S = \frac{1}{V}\right] > \left(\frac{\partial P}{\partial H}\right)_V.$$

When $\alpha_P = 1/T$, the isotherms present a vertical tangent, the locus
of the *Joule–Thomson effect* inversion point:

$$\alpha_P > \frac{1}{T} \Rightarrow \left(\frac{\partial P}{\partial H}\right)_T < 0.$$

On an enthalpy–pressure diagram, branches corresponding to the
liquid and the vapour in equilibrium meet at the critical point with a
horizontal tangent. The branch of the saturated vapour presents a
point with a vertical tangent, corresponding to a maximum of the H
function, a maximum also found on an enthalpy–entropy diagram.

The differential $dH = T\,dS + V\,dP$ and the expression of $(\partial H/\partial S)_x$ (see *enthalpy–entropy diagram*) allow a straightforward calculation:

$$\left(\frac{\partial P}{\partial H}\right)_x = \frac{(\partial P/\partial T)x}{c_x + V(\partial P/\partial T)_x}$$

with:

$$c_x = T\left(\frac{\partial S}{\partial T}\right)_x = c_P - T\left(\frac{\partial V}{\partial T}\right)_P\left(\frac{\partial V}{\partial T}\right)_x .$$

The subscript x signifies equilibrium between two phases; c_x represents the *heat capacity* along the liquid–vapour equilibrium curve; $(\partial P/\partial T)_x$, the slope of the saturated vapour pressure curve, is given by the *Clapeyron equation*: $(\partial P/\partial T)_x = \Lambda_{lu} H/T \Lambda_{tr} V$.

Entropology—'Discussion of *entropy*'. A name sometimes given to thermodynamics without differential equations. Here is a nice example of it: 'Taken as a whole, therefore, civilization can be described as a prodigiously complicated mechanism: tempting as it would be to regard it as our universe's best hope of survival, its true function is to produce what physicists call entropy· inertia, that is to say. Every scrap of conversation, every line set up in type, establishes a communication between two interlocutors, levelling what had previously existed on two different planes and had had, for that reason, a greater degree of organization. "Entropology", not anthropology, should be the word for the discipline that devotes itself to the study of this process of disintegration in its most highly evolved forms'. (*A World on the Wane*, London, 1961, p. 397; translation by John Russell of *Tristes Tropiques* by Claude Lévi-Strauss, French anthropologist, born 1908.)

Entropy—A term proposed by Clausius in 1868, from the Greek root εντροπη, the act of turning round (τροπη, change of direction), implying the idea of *reversibility*. The entropy S is a *state function*, defined, following the *second law, corollary 7*, by the relation:

$$dS \equiv \beta(dE - dF).$$

The *energy* E and the *Helmholtz energy* F being state functions, their difference is also a state function. The proportionality coefficient β defines the *thermodynamic temperature* T of the *reservoir* with which the *system* exchanges *heat* and *work* ($\beta = 1/T$). With corollary 11, the expression of dS is obtained in a more useful form:

$$dS = dQ_{rev}/T.$$

Remark 1: It is possible to put directly $dS \equiv dQ_{rev}/T$, by definition, without reference to the second law. However, dQ is not generally an *exact differential*. It is thus necessary to show that the function S thus defined is really a state function, in other words, that dS is an exact differential.

With this aim, let us consider a system following a reversible cycle. It is always possible to decompose the cycle into an infinity of *Carnot cycles*. It is verified that, over a Carnot cycle, $\oint (dQ/T) = 0$. The equation is also valid for the reversible cycle obtained by summing all the elementary Carnot cycles. The function defined by $dS = (dQ_{rev}/T)$ is thus a state function for the system.

Mathematicians demonstrate this result in another way. Let us take the expression of dQ_{rev} from the *calorimetric coefficients*:

$$dQ_{rev} = c_P \, dT + \hbar \, dP = c_P \, dT - (c_P - c_V)(\partial T/\partial P)_V \, dP.$$

Like every differential that is a function of two variables, dQ_{rev} is an *integrable differential* and it is easily verified, in the case of an ideal gas, that the integrating factor is $1/T$, in other words, that the differential (dQ_{rev}/T) is exact.

If (dQ_{rev}/T) is an exact differential for ideal gases, it is also exact for any system. If we consider two systems A (ideal gas) and B (ordinary) at equilibrium, the assembly (A + B) being *isolated*:

$$dQ_A + dQ_B = 0 \quad \text{and} \quad \oint [(dQ_A/T) + (dQ_B/T)]_{rev} = 0.$$

Now, we have just verified that $\oint (dQ_A/T)_{rev} = 0$ if A is an ideal gas. Then: $\oint (dQ_B/T)_{rev} = 0$ for any system B. The function S, defined by $dS = (dQ/T)_{rev}$, is then a state function.

This proof, given by Max Planck, has been criticized by Tatiana Ehrenfest with the following argument: it is not obvious that system B, of which nothing is known, loops its cycle together with system A, in which case, it would be necessary to write:

$$\oint (dQ_A/T)_{rev} + \int (dQ_B/T)_{rev} = 0$$

and the proof falls down! Actually, if the transformation is cyclic for the system A $[\oint (dQ_A/T)_{rev} = 0]$, it must also be cyclic for B, but the proof of this statement implies the use of the second law.

Remark 2: Entropy, like every *extensity*, is conservative for reversible transformations. It is easy to verify with some simple examples that it is not conserved for *irreversible* transformations, as opposed to other extensities such as volume or enthalpy. Let us consider for instance the transformation:

(1 L water at T_1) + (1 L water at T_2) → (2 L water at T_m).

The heat capacity of water being supposed constant, we calculate:

$$\Delta H = \int_{T_1}^{T_m} c_P\, dT + \int_{T_2}^{T_m} c_P\, dT = 0$$

$$\Delta S = \int_{T_1}^{T_m} \frac{c_P}{T}\, dT + \int_{T_2}^{T_m} \frac{c_P}{T}\, dT = c_P \ln \frac{T_m^2}{T_1 T_2} > 0,$$

since $T_m = (T_1 + T_2)/2 > \sqrt{T_1 T_2}$.

This observation may be generalized: 'When an isolated system evolves spontaneously, its entropy can only increase'. This statement, deduced from the *second law*, *corollary 10*, may be posed directly as an expression of the second law.

Remark 3: When the system is not isolated, it is possible to apply the above statement to the *universe*:

$$dS_{system} + dS_{surroundings} = dS_{universe} \geq 0.$$

$dS_{surroundings} = dQ/T$, because it is always possible to find a reversible transformation by which the surroundings exchange $-dQ$ with its surroundings, in other words, with the system!

$dS_{universe} = d\sigma \geq 0$ is the entropy created by the irreversibility of the transformation. We obtain the fundamental relationship (*Second law*, *Remark 2*):

$$dS = \frac{dQ}{T} + d\sigma.$$

$d\sigma > 0$ for an irreversible transformation, which is a consequence of the second law, and $d\sigma = 0$ for a reversible transformation, which comes from the definition of entropy.

Remark 4: At the risk of breaking down an open door, let us recall the good reflexes necessary for calculating a change in entropy:

- $\Delta S = \int_i^f dS = S_f - S_i$, independent of the path followed!
- $\Delta S = \int_i^f (dQ_{rev}/T)$ only for a reversible path!
- If the transformation is irreversible, one must find a reversible transformation, which leads the system from the same initial state i to the same final state f.
- If it is impossible to find a reversible path, one must then use the general relationship: $dS = (dQ/T) + d\sigma$; however, the situation may

seem to be a dead end, because this relation is generally used to calculate $d\sigma$!

- Actually, there are no hopeless situations, because it is always possible, thanks to the *third law*, which allows an absolute value to be assigned to the entropy, to calculate separately S_i and S_f, then to evaluate the difference: $\Delta S = S_f - S_i$.

Entropy creation—See *Entropy production*.

Entropy of mixing—See also: *Quantity of mixing, Integral molar quantity, Partial molar quantity*. The integral entropy of mixing $\Delta_{mix}S$ is the variation of entropy during the transformation:

$$\sum n_i \langle A_i \rangle \rightarrow \langle \text{Mixture} \rangle.$$

The integral molar entropy of mixing is the integral entropy of mixing referred to one mole:

$$\sum x_i \langle A_i \rangle \rightarrow \langle \text{Mixture} \rangle \qquad (\text{with } \sum x_i = 1).$$

The partial molar entropy of mixing of i, or partial molar entropy of dissolution of i, represented by Δs_i, is the difference between s_i, the partial molar entropy of i in the mixture, and s_i°, the molar entropy of i outside the mixture. δS, the entropy variation during the transformation:

$$\langle \text{Mixture} \rangle + \delta n_i \langle A_i \rangle \rightarrow \langle \text{Mixture} \rangle$$

is given by:

$$\delta S = (s_i - s_i^\circ)\delta n_i = \Delta s_i \, \delta n_i.$$

Between partial molar entropies of dissolution and the integral entropy of mixing there is:

- the definition relationship: $\Delta s_i = (\partial \Delta_{mix} S / \partial n_i)_{T,P,n_{j \neq i}}$;
- the *Euler identity*: $\Delta_{mix}S = \sum n_i \Delta s_i$;
- the *Gibbs–Duhem equation*: $\sum n_i \, d\Delta s_i = 0$ (T and P constant).

EXPRESSION OF ENTROPIES OF MIXING: Entropies of mixing derive from *Gibbs energies of mixing*:

$$\Delta s_i = -(\partial \Delta \mu_i / \partial T)_P = -R \ln a_i - RT (\partial \ln a_i / \partial T)_P$$

$$\Delta_{mix}S = -(\partial \Delta_{mix} G / \partial T)_P = -R \sum n_i \ln a_i - RT \sum n_i (\partial \ln a_i / \partial T)_P.$$

In practice, it is wise not to be too confident about the precision obtained by deriving experimental curves. Moreover, it is generally hard to obtain entropies of mixing by direct experimental measurements. They are more often deduced from the relation:

$$\Delta_{mix}G = \Delta_{mix}H - T\Delta_{mix}S.$$

For an ideal solution: $a_i = x_i$ when the *standard state* is the pure constituent i. Entropies of mixing are then expressed by:

$$\Delta s_i^{id} = -(\partial \Delta \mu_i^{id}/\partial T)_P = -R \ln x_i$$

$$\Delta_{mix}S^{id} = -(\partial \Delta_{mix}G^{id}/\partial T)_P = -R\sum n_i \ln x_i.$$

Entropy of reaction—The entropy of a *chemical reaction* $\sum v_i M_i = 0$ is given, from the *second law*, by:

$$\Delta_r S = \frac{\Delta_r Q}{T} + \Delta_r \sigma.$$

This expression is hard to use, because of the term $\Delta_r \sigma$, the entropy created by the *irreversibility* of the transformation, which is a function of the path followed. In practice, $\Delta_r S$ is directly calculated by:

$$\Delta_r S = S_f - S_i = \sum v_i S_i.$$

Tables of thermodynamic data allow the calculation of $S_i^\circ(T)$, the *standard entropy* of the constituent M_i at temperature T. When the constituents M_i which take part in the reaction are not in their *standard state*, it is necessary to make a correction. The entropy $\Delta_r S$ of the reaction is then given by:

$$\Delta_r S(P,T) = \Delta_r S^\circ(T) + \sum v_i \int_{p_i^\circ}^{p_i} - \left(\frac{\partial v_i}{\partial T}\right)_P dP.$$

When the constituents M_i are *solutes* at a concentration c_i, the standard concentration being c_i°:

$$\Delta_r S(T, c_i) = \Delta_r S^\circ(T) + \sum_i v_i (s_i - s_i^\circ).$$

$\Delta s_i = s_i - s_i^\circ$: partial molar entropy of dissolution of the solute i taken in its standard state (concentration c_i°) in the solution (concentration c_i).

Remark: As opposed to what is often observed with the *enthalpy of reaction*, where it is possible to let $\Delta_r H \approx \Delta_r H^\circ$, the correction allowing a standard entropy of reaction to be converted to an entropy of reaction is generally not negligible, because enthalpies and entropies have a different behaviour with respect to pressure or dilution. For instance, the enthalpy of an ideal gas is independent of pressure whereas its entropy varies as $R\ln(P^\circ/P)$ when its pressure goes from P° to P. Likewise, the partial molar enthalpy of a constituent i in an *ideal solution* is independent of the dilution whereas its partial molar entropy varies as $R\ln(c_i^\circ/c_i)$ when its concentration goes from c_i° to c_i.

Entropy production—Entropy being a non-conservative quantity, the *second law* states that, for an *isolated* system, every spontaneous, and hence *irreversible* transformation is characterized by an increase in entropy. The entropy production is the entropy created by the irreversibility of the transformation during unit time:

$$\dot\sigma \equiv \frac{d\sigma}{dt} > 0.$$

Let us examine the main expressions for entropy production:

HEAT TRANSFER: If two systems A and B in thermal contact, the assembly (A + B) being isolated, exchange a heat quantity $dQ = dQ_B = -dQ_A$, the entropy production is:

$$\dot\sigma \equiv \frac{d\sigma}{dt} = \frac{dQ}{dt}\left(\frac{1}{T_B} - \frac{1}{T_A}\right) > 0,$$

where T_A and T_B are the temperatures of the two systems A and B. If $T_B > T_A$, $dQ < 0$: heat flows spontaneously from the warm body towards the cold one. The contrary would be astonishing; however, isn't it always advisable to do such elementary checking?

MASS TRANSFER: If two open systems A and B, the assembly (A + B) being isolated, exchange a quantity of i: $dn_i = dn_{iB} = -dn_{iA}$, the *work* exchanged during the transfer is expressed by:

$$dW = (\mu_{iB} - \mu_{iA})\, dn_i.$$

Since the system (A + B) is isolated:

$$dU = T\, dS + dW = 0.$$

The entropy variation is the entropy created by the irreversibility of the transformation, whence the entropy production is:

$$\dot\sigma \equiv \frac{d\sigma}{dt} = -\frac{dn_i}{dt}\left(\frac{\mu_{iB}}{T_B} - \frac{\mu_{iA}}{T_A}\right) > 0.$$

At thermal equilibrium $(T_B = T_A)$, $\mu_{iB} > \mu_{iA} \Rightarrow dn_i = dn_{iB} < 0$: the constituent i migrates in the direction of decreasing chemical potentials. The expression of the entropy production may be generalized on the assumption of thermal equilibrium:

$$\dot{\sigma} \equiv \frac{d\sigma}{dt} = -\frac{1}{T}\sum_i \frac{dn_{iB}}{dt}(\mu_{iB} - \mu_{iA}) > 0.$$

Entropy production must be globally positive; however, it is not forbidden for one chemical species, to diffuse against its chemical potential gradient.

E

TRANSFER OF ELECTRICITY: If $dn_{iB} = -dn_{iA} = dn_i$ moles of ions with charge z_i are transferred from a *Galvani potential* Φ_A towards a Galvani potential Φ_B, the electrical work exchanged is expressed by:

$$dW = z_i \mathscr{F}(\Phi_B - \Phi_A)dn_i.$$

When several ionic species are transferred from A to B, the entropy production is given by:

$$\dot{\sigma} \equiv \frac{d\sigma}{dt} = -\frac{1}{T}\sum_i \frac{dn_{iB}}{dt} z_i \mathscr{F}(\Phi_B - \Phi_A) > 0.$$

Cations $(z_i > 0)$ will travel spontaneously in the direction of decreasing electric potential $(\Phi_B > \Phi_A \Rightarrow dn_{iB} < 0)$. If the systems A and B are characterized by a chemical potential and by an electric potential which differ, the transfer of dn_i moles from A towards B is given by:

$$dE = (\mu_{iB} - \mu_{iA})dn_i + z_i \mathscr{F}(\Phi_B - \Phi_A)dn_i,$$

with $\tilde{\mu}_i = \mu_i + z_i \mathscr{F}\Phi$, the *electrochemical potential* of the ion i:

$$\dot{\sigma} \equiv \frac{d\sigma}{dt} = -\frac{1}{T}\sum_i \frac{dn_{iB}}{dt}(\tilde{\mu}_{iB} - \tilde{\mu}_{iA}) > 0.$$

Ions will travel spontaneously from the higher electrochemical potential towards the lower one $(\tilde{\mu}_{iB} > \tilde{\mu}_{iA} \Rightarrow dn_{iB} < 0)$.

CHEMICAL REACTIONS: If the *extent* ξ of a chemical reaction $\sum \nu_i M_i = 0$ varies by $d\xi$ (at T and P constant), its *affinity* is given by:

$$\mathscr{A} = T(d\sigma/d\xi)_{T,P} = -\sum \nu_i \mu_i = -\Delta G.$$

The entropy production is then expressed by:

$$\dot{\sigma} \equiv \frac{d\sigma}{dt} = \frac{\mathscr{A}}{T}\cdot\frac{d\xi}{dt} = \frac{\mathscr{A}}{T}\upsilon > 0,$$

where $\upsilon = d\xi/dt$ is the *reaction rate*. Rate and affinity have the same sign: a reaction is carried out in the forward direction ($d\xi > 0$) when $\mathscr{A} > 0$. With several coupled reactions:

$$\dot{\sigma} \equiv \frac{d\sigma}{dt} = \frac{1}{T} \sum_j \mathscr{A}_j \upsilon_j > 0.$$

The entropy production is the sum of the entropy productions of the various reactions. It must be globally positive. However, it is not forbidden for one reaction to occur with a decrease in entropy: the *coupling* of several reactions allows a particular reaction to advance in the direction opposite to that given by its affinity.

ELECTROCHEMICAL REACTIONS: To the expression of the *Gibbs energy* of the reaction is added a term corresponding to the change in the electrical energy:

$$\Delta \tilde{G} = \sum v_i (\mu_{iB} - \mu_{iA}) + \sum v_i z_i \mathscr{F} (\Phi_B - \Phi_A).$$

$\Delta \tilde{G}$ is the change in Gibbs energy of the reaction, called the electrochemical Gibbs energy because of the presence of electric terms; the first term on the right-hand side represents the pure chemical contribution whereas the second term represents the electrical contribution. By introducing the *electrochemical affinity* ($\tilde{\mathscr{A}} = -\Delta \tilde{G}$):

$$\tilde{\mathscr{A}} = -\sum v_i [(\mu_{iB} + z_i \mathscr{F} \Phi_B) - (\mu_{iA} + z_i \mathscr{F} \Phi_A)]$$

$$\tilde{\mathscr{A}} = -\sum v_i (\tilde{\mu}_{iB} - \tilde{\mu}_{iA}),$$

we obtain the expression for the entropy production:

$$\dot{\sigma} \equiv \frac{d\sigma}{dt} = \frac{\tilde{\mathscr{A}}}{T} \cdot \frac{d\xi}{dt} = \frac{\tilde{\mathscr{A}}}{T} \upsilon > 0,$$

where $\upsilon = d\xi/dt$ is the rate of the electrochemical reaction. In a galvanic cell, a reaction advances in the forward direction ($d\xi > 0$) when $\tilde{\mathscr{A}} > 0$.

Environment—See *Surroundings*.

Equation of state—A relation between state variables. Equations of state most often encountered are those of *ideal gases* or *real gases*. For condensed matter, it is generally more convenient to use expansivity α_P or compressibility χ_T instead of an equation of state:

$$\alpha_P = \frac{1}{V} \left(\frac{\partial V}{\partial T} \right)_P \qquad \chi_T = -\frac{1}{V} \left(\frac{\partial V}{\partial P} \right)_T.$$

Approximate values for liquids are $\alpha_P \approx 10^{-4}\,\mathrm{K}^{-1}$; $\chi_T \approx 10^{-10}\,\mathrm{Pa}^{-1}$ (or $\approx 10^{-5}\,\mathrm{atm}^{-1}$). When limited to a first-order development, it is however possible to write: $V \approx V_{\mathrm{o}}(1 + \alpha_P T - \chi_T P)$. There are equations of state for various areas, for instance:

TENSION IN A WIRE: Hooke's law (English astronomer, 1635–1703) states that, in the elastic domain, the tension is proportional to the lengthening of a wire: $\mathscr{F} = A(T)(\lambda - \lambda_{\mathrm{o}})$.

SURFACE TENSION: For a liquid in equilibrium with its vapour:

$$\Gamma = \Gamma_{\mathrm{o}}(1 - T/T')^n,$$

where n is an experimental constant, between 1 and 2, and T' is an empirical temperature, close to the critical temperature. This law is no longer true around the critical temperature; however, $\Gamma \to 0$ when $T \to T_{\mathrm{c}}$.

ELECTRIC POLARIZATION: This is related to the *electric field E* and *electric excitation D* by: $D = \varepsilon_{\mathrm{o}} E + P$. A typical equation of state for a homogeneous dielectric is: $P = (a + b/T)E$.

MAGNETIC MOMENT: The Curie law gives an equation of state for a paramagnetic substance:

$$\mathscr{M} = (nC/T)\mathscr{H},$$

where n is the amount of substance, C an experimental constant and $\mathscr{M}$ the magnetization (magnetic moment per volume unit) of the material in a magnetic field $\mathscr{B}$. $\mathscr{M} - \mu_{\mathrm{o}}^{-1}\mathscr{B} - \mathscr{H}$.

Equilibrium—A *system* is in equilibrium when neither its state nor that of the *surroundings* evolves with time.

An equilibrium may be stable, metastable, undifferentiated, or unstable. An equilibrium state is obtained by seeking the extrema of a *thermodynamic potential* function, whose nature depends on the constraints imposed on the system.

The absolute minimum of the thermodynamic potential corresponds to the only *stable equilibrium*; other minima correspond to states of *metastable equilibria*. Maxima correspond to states of *unstable equilibria. Undifferentiated equilibria* are encountered when it is possible to modify a state variable without a change in the surroundings.

Outside the extrema, there is no possible equilibrium and the states are merely unstable: the system evolves spontaneously towards an equilibrium, stable or metastable, although the evolution may be very slow.

100 Equilibrium constant

Equilibrium constant—The *Gibbs energy* of a *chemical reaction* $\sum v_i M_i = 0$, which occurs in a mixture formed of n_i moles of M_i, is given by the expression:

$$\Delta_r G = \Delta_r G^\circ + RT \ln \prod_i a_i^{v_i},$$

a_i being the *activity* of the constituent M_i in the mixture.

For the sake of homogeneity, it is convenient to state:

BY DEFINITION: $K \equiv \exp(-\Delta_r G^\circ / RT)$.

I should add, for the inattentive reader:

BY DEFINITION: $\Delta_r G^\circ \equiv -RT \ln K$.

K is called the equilibrium constant of the chemical reaction.

Remark 1: One must be convinced of the absolute identity of these two concepts: the equilibrium constant and *standard Gibbs energy of reaction* represent the same thermodynamic animal. It is immaterial in expressing an equilibrium condition, whether one says that the Gibbs energy of the reaction is zero, or that the *mass action product* is equal to the equilibrium constant.

Remark 2: This 'constant' is ill-named, because it depends on the temperature, on the balance equation and on the choice of the *standard state* for each constituent. In the particular cases where the standard pressure is not defined, the 'constant' should also depend on the pressure!

Remark 3: The equilibrium constant is sometimes denoted K° and called the 'standard equilibrium constant'. This good sense, consistent with logic $[K^\circ \equiv \exp(-\Delta_r G^\circ / RT)]$, presents an advantage, that of 'hammering home the nail', but also a drawback, by letting it be supposed that a mere 'equilibrium constant' that is not standard exists, which is impossible. The expression $\exp(-\Delta_r G / RT)$, a function of the pressure, temperature, composition, and extent of the reaction, contains nothing that allows any confusion with an equilibrium constant!

Remark 4: An equilibrium constant is a number, without any dimensions. When students write the equilibrium condition for the reaction $N_2 + 3H_2 \rightarrow 2NH_3$ in the form:

$$K_P(T) = P_{NH_3}^2 / P_{N_2} \cdot P_{H_2}^3,$$

they are generally fully aware of the approximation made: for real gases, *fugacities* must be used instead of pressures, but it is more often forgotten that the only correct expression is that using *activities*:

$$K(T) = a_{NH_3}^2/a_{N_2} \cdot a_{H_2}^3 \quad \text{with} \quad a_i = f_i/f_i^\circ \approx p_i/p_i^\circ.$$

Naturally, it is possible for the sake of simplicity to say that $a_i = f_i \approx p_i$, but pressures have to be expressed in *bars* if the standard pressure selected is $p_i^\circ = 1$ bar for every gas. The student who thinks to act correctly with pressures expressed in *pascals* would be mistaken, in this particular case, by a factor of 10^{10}!

Remark 5: It is possible to encounter (in exercises set for undergraduate students rather than in current practice!) various expressions of the equilibrium condition:

$$K_P(T) = P_{NH_3}^2/P_{N_2} \cdot P_{H_2}^3; \qquad K_c(T) = c_{NH_3}^2/c_{N_2} \cdot c_{H_2}^3;$$

$$K_x(T, P) = x_{NH_3}^2/x_{N_2} \cdot x_{H_2}^3.$$

P_i, c_i, and x_i represent respectively pressures (in bars), *molarities* (in mol L^{-1}), and *mole fractions* (dimensionless) of the species i. It is easy to obtain, by assuming ideal gases, the relations:

$$K_P = K_c(RT)^{\sum v_i} = K_x P^{\sum v_i},$$

with $\sum v_i = -2$ and $R = 0.0831451$ because P is in bars and V in litres.

It is interesting to note that the transition from K_P to K_c or K_x expresses only a change of standard state. To use K_c amounts to choosing a standard pressure $P^\circ = RT/V^\circ$ with $V^\circ = 1$ litre. To use K_x means choosing the pure gas i as standard state, that is, fixing no standard pressure. In such a case, the constant K_x also depends on the pressure. In practice, it is strongly recommended to work only with K_P, that is, to let $P^\circ = 1$ bar.

Remark 6: The equilibrium constant has been introduced starting from the *standard Gibbs energy of reaction*. From another aspect, the function G represents the *thermodynamic potential* of a constrained system with T and P constant, but these two results are independent. Indeed, it is clear that the equilibrium condition: $\prod_i a_i^{v_i} = \exp(\Delta_r G^\circ/RT)$ gives the state of the system at equilibrium (!), but it is obvious that this state depends neither on the path followed by the system before reaching equilibrium, nor on the choice of a standard state, a choice arbitrarily defined by the thermodynamicist.

102 Equilibrium constant

Let us consider a system consisting of n_i moles of species M_i and, between them, the possible reaction $\sum v_i M_i = 0$. The differentials of the functions F and G are given by:

$$dF = -S\,dT - P\,dV + \sum \mu_i\,dn_i$$

$$dG = -S\,dT + V\,dP + \sum \mu_i\,dn_i.$$

At constant temperature and volume, $dF = \sum \mu_i\,dn_i$, whereas at constant temperature and pressure, $dG = \sum \mu_i\,dn_i$, which may be expressed by introducing the *extent* of the reaction $d\xi = dn_i/v_i$, as:

- at constant T and V: $dF = \sum v_i\mu_i\,d\xi = \Delta_r G\,d\xi$;
- at constant T and P: $dG = \sum v_i\mu_i\,d\xi = \Delta_r G\,d\xi$.

Indeed, μ_i, the *chemical potential* of i in the mixture, represents the partial molar *Gibbs energy* of i in the mixture, and not the partial molar *Helmholtz energy* of i in the mixture (see *Partial molar quantity*). $\sum v_i\mu_i$ represents therefore the Gibbs energy and not the Helmholtz energy of the reaction:

$$\left(\frac{\partial F}{\partial \xi}\right)_{T,V} = \left(\frac{\partial G}{\partial \xi}\right)_{T,P} = \sum v_i\mu_i = \Delta_r G.$$

In both cases, the equilibrium condition is written, because $d\xi$ has an arbitrary value:

$$\sum v_i\mu_i = \Delta_r G = 0.$$

When developing the expression for the chemical potential:

$$\mu_i = \mu_i^\circ + RT\ln a_i,$$

the equilibrium condition is again obtained:

$$\Delta_r G = \Delta_r G^\circ + RT\ln \prod_i a_i^{v_i} = 0.$$

Remark 7: The equilibrium constant K and the *mass action product* Π must not be confused:

$$\Delta_r G = \Delta_r G^\circ + RT\ln \prod_i a_i^{v_i} = -RT\ln K + RT\ln \Pi.$$

The equilibrium constant can be calculated from thermodynamic data tables independently of any actual system, whereas the mass action product Π is explicitly a function of the state of the actual system under study. The equality $K = \Pi$, implying $\Delta G = 0$, is verified only at equilibrium!

Remark 8: In statistical thermodynamics, when using the relation:

$$K = \exp(-\Delta_r G^\circ / RT),$$

it must be kept in mind that the expression $G = -NkT \ln(Z/N)$ (see *partition function*) gives an absolute value for G, because it leads to $G = 0$ at $T = 0$! Actually, $T = 0 \Rightarrow \Delta_r S = 0$ and $\Delta_r G = \Delta_r H \neq 0$. The equilibrium constant of a chemical reaction $\sum \nu_i M_i = 0$ is then given by:

$$K = \left[\prod_i \left(\frac{Z_i^\circ}{N} \right)^{\nu_i} \right] \times \exp\left[-\frac{\Delta_r H^\circ(0)}{RT} \right],$$

where $\Delta_r H^\circ(0)$ is the standard enthalpy of the reaction at $0\,\mathrm{K}$.

Ergodic—The 'ergodic hypothesis' was first formulated by Boltzmann in 1887. To the question: 'Given a system characterized by its energy E, hence constrained to move on a surface $E - H(q_1, \ldots, q_n, p_1, \ldots, p_n)$, q_i representing the coordinates of position and p_i those of velocity, which points of the surface are reached during the course of time?', Boltzmann answered: 'During the course of time, the path goes through every point of the surface'.

Such a rigid situation is unbearable for mathematicians; that is why Paul Ehrenfest in 1911 stated the quasi-ergodic hypothesis: 'During the course of time, the path goes as near as desired to any point of the surface'. If we consider a constrained system, the quasi-ergodic hypothesis leads to the postulate that if every state available to the system, taking into account the imposed constraints, has the same probability, every system spends, over a long period, the same time in each state.

The ergodic hypothesis is interesting because it allows a difficult problem: the calculation of the mean value of a physical quantity over time, to be replaced by a (slightly) easier problem: the calculation of the mean value, at a given time, of this quantity over a great number of systems. Ergodicity nevertheless remains a property hard to prove.

Ericsson cycle—Very similar to the *Stirling cycle*, it works between two isobars and two isotherms. It was proposed in 1833 by the Swedish engineer John Ericsson (1803–1889).

Euler, Leonhard (1707–1783)—Swiss mathematician whose considerable works cover every mathematical branch developed during the

104 Euler identity

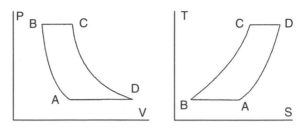

Ericsson cycle: P–V (left) and T–S (right) diagrams.

eighteenth century. The *Euler identity*, relating to first-degree homogeneous equations, finds application in the thermodynamics of mixtures.

Euler identity—Let Y be an integral quantity, homogeneous and of the first degree as a function of the *amount of substance*:

$$Y(T, P, \lambda n_1, \lambda n_2, ..., \lambda n_i, ...) = \lambda \cdot Y(T, P, n_1, n_2, ..., n_i, ...).$$

On differentiating both sides with respect to λ, it follows that:

$$\sum \left(\frac{\partial Y}{\partial(\lambda n_i)}\right)\left(\frac{\partial(\lambda n_i)}{\partial \lambda}\right) = Y.$$

The introduced parameter is eliminated by putting $\lambda = 1$. The relation so obtained is the Euler identity:

$$Y = \sum n_i \left(\frac{\partial Y}{\partial n_i}\right)_{T,P,n_{j \neq i}}.$$

$y_i = (\partial Y/\partial n_i)_{T,P,n_{j \neq i}}$ is the *partial molar quantity* of i in the system.

Physically, the Euler identity $Y = \sum n_i y_i$ means that the total quantity (the integral quantity Y) is the sum of its parts (each mole of i contributes a part y_i to the value of Y). It also applies to quantities of mixing:

$$\Delta_{mix} Y = \sum n_i \Delta y_i.$$

$\Delta_{mix} Y = Y - Y^\circ$: integral quantity of mixing.
$\Delta y_i = y_i - y_i^\circ$: partial molar quantity of mixing.

Remark: It is possible to express $Y(T, V, n_1, n_2, ..., n_i, ...)$ and then to calculate $(\partial Y/\partial n_i)_{T,V,n_{j \neq i}}$. The result is not a partial molar quantity, and thus does not verify the Euler identity, because $Y(T, V, n_1, n_2, ..., n_i, ...)$ is no longer a homogeneous function of the first degree in n_i!

Eutectic—From the Greek ευτηκτος, easily melted. Some metallurgists, captivated by the beautiful structures observed under the microscope, have wrongly seen in this word the root ευ-τεκτον, fine construction, from τεκτον, carpenter. This term, which appeared in 1906, describes the transformation during heating:

$$\langle\text{Solid }\alpha\rangle + \langle\text{Solid }\beta\rangle \rightarrow (\text{Liquid}).$$

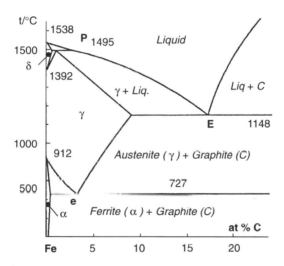

Eutectic: The iron carbon diagram shows a eutectic E, a eutectoid e and a peritectic P. The α and δ forms of iron, body-centred cubic, are identical; the γ form is face-centred cubic.

A eutectic mixture, characterized by its overall composition (eutectic composition), by that of the two phases α and β, and by its fusion temperature (eutectic temperature), behaves on melting like a pure substance. The coordinates of the eutectic point are obtained by taking the point where the *liquidus* curves of the pure compounds α and β meet. The word 'eutectoid', coined in 1922 from *eutectic*, describes the transformation on heating: $\langle\text{Solid }\alpha\rangle + \langle\text{Solid }\beta\rangle \rightarrow \langle\text{Solid }\gamma\rangle$.

Exact differential—Sometimes called 'total differential'. To speak of a 'total exact differential' is a pleonasm! It is always possible to calculate the *differential* dQ of a given function $Q(x, y, z)$. However, if dQ is arbitrarily given, the function Q generally does not exist.

A differential dQ is said to be 'exact' if the function Q exists. A differential $dQ = A(x)dx$, a function of only one variable, is always exact. In order that a differential $dQ = A(x, y)dx + B(x, y)dy$, a function of two variables, be exact, it is necessary and sufficient that between A and B there exists the relation:

$$(\partial A/\partial y)_x = (\partial B/\partial x)_y.$$

A differential $dQ = A(x, y, z)dx + B(x, y, z)dy + C(x, y, z)dz$ is exact if between the functions A, B and C there exist the relations:

$$\left(\frac{\partial B}{\partial z}\right)_{x,y} = \left(\frac{\partial C}{\partial y}\right)_{x,z} ; \quad \left(\frac{\partial C}{\partial x}\right)_{y,z} = \left(\frac{\partial A}{\partial z}\right)_{x,y} ; \quad \left(\frac{\partial A}{\partial y}\right)_{x,z} = \left(\frac{\partial B}{\partial x}\right)_{y,z}.$$

These conditions, which are easy to generalize, arise from the independence of the order of differentions in the calculation of the second derivatives. So, in order for a differential dQ, that is a function of four variables to be an exact differential, there are six conditions to satisfy.

In summary, when a differential dQ is exact:

- the function Q exists;
- $\int_i^f dQ = Q(f) - Q(i)$, independent of the path followed.

In thermodynamics, when dQ is exact, the function Q is a *state function* of the system. The thermodynamic functions U, S, H, F and G are state functions. Generally, neither *work* nor *heat* is a state function.

Excess Gibbs energy of mixing—Like all *excess quantities*, excess Gibbs energy of mixing is defined by: $\Delta_{mix} G^{xs} \equiv \Delta_{mix} G - \Delta_{mix} G^{id}$,

$$\Delta_{mix} G^{xs} = RT \sum n_i \ln a_i - RT \sum n_i \ln x_i.$$

It follows, on introducing the *activity coefficient* $\gamma_i = a_i/x_i$, that:

$$\Delta_{mix} G^{xs} = RT \sum n_i \ln \gamma_i = \sum n_i \Delta \mu_i^{xs},$$

with $\Delta \mu_i^{xs} = RT \ln \gamma_i$. The *Euler identity* is again obtained.

The excess Gibbs energy of mixing allows an indeterminacy to be removed. An activity may tend towards zero (hence its logarithm towards $-\infty$), but an activity coefficient always retains a finite, non-zero value.

Excess quantity—The difference between a measured quantity for a real solution and the quantity calculated for an *ideal solution*:

$$Y^{xs} = Y - Y^{id}.$$

- The excess quantities may be integral, for instance:

$$\Delta_{mix}H^{xs} \equiv \Delta_{mix}H \quad \text{because } \Delta_{mix}H^{id} = 0$$

$$\Delta_{mix}V^{xs} \equiv \Delta_{mix}V \quad \text{because } \Delta_{mix}V^{id} = 0$$

$$\Delta_{mix}G^{xs} = \Delta_{mix}G - \Delta_{mix}G^{id} = RT\sum n_i \ln \gamma_i$$

$$\Delta_{mix}S^{xs} = \Delta_{mix}S - \Delta_{mix}S^{id} = -RT\sum n_i(\partial \ln \gamma_i/\partial T)_P.$$

E

- They may be partial molar, for instance:

$$\Delta h_i^{xs} \equiv \Delta h_i \quad \text{because } \Delta h_i^{id} = 0$$

$$\Delta v_i^{xs} \equiv \Delta v_i \quad \text{because } \Delta v_i^{id} = 0$$

$$\Delta \mu_i^{xs} = \Delta \mu_i - \Delta \mu_i^{id} = RT \ln \gamma_i$$

$$\Delta s_i^{xs} = \Delta s_i - \Delta s_i^{id} = -RT(\partial \ln \gamma_i/\partial T)_P.$$

Between excess partial molar quantities y_i^{xs} and excess integral quantities Y^{xs}, there exist:

- the *definition* relationship: $y_i^{xs} = (\partial Y^{xs}/\partial n_i)_{T,P,n_{j\neq i}}$
- the *Euler identity*: $Y^{xs} = \sum n_i y_i^{xs}$
- the *Gibbs–Duhem equation*: $\sum n_i \, dy_i^{xs} = 0$ (*T* and *P* constant).

Exergy—A thermodynamicist is concerned with *energy* defined for a system from the *first law*, whereas an engineer looks at the *work* that can be done by a system or extracted from a *transformation*. The word 'exergy', coined by Rant in 1956, rapidly became very popular. However, the concept has existed since Gibbs ('available energy of the body and medium', 1873) and Gouy ('possibility of obtaining work', 1889).

By definition, the exergy Ξ of a system with respect to a *reservoir* is the maximum work done by the system during a transformation which brings it into equilibrium with the reservoir. By convention, exergy is positive when putting the system into equilibrium with the reservoir results in doing work on the *surroundings*.

From the *second law, corollary 5*, in order that this work be a maximum, it is necessary that the transformation be *reversible*. It is then related, from *corollary 6*, to the change in the *Helmholtz energy* of the system:

$$\Delta F = F_2 - F_1 = \Delta W_{rev}.$$

If F_0 is the Helmholtz energy of the system in equilibrium with the reservoir:

$$\Delta F = F_2 - F_1 = (F_2 - F_0) - (F_1 - F_0) = \Xi_1 - \Xi_2.$$

Exothermic—From the Greek $\varepsilon\xi\omega$, outside, and $\theta\varepsilon\rho\mu\omega$, to heat. It is said of a transformation in which a *system* gives heat to the *surroundings*. If the transformation occurs at constant pressure, $\Delta H < 0$; if it occurs at constant volume, $\Delta U < 0$. If the system undergoes a transformation which is both *exothermic* and *adiabatic*, its temperature increases.

Expansivity—The linear expansivity, or linear expansion coefficient, is the ratio, reduced to unit length, of the elongation divided by the temperature change. It is necessary to specify the conditions of the transformation. For instance, at constant *stress*:

$$\kappa_{\mathscr{F}} \equiv \frac{1}{l}\left(\frac{\partial l}{\partial T}\right)_{\mathscr{F}}.$$

The volumetric expansivity, or volumetric expansion coefficient, is the ratio, reduced to unit volume, of the volume change divided by the temperature change. It is possible to define:

- the isobaric volumetric expansivity: $\alpha_P \equiv \dfrac{1}{V}\left(\dfrac{\partial V}{\partial T}\right)_P$;

- the isentropic volumetric expansivity: $\alpha_S \equiv \dfrac{1}{V}\left(\dfrac{\partial V}{\partial T}\right)_S$.

Explicit—From Latin *explicitus*, past participle of *explicare*. It signifies that which is clearly expressed. It is sometimes better to be explicit, at the risk of appearing dull.

Extensive quantity—An extensive quantity, or extensity, is a quantity proportional to the dimension of the system. The dimension may be 1 (a length), 2 (a surface), or most often 3 (a volume, an amount of electricity, an amount of substance, an entropy, the functions U, H, F, G, etc.).

An extensive quantity $Y(T, P, n_1, n_2, \ldots, n_i, \ldots)$, proportional to the amount of substance, is expressed by a homogeneous function of the first degree in number of moles:

$$Y(T, P, \lambda n_1, \lambda n_2, \ldots, \lambda n_i, \ldots) = \lambda \times Y(T, P, n_1, n_2, \ldots, n_i, \ldots).$$

Extensive quantities satisfy the *Euler identity*:

$$Y = \sum n_i \left(\frac{\partial Y}{\partial n_i} \right)_{T,P,n_{j \neq i}} = \sum n_i y_i.$$

Theorem: For a *reversible* transformation, the extensity is conservative. Let us consider two systems A and B at equilibrium, which may exchange extensity, the assembly $(A + B)$ remaining isolated; the change in *internal energy* of the whole system is:

$$dU = dU_A + dU_B = \sum X_{iA}\, dx_{iA} + \sum X_{iB}\, dx_{iB},$$

with $dU = 0$ because the assembly $(A + B)$ is isolated and, $X_{iA} = X_{iB} = X_i$ (equilibrium expressed by equating the tensions) because the exchange of extensity between A and B is reversible. As a consequence: $\sum X_i(dx_{iA} + dx_{iB}) = 0$, whatever i. At equilibrium, extensity gained by A is lost by B, and conversely.

Remark 1: This theorem, established for a reversible transformation, is often applicable to *irreversible* transformations. A volume or a quantity of electricity may neither be created nor disappear! *Entropy* has a distinctive behaviour because, for an irreversible transformation, there is no conservation, only creation, of entropy. This behaviour lies at the basis of the *second law* of thermodynamics.

Remark 2: The extensities most often encountered, whose dimension is 3, are homogeneous functions and of the first degree in amount of substance provided that the quantities held constant are intensive quantities (T and P for instance). This would no longer be true if an extensive quantity, such as volume or entropy, were held constant.

Remark 3: The above remark also applies to extensive quantities whose dimension is 2, provided that the only substance taken into account is the matter situated on the surface. The amount of substance at an interface is of course proportional neither to the volume nor to the mass of the system.

Extent—Let us consider a *closed* system, containing n_i moles of M_i, between which a *chemical reaction* $\sum v_i M_i = 0$ may proceed. The extent ξ of this reaction is defined by: $d\xi = dn_i / v_i$, an expression independent of the constituent M_i selected. An extent has the dimensions of amount of substance and is expressed in *moles*.

If, at constant T and P, the extent of a reaction changes by $d\xi$, the amount of M_i will change by $v_i d\xi$ and the *Gibbs energy* of the mixture by:

$$\mathrm{d}G = \left(\sum v_i \mu_i\right) \mathrm{d}\xi = -\mathscr{A}\,\mathrm{d}\xi.$$

$\mathscr{A} = -\sum v_i \mu_i$ is the *affinity* of the chemical reaction. A reaction proceeds spontaneously in the forward direction ($\mathrm{d}\xi > 0$) if the Gibbs energy G of the mixture decreases, that is, if its affinity is positive.

External energy—The name given to the part of the *energy* of a system which is a function only of its macroscopic parameters and of its coordinates in an external field. The external energy may be divided into two contributions: *kinetic energy* and *potential energy*. The external energy is a state function of the system.

   Although the potential energy of a mass m in a gravitational field is clearly an external energy, one may question the status of a material placed in an *electric field* or a *magnetic field*. Briefly, are *electric energy* and *magnetic energy* internal energy or external energy? Authors generally avoid taking a position; those who come to a decision class these energies as internal energy, without too much justification.

   Actually, an acute analysis of the situation shows that electric and magnetic energies have an internal contribution and an external contribution. The fundamental reason proceeds from the fact that the presence of matter in an electric or magnetic field modifies these fields, which is not the case for a gravitational field. It is possible to show that the external contribution to electric energy, due to fixed charges, is potential energy, whereas the external contribution to magnetic energy, due to electric current, is kinetic energy.

F

Factor—The term 'factor' is reserved for the proportionality coefficient k linking two physical quantities A and B whose dimensions are the same: $A = kB$. A factor is therefore dimensionless.

Falsifiable—A proposition is said to be falsifiable, or refutable, in the meaning of *Karl Popper*, if it is possible to imagine an experiment capable of disproving it. According to Popper, a hypothesis or proposition may be called 'scientific' only if it is falsifiable, such as all the principles of thermodynamics. The sentence 'Every student likes thermodynamics' is easily falsifiable, because it can be confirmed or refuted experimentally.

Farad—The unit of electric *capacity* in the SI system (symbol F), named from Michael Faraday, English physicist and chemist (1791–1867):

$$1\,F = 1\,C\,V^{-1} = 1\,A^2\,m^{-2}\,kg^{-1}\,s^4.$$

F

The farad must not be confused with the faraday. The faraday, an obsolete unit of amount of electricity, represents the electric charge of one mole of electrons, that is to say, $96\,485.309\,C$.

Fermi, Enrico (1901–1954)—Italian physicist, who was awarded the Nobel prize in physics in 1938 for having achieved the first fission of uranium. With Paul Dirac in 1926, he perfected the statistics describing the behaviour of *fermions*.

Fermi–Dirac distribution—The name given to the energy *distribution* in an assembly of N particles obeying *Fermi–Dirac statistics* and subjected to the *constraints* that N (number of particles) and E (energy) are constant. By seeking the maximum of the *entropy* function:

$$S = k\ln\Omega = k\ln\prod_i \frac{g_i!}{N_i!(g_i - N_i)!},$$

and by introducing the *Lagrange multipliers* α and β to take into account the constraints (see *Microcanonical ensemble*), we obtain:

$$N_i = \frac{g_i}{\exp(\alpha)\cdot\exp(\beta\varepsilon_i) + 1}$$

$$N_i = \frac{g_i}{\exp(-G/NkT)\cdot\exp(+\varepsilon_i/kT) + 1}.$$

This distribution differs from the *Boltzmann distribution* only in the presence of the term $+1$ in the denominator. When $g_i \gg N_i$, which is often verified, except at low temperatures, the Fermi–Dirac distribution is reduced to that of Boltzmann. By comparison of the above relation in the case $g_i \gg N_i$ with the Boltzmann distribution:

$$N_i = \frac{N}{Z}g_i\exp\left(-\frac{\varepsilon_i}{kT}\right),$$

we again find the expression for the *Gibbs energy* G for N particles (see *Partition function*): $G = -NkT \ln(Z/N)$.

Fermi–Dirac statistics—The name given to the statistics followed by *fermions*, particles whose spin is half-integer. Fermions are characterized by the fact that a *quantum state* cannot be occupied by more than one particle. If N_i indistinguishable particles occupy the energy level ε_i of *degeneracy* g_i, the number of possible permutations is given by:

$$\Omega_i = \frac{g_i!}{N_i!(g_i - N_i)!}.$$

This expression represents the number of permutations between N_i cells containing one fermion and $(g_i - N_i)$ empty cells, that is to say, the number of combinations between g_i objects taken N_i by N_i. By forming the product at all energy levels ε_i, we get the fundamental relationship giving the number of microscopic states in Fermi–Dirac statistics:

$$\Omega_{\mathrm{FD}} = \prod_i \frac{g_i!}{N_i!(g_i - N_i)!}.$$

By using the *Stirling formula* $(\ln N! \approx N \ln N - N)$, we obtain the following expression for the entropy:

$$S = k \ln \Omega_{\mathrm{FD}} = k \sum_i \left[N_i \ln\left(\frac{g_i}{N_i} - 1\right) - g_i \ln\left(1 - \frac{N_i}{g_i}\right) \right].$$

Fermion—A particle whose spin is half-integer. A fermion obeys the Pauli exclusion principle, which stipulates that a quantum state can be empty or occupied by only one particle. A fermion follows *Fermi–Dirac statistics*. Protons, neutrons, electrons (spin = 1/2), helium-3 nuclei (spin = 3/2), etc. are fermions.

Field—A vector field in a system of axes $(\boldsymbol{i}, \boldsymbol{j}, \boldsymbol{k})$ is defined by:

$$\boldsymbol{F} = A(x, y, z) \cdot \boldsymbol{i} + B(x, y, z) \cdot \boldsymbol{j} + C(x, y, z) \cdot \boldsymbol{k}.$$

This field derives from a potential if there is a function V such that:

$$A = (\partial V / \partial x)_{y,z} \quad B = (\partial V / \partial y)_{x,z} \quad C = (\partial V / \partial z)_{x,y},$$

which may be expressed by: $\boldsymbol{F} = \mathrm{grad}\, V$ or, when the force is directed towards decreasing potential: $\boldsymbol{F} = -\mathrm{grad}\, V$.

If a field derives from a potential, the circulation of the vector field from a point i to a point f is independent of the path followed:

$$\int_i^f \boldsymbol{F} \cdot \mathbf{d}\boldsymbol{r} = V(\mathrm{f}) - V(\mathrm{i}).$$

Not all force fields derive from a potential. A necessary and sufficient condition for a vector field $\boldsymbol{F}$ to be the gradient of a scalar function V is that $\mathrm{curl}\, \boldsymbol{F} = 0$, that is to say:

$$\left(\frac{\partial B}{\partial x}\right)_{y,z} = \left(\frac{\partial A}{\partial y}\right)_{x,z}; \quad \left(\frac{\partial A}{\partial z}\right)_{x,y} = \left(\frac{\partial C}{\partial x}\right)_{y,z}; \quad \left(\frac{\partial C}{\partial y}\right)_{x,z} = \left(\frac{\partial B}{\partial z}\right)_{x,y}.$$

In this case the field is said to be irrotational.

If $\boldsymbol{F}$ is not irrotational, there may exist a scalar function $\lambda(x, y, z)$ such that $\mathrm{curl}(\lambda \boldsymbol{F}) = 0$, which means that the differential:

$$\mathrm{d}Q = \boldsymbol{F} \cdot \mathbf{d}\boldsymbol{r} = A\,\mathrm{d}x + B\,\mathrm{d}y + C\,\mathrm{d}z$$

is integrable, the scalar λ being the *integrating factor*. Moreover:

$$\mathrm{curl}(\lambda \cdot \boldsymbol{F}) = \lambda \cdot \mathrm{curl}\, \boldsymbol{F} - \boldsymbol{F} \cdot \mathrm{grad}\, \lambda.$$

By taking the scalar product of the two members by $\boldsymbol{F}$, we get:

$$\boldsymbol{F} \cdot \mathrm{curl}(\lambda \cdot \boldsymbol{F}) = \lambda \cdot \boldsymbol{F} \cdot \mathrm{curl}\, \boldsymbol{F}.$$

The existence of an integrating factor, that is to say the condition $\mathrm{curl}(\lambda \boldsymbol{F}) = 0$, yields: $\boldsymbol{F} \cdot \mathrm{curl}\, \boldsymbol{F} = 0$, in other words:

$$A\left[\left(\frac{\partial C}{\partial y}\right)_{x,z} - \left(\frac{\partial B}{\partial z}\right)_{x,y}\right] + B\left[\left(\frac{\partial A}{\partial z}\right)_{x,y} - \left(\frac{\partial C}{\partial x}\right)_{y,z}\right]$$
$$+ C\left[\left(\frac{\partial B}{\partial x}\right)_{y,z} - \left(\frac{\partial A}{\partial y}\right)_{x,z}\right] = 0.$$

When $\boldsymbol{F} \cdot \mathrm{curl}\, \boldsymbol{F} \neq 0$, the differential $\mathrm{d}Q$ is not integrable. However, if $\boldsymbol{F}$ is dependent on only two variables, the condition $\boldsymbol{F} \cdot \mathrm{curl}\, \boldsymbol{F} = 0$ is satisfied and there is always an integrating factor.

First law—Also known as the equivalence principle or energy conservation principle, its first explicit statement was given by Clausius in 1850:

There is a state function E, called 'energy', whose differential equals the work exchanged with the surroundings during an adiabatic process.

Actually, this statement may be deduced from the *stable equilibrium principle*.

Corollary 1: To make a *system* evolve from a well-defined initial state to a well-defined final state by an *adiabatic process*, we must exchange with the *surroundings* a quantity of *work* which depends neither on the path followed nor on the nature (mechanical, electrical, etc.) of the work exchanged:

$$\Delta E = E_f - E_i = \int_i^f \sum dW_{ad}.$$

Remark 1: The state function E introduced here under the name of energy is sometimes improperly called *internal energy U*. In reality, energy may be decomposed into an internal energy, a function of characteristic parameters of the system, and an *external energy*, a function of the imposed *fields*. External energy may be separated into *kinetic energy* and *potential energy*.

Corollary 2: There are processes, called *diabatic*, for which the above relationship is not satisfied. For these processes, the energy transferred equals the work exchanged with the surroundings during an adiabatic process in which the system evolves from the same initial state to the same final state.

Remark 2: Experience shows that it is not always possible to find an adiabatic process in which a system evolves from a given initial state to a given final state. However, it is always possible to find an adiabatic process in which the reverse transformation is allowed. For instance, it is impossible to cool water, but it is quite possible to heat water by means of an adiabatic process.

Corollary 3: By calling *heat*, dQ, the difference between the energy change dE of a system and the work exchanged with the surroundings, $dQ = dE - \sum dW$, the first law may be expressed in the form: 'the energy change of a system during a *transformation* equals the sum of the heat and work exchanged with the surroundings': $dE = dQ + dW$.

Remark 3: The above corollary, which states the energy conservation principle, is often presented as the first law. This presentation is quite legitimate (once this third corollary is admitted, the first two, as well as the Clausius statement, are deduced), but it supposes that heat has been defined beforehand, independently of the first law. Finally, it must be pointed out that the first law consists not in the expression

'$dE = dQ + dW$', but in the fact that E is a state function, that is dE is an exact differential, whereas neither Q nor W is a state function.

Flux—It is possible to characterize the flux of a vector field or the flux of an extensity.

ELECTROMAGNETISM: The flux Φ of a vector field $\mathscr{B}$ through a surface $\mathscr{S}$ is defined by its differential: $d\Phi = \mathscr{B} \cdot d\mathscr{S}$, $d\mathscr{S}$, representing a vector perpendicular to the surface element $d\mathscr{S}$. If the surface $\mathscr{S}$ is closed, the vector $d\mathscr{S}$ is directed towards the outside of the surface. By integration:

$$\Phi = \iint_{\mathscr{S}} \mathscr{B} \cdot d\mathscr{S}.$$

If $\mathscr{B}_x$, $\mathscr{B}_y$, and $\mathscr{B}_z$ are the components of the vector $\mathscr{B}$, the total flux Φ_t out of a volume element dV is given by:

$$d\Phi_t = \left[\left(\frac{\partial \mathscr{B}_x}{\partial x} \right) + \left(\frac{\partial \mathscr{B}_y}{\partial y} \right) + \left(\frac{\partial \mathscr{B}_z}{\partial z} \right) \right] dV = \operatorname{div} \mathscr{B} \cdot dV.$$

The divergence of a vector field represents the limit of the flux per unit volume through a closed surface when $V \to 0$:

$$\Phi = \iint_{\mathscr{S}} \mathscr{B} \cdot d\mathscr{S} = \iiint_V \operatorname{div} \mathscr{B} \cdot dV \quad \text{(Green's formula)}.$$

- A flux of magnetic field, or simply a magnetic flux, is expressed in webers:
$$1 \, \text{Wb} = 1 \, \text{m}^2 \text{kg s}^{-2} \text{A}^{-1}.$$

- A flux of electric field is expressed in V m:
$$1 \, \text{V m} = 1 \, \text{m}^3 \text{kg s}^{-3} \text{A}^{-1}.$$

Flux being thus defined, a flux density is then a flux per unit surface, that is to say, a field.

- A magnetic flux density, or *magnetic field*, is expressed in teslas:
$$1 \, \text{T} = 1 \, \text{Wb m}^{-2} = 1 \, \text{kg s}^{-2} \text{A}^{-1}.$$

- An electric flux density, or *electric field*, is expressed in V m^{-1}:
$$1 \, \text{V m}^{-1} = 1 \, \text{m kg s}^{-3} \text{A}^{-1}.$$

IRREVERSIBLE PROCESSES: In this domain there is usually much confusion in terminology, so it is necessary to pay great attention. The flux of an extensive quantity x through a surface is most often defined as a current density. It then represents the quantity x that crosses unit surface during unit time. It is a vector quantity:

- a flux of heat J_Q is expressed in $J\,m^{-2}s^{-1}$ (or $W\,m^{-2}$);
- a flux of matter J_i is expressed in $mol\,m^{-2}s^{-1}$;
- a flux of electricity J_e is expressed in $C\,m^{-2}s^{-1}$ (or $A\,m^{-2}$).

If a volume V is neither a *source* nor a *sink* for the quantity x, the flux of x through the surface delimiting this volume equals zero. On the other hand, if the quantity x may be created or disappear in the volume V, the flux of x through the surface no longer equals zero and it may be convenient to reduce the flux to unit volume instead of unit surface. The flux thus defined loses its vector character, becoming a scalar quantity:

- a flux of heat J_Q is expressed in $J\,m^{-3}s^{-1}$ (or $W\,m^{-3}$);
- a flux of matter J_i is expressed in $mol\,m^{-3}s^{-1}$;
- a flux of electricity J_e is expressed in $C\,m^{-3}s^{-1}$ (or $A\,m^{-3}$);
- a flux of *entropy* is expressed in $J\,K^{-1}m^{-3}s^{-1}$.

Corresponding 'currents' are naturally expressed in W (heat current), in $mol\,s^{-1}$ (current of matter), in A (electric current), or in $W\,K^{-1}$ (entropy current).

Remark: When expressing the *entropy production* by $\dot{\sigma} = \sum J_i X_i$, J_i represents either a generalized current when $\dot{\sigma}$ is the entropy created during unit time, or a flux when $\dot{\sigma}$ is the entropy created per unit volume during unit time. X_i then represents a *generalized force*, or a *generalized affinity*. It is an intensive quantity whose value does not depend on whether or not $\dot{\sigma}$ is reduced to unit volume or unit amount of substance.

A flux being a current density, the expression 'flux density' applied to the transfer of heat, matter, electricity, or entropy must be outlawed. It is confined to those who assimilate (wrongly) flux and current. In electromagnetism, a flux density is a field.

Force—A force is the derivative of a momentum with respect to time: $F = d(mv)/dt$. If the mass is constant, force becomes the product of mass times acceleration. The unit of force in the SI system is the newton: $1\,N = 1\,m\,kg\,s^{-2}$. Force is a *tension* whose conjugate *extensity* is length.

Fountain effect—This effect, discovered in 1938 by Allen and Jones, is one of the most spectacular manifestations of the *thermomolecular effect* in superfluid helium-4 below 2.19 K. If we dip into liquid helium (system A) a bent tube filled with a fine powder and surmounted by a capillary (system B), and irradiate the bend, we observe the superfluid helium move through the powder, falling as rain at the free end of the capillary. The two systems A and B filled with helium, the assembly (A + B) being *isolated*, are at different temperatures, but lower than 2.19 K. Only the superfluid helium moves through the capillary. As the capillary does not allow any transfer of entropy or heat, the temperature and the entropy of each system must remain constant. The equilibrium state of a system evolving at constant S and V is obtained by seeking the minimum of the *internal energy* function.

For each closed system E (E = A, B): $dU = T_E \, dS_E - P_E \, dV_E + \mu_E \, dn_E$.

Assuming $dS_E = dV_E = 0$ and taking into account conservation of matter ($dn = dn_B = -dn_A$), the equilibrium condition is reduced to:

$$dU - dU_A + dU_B = (\mu_B - \mu_A)\,dn = 0,$$

that is, to equality of the *chemical potential* of helium in each system. If there is a temperature difference between the two compartments, a pressure difference must develop at equilibrium:

$$d\mu = -s\,dT + v\,dP = 0,$$

whence: $dP/dT = s/v$, where s and v are respectively the molar entropy and the molar volume of helium. If, initially, $T_B > T_A$ and $P_B = P_A$, helium will flow spontaneously from the colder compartment A towards the warmer B, until the relation $dP/dT = s/v$ is obeyed. This does not contradict the *second law*, because only superfluid helium, which does not carry entropy, goes through the capillary.

Fourth law—Since the *third law* postulates the existence of an *absolute zero*, the lower limit of temperature, it seems natural to raise the question of possible upper limit. Statements encountered in the literature under the name of 'fourth law' postulate the existence of such a limit. Here, the designation 'principle' is excessive: at most it is a conjecture.

There are two kinds of reasoning, which both lead to placing the upper limit of temperature between 10^{11} and 10^{12} K.

The first argument, developed by Yves Rocard in 1952, is based on *electric field* calculations. By taking the electron radius at 2×10^{-15} m, one calculates an electric field at its surface of:

$$E = q/4\pi\varepsilon_0 r^2 \approx 3.6 \times 10^{20}\,\mathrm{V\,m^{-1}},$$

and hence an energy density of the electric field of:

$$dW/dV = \varepsilon_0 E^2/2 \approx 1.15 \times 10^{30} \, \text{J m}^{-3}.$$

Now, the energy density of *radiation* is given by Stefan's law:

$$dW/dV = aT^4 = 7.56 \times 10^{-16} T^4.$$

Identifying the two expressions of dW/dV leads to $T \approx 2 \times 10^{11}$ K. As we do not know of any physical means of obtaining an electric field higher than 3.6×10^{20} V m^{-1}, it is reasonable to think that 2×10^{11} K represents a temperature limit for radiation.

The second argument, due to D. C. Kelly in 1973, comes from the observation that adding energy at constant volume may lead to an increase in temperature or to a creation of new particles. Heating hydrogen, for instance, produces the following steps:

- dissociation of molecules into atoms (10^3–10^4 K);
- ionization of atoms (10^4–10^5 K);
- creation, by collision, of electron–positron pairs ($\sim 10^9$ K);
- creation of pions by proton–proton collisions ($\sim 10^{11}$ K).

Above a limit around 10^{11}–10^{12} K, the kinetic energy of protons becomes higher than the rest mass of pions. Adding energy at constant volume does not increase the temperature further, but does increase the number and the variety of particles present.

Fraction—When two physical quantities A and B are proportional, with the same dimensions, $A = kB$, the dimensionless number k is called a fraction if it is lower than unity, for instance if it represents the quotient of a part over the whole. In the general case, one speaks of a *ratio*. The most commonly encountered fractions are the following:

MASS FRACTION, symbolized by w_i, the quotient of the mass of the constituent i over the whole mass of the mixture. The mass percentage represents 100 times the mass fraction.

MOLE FRACTION, symbolized by x_i in a condensed phase and by y_i in a gaseous one, the quotient of the amount of constituent i in a mixture over the total number of moles: $x_i = n_i/\sum n_i$. If the pure constituent i ($x_i = 1$) is selected as the *standard state*, compositions in the mixture are commonly expressed in mole fractions.

VOLUME FRACTION, symbolized by φ_i, the quotient of the volume of i in a mixture over the total volume of the mixture. In a single-phase mixture, this concept is a dangerous one, because the volume of one mole of i in the mixture (*partial molar volume*) generally differs from the molar volume of i out of the mixture. The partial molar volume may even be negative, which removes any meaning from volume fraction. In a multiphase mixture, it is possible to speak of the volume fraction of a phase, the ratio of the volume occupied by this phase to the total volume. Volume fractions are of interest for polymer blends, for which the notion of mole fraction loses its meaning.

Free energy—See *Helmholtz energy*, defined by $F \equiv E - TS$, and *Gibbs energy*, defined by $G \equiv H - TS$. Here, the word 'free' has to be taken as meaning 'available in the form of useful work'. Let ΔF be the Helmholtz free energy change of a system for a given transformation. ΔF represents the minimum work exchanged by the system with the surroundings, a minimum observed when the transformation is reversible. If $\Delta F > 0$, ΔF represents the minimum work which must be done on the system; if $\Delta F < 0$, $|\Delta F|$ then represents the maximum work which can be done by the system on the surroundings. The Gibbs free energy variation ΔG represents the minimum work exchanged by the system with the surroundings without taking into account the work of pressure forces. As for ΔF, the minimum is observed when the transformation is reversible. For instance, when a chemical reaction proceeds in a galvanic cell, ΔG represents the minimum electric work exchanged by the cell with the surroundings.

Free enthalpy— A name sometimes given to the *Gibbs energy* function, $G = H - TS$, by people whose mother tongue is not Shakespeare's language.

Fugacity—A physical quantity introduced by Gilbert N. Lewis in 1901 and whose dimensions are those of pressure; it expresses the tendency for a constituent to escape from a medium. The fugacity of i in a condensed phase is defined by the fugacity of its vapour at equilibrium:

$$d\mu_i = -s_i dT + v_i dP.$$

By integrating the expression for $d\mu_i$ at constant temperature, we obtain:

$$\mu_i(T, p_i) - \mu_i(T, p_i^\circ) = \int_{p_i^\circ}^{p_i} v_i dP.$$

If the constituent i is an ideal gas: $v_i = RT/p_i$, whence:

$$\mu_i(T, p_i) - \mu_i(T, p_i^\circ) = RT \ln \frac{p_i}{p_i^\circ}.$$

If the constituent i is a real gas, the calculation of the integral requires knowledge of an *equation of state*. To avoid this difficulty, the problem is assumed solved and, by analogy, an equation similar to the above equation is written for real gases:

$$\mu_i(T, p_i) - \mu_i(T, p_i^\circ) = RT \ln \frac{f_i}{f_i^\circ},$$

where f_i is the fugacity of i in the mixture and f_i° the *standard fugacity*. Such a relationship defines the fugacity only with a multiplicative constant. To fix this constant, it is possible to take into account the behaviour of every real gas which tends towards that of ideal gas when the pressure tends towards zero: $f_i/p_i \to 1$ when $p_i \to 0$, whence we obtain an unequivocal relationship between pressure and fugacity:

$$RT \ln \frac{f_i}{f_i^\circ} = \int_{p_i^\circ}^{p_i} v_i \, dp \quad \text{with} \quad \lim_{p_i \to 0} \left(\frac{f_i}{p_i} \right) = 1.$$

With $p_i^\circ \to 0$ (then $f_i^\circ/p_i^\circ \to 1$), it follows that:

$$RT \ln \frac{f_i}{p_i} = \int_0^{p_i} \left(v_i - \frac{RT}{p} \right) dp.$$

This relationship between pressure and fugacity is rigorous, whether gas i be pure or not; however, if i is in a mixture, the relation is hard to apply because the equation of state $f(p_i, v_i, T) = 0$ is generally unknown.

In practice, it is possible to assimilate pressure and fugacity below 1 bar. Near ambient temperature, the error is of the order of 0.01% for nitrogen, 0.1% for CO_2, CH_4, or H_2O, and 1% for more complex compounds such as pentane, benzene, methanol, etc.

Finally, note that the integral $\int_0^{p_i}(v_i - (RT/p)) dp$ is always finite: If $p \to 0$, $v_i \to \infty$ and $RT/p \to \infty$, but the difference $(v_i - RT/p)$ always tends towards B, the second virial coefficient.

Fugacity coefficient—This is defined by:

$$\gamma_i \equiv f_i/p_i,$$

the ratio of the *fugacity* of a constituent i to its partial pressure. The fugacity coefficient identifies with the *activity coefficient* when pure i under 1 *bar* pressure is chosen as standard state. The advantage of a fugacity coefficient lies in the fact that it remains finite at low pressures. For a pure gas i, $\gamma_i \to 1$ when $p_i \to 0$, which is not necessarily true for a gas i in a mixture! (See *Interaction parameter*.)

Fusion—Fusion, or melting, is the transformation from the solid state to the liquid state; it is an *endothermic* phenomenon. A noteworthy exception is that of helium-3 below 0.35 K. The slope of the solid–liquid equilibrium curve is given by the *Clapeyron equation*:

$$(\mathrm{d}P/\mathrm{d}T)_{eq} = \Delta_{fus}H/T \cdot (v_{liq} - v_{sol}).$$

Generally, the slope is positive (because $\Delta_{fus}H > 0$ and $v_{liq} > v_{sol}$). Water and bismuth, for which $v_{liq} < v_{sol}$, have an exceptional behaviour.

G

Galvanic cell—A device able to turn chemical energy into electrical energy. The first galvanic cell, demonstrated by Volta in 1800, was made by piling up alternate copper and zinc sheets separated by cloth dipped in acid. It is possible to construct a galvanic cell around any redox reaction separated into two half-reactions. The oxidation, $Red_1 \to Ox_1 + ne^-$, occurs in a compartment containing the *anode*; the reduction, $Ox_2 + ne^- \to Red_2$, occurs in a compartment containing the *cathode*. The two compartments are separated by an ion-conductive membrane.

Consider for instance a fuel cell in which the following reaction occurs:

$$CH_4 + 2O_2 \to CO_2 + 2H_2O.$$

The combustion may be broken down into two half-reactions:

- at the anode: $CH_4 + 2H_2O \to CO_2 + 8H^+ + 8e^-$;
- at the cathode: $2O_2 + 8H^+ + 8e^- \to 4H_2O.$

122 Galvanic cell

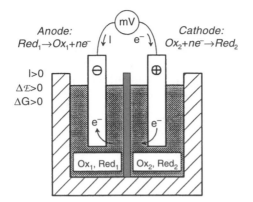

Anode:
$Red_1 \rightarrow Ox_1 + ne^-$

Cathode:
$Ox_2 + ne^- \rightarrow Red_2$

$I > 0$
$\Delta \mathscr{E} > 0$
$\Delta G > 0$

Ox$_1$, Red$_1$ | Ox$_2$, Red$_2$

Galvanic cell: *Outside the cell, current travels from the positive pole to the negative one, whereas electrons travel in the opposite direction. In the above cell, the overall reaction proceeds in the direction:*

$$Ox_2 + Red_1 \rightarrow Ox_1 + Red_2.$$

Somewhat different from galvanic cells are 'concentration cells', in which the reaction occurring at the anode is the reverse of the reaction occurring at the cathode. The two compartments differ only in the concentrations of the reactants.

By convention, the electromotive force of a cell equals the difference between the potential of the right-hand electrode and that of the left-hand electrode: $\Delta \mathscr{E} = \mathscr{E}_2 - \mathscr{E}_1$. Consequently, the current circulating outside the cell, from electrode 2 to electrode 1, will be considered as positive. Under these conditions, the reactions associated with the electrodes are those which make the electrons travel, outside the cell, from electrode 1 to electrode 2, which corresponds to oxidation in compartment 1 and reduction in compartment 2. In summary: $\Delta \mathscr{E} = \mathscr{E}_2 - \mathscr{E}_1 > 0 \Rightarrow I > 0$: the reaction proceeds in the direction:

$$Red_1 + Ox_2 \rightarrow Ox_1 + Red_2.$$

If the cell works at constant temperature and pressure, the *Gibbs energy of reaction* is related to the electromotive force by:

$$\Delta_r G = -n \mathscr{F} \Delta \mathscr{E},$$

$\mathscr{F}$ being the Faraday constant and n the number of electrons exchanged during the reaction. $\Delta_r G$, the Gibbs energy of the reaction, represents the available energy under the condition of *reversibility*, excluding the work of pressure forces. When the redox reaction takes place once, the cell turns the available energy into electrical work, that is, '$n \mathscr{F} \Delta \mathscr{E}$' if n electrons are exchanged. The minus sign arises from the sign convention for $\Delta_r G$ and $\Delta \mathscr{E}$. The entropy and enthalpy of the reaction are calculated by means of classical differentials:

$$\Delta_r S = n\mathscr{F}\left(\frac{\partial \Delta\mathscr{E}}{\partial T}\right)_P \qquad \Delta_r H = n\mathscr{F}T^2\frac{\partial}{\partial T}\left(\frac{\Delta\mathscr{E}}{T}\right)_P.$$

Galvani potential—The electric potential difference between a point situated inside a metal and a point situated in a solution in contact with the metal. This potential, also called 'inner potential' is not measurable. The difference between the inner potential φ and the outer potential ψ (or *Volta potential*) represents χ, the surface potential: $\varphi = \psi + \chi$.

G

Gamma—A name sometimes given to the nanotesla ($1\gamma = 10^{-9}\,T$).

Gas—From the Greek $\chi\alpha o\varsigma$, chaos. This word was proposed by Jan Baptist van Helmont, Flemish physician and chemist (1577–1644), who was the first to identify carbon dioxide. The gaseous state is characterized by very weak interactions between particles, which leads to high *compressibility*, and by a marked tendency to occupy the available volume. On increasing the pressure, the mean distance between the particles decreases, the interactions increase, and the compressibility decreases. At high pressure, there is no sharp boundary between the *liquid* and gaseous states. A fluid is considered as gaseous when the kinetic energy of a particle is high in comparison with the interaction potential energy between two particles.

Gas mixture—A mixture of *ideal gases* obeys the equation of state $PV = \sum_i n_i RT$. Consequences are Amagat's and Dalton's laws:

• Amagat's law (1893) concerns *partial volumes*: the volume of a mixture equals the sum of the partial volumes of the constituents. The partial volume of a constituent i equals the volume occupied by this constituent alone if it were at the total pressure of the mixture. Amagat's law is expressed by the equation:

$$PV = P\sum V_i = RT\sum n_i.$$

• Dalton's law (1805) concerns *partial pressures*: the pressure of a mixture equals the sum of the partial pressures of the constituents. The partial pressure of a constituent i equals the pressure exerted by this constituent if it were alone in the whole volume available to the mixture. Dalton's law is expressed by the equation:

$$PV = V \sum P_i = RT \sum n_i.$$

A mixture of real gases obeys neither one nor the other. Sometimes, mainly at low pressure, it may follow one of them, more often that of Amagat; the other is then not satisfied. It is possible to expand the equation of state of a real gas mixture with a virial form:

$$P = \left(\frac{n}{V}\right)RT + B\left(\frac{n}{V}\right)^2 + C\left(\frac{n}{V}\right)^3 + \cdots$$

For a binary mixture, B, the second virial coefficient may be represented by a sum of terms characterizing interactions between pairs 1–1, 1–2 and 2–2:

$$B = B_{11}y_1^2 + 2B_{12}y_1y_2 + B_{22}y_2^2,$$

where y_i is the mole fraction of constituent i in the gaseous mixture. The expression may be generalized for any mixture:

$$B = \sum_i \sum_j B_{ij}y_iy_j,$$

with, naturally, $B_{ij} = B_{ji}$. The coefficient B has the characteristics of an *integral molar quantity* and may be expressed by:

$$B = \sum b_iy_i,$$

where b_i is the *partial molar quantity* of constituent i:

$$b_i = B + \sum_{j=2}^{n} (\delta_{ij} - y_j)\left(\frac{\partial B}{\partial y_j}\right)_{y_{j \neq i}}.$$

By introducing $D_{12} = 2B_{12} - B_{11} - B_{22}$, we obtain, for a binary mixture:

$$b_1 = B_{11} + D_{12}(1 - y_1)^2$$
$$b_2 = B_{22} + D_{12}(1 - y_2)^2.$$

Gauss—A former cgs unit of *magnetic field*. Its use is now discouraged ($1G = 10^{-4} T$).

Gaussian distribution—See *Normal distribution*. The well-known 'Gaussian curve' was actually first put forward by Jacques I. Bernoulli (1654–1705) in his posthumous treatise *Ars conjectandi* published in 1713.

Generalized affinity—In the thermodynamics of irreversible processes, *entropy production* is given by:

$$\frac{d\sigma}{dt} \equiv \dot{\sigma} = \sum J_i X_i.$$

J_i represents a *flux* or *generalized current*, depending on whether or not the entropy production is reduced to unit volume.

X_i represents a generalized affinity (or force).

Example 1: When an amount of heat dQ is transferred from a system A to a system B whose respective temperatures are T_A and T_B, the entropy production is given by:

$$\dot{\sigma} = \frac{dQ}{dt}\left(\frac{1}{T_B} - \frac{1}{T_A}\right).$$

The generalized affinity is then: $[(1/T_B) - (1/T_A)]$.

Example 2: When the *extent* of a chemical reaction varies by $d\xi$, the entropy production at constant T and P is given by:

$$\dot{\sigma} = -\frac{1}{T}\sum v_i \mu_i \frac{d\xi}{dt} = \frac{\mathscr{A}}{T}\frac{d\xi}{dt}.$$

$\mathscr{A}$ is the *affinity* of the reaction as defined by De Donder.
$\mathscr{A}/T$ is the generalized affinity of the chemical reaction.

Example 3: When, in an electrochemical system, a constituent i whose charge is z_i is transferred from a potential Φ_A to a potential Φ_B, the extent of the transfer may be defined by $dn = dn_B = -dn_A$. The entropy production is then given by:

$$\dot{\sigma} = -\left(\frac{\tilde{\mu}_A}{T}\frac{dn_A}{dt} + \frac{\tilde{\mu}_B}{T}\frac{dn_B}{dt}\right) = \frac{\tilde{\mathscr{A}}}{T}\cdot\frac{d\xi}{dt}.$$

$\tilde{\mathscr{A}}$ is the electrochemical affinity corresponding to the transfer of i:

$$\tilde{\mathscr{A}} = \tilde{\mu}_A - \tilde{\mu}_B = (\mu_A + z_i \mathscr{F}\Phi_A) - (\mu_B + z_i \mathscr{F}\Phi_B).$$

$\tilde{\mathscr{A}}/T$ is the generalized affinity corresponding to the transfer of i.

Generalized current—In the thermodynamics of irreversible processes, the expression 'generalized current' is reserved for an extensive quantity exchanged in unit of time. For instance, a current of matter is expressed in $mol\,s^{-1}$; a current of energy (heat or work) is expressed in W, or $J\,s^{-1}$. This definition of a current is coherent with that of an electric current, which is expressed in A, or Cs^{-1}. When the *entropy production* is expressed by $\dot{\sigma} = \sum J_i X_i$, the term J_i may represent a generalized current, or a *flux*, that is, a current density.

Generalized force—The concept of generalized force must be used cautiously, because there are two points of view that differ but are legitimate. It is indeed possible to consider that a *force* whose point of application moves produces *work* or motion.

IN CLASSICAL THERMODYNAMICS, which takes the standpoint of work, the change in *internal energy* is given by:

$$dU = \sum X_i \, dx_i.$$

X_i represents a *tension* and x_i an *extensive quantity*. The collective name of 'generalized forces' is sometimes given to tensions, which allows use of the terms 'surface tension forces', 'pressure forces', 'electric field forces', 'magnetic field forces'; under these conditions the associated extensities become 'generalized displacements'. Such expressions do not clarify anything, generate confusion and make double use of the concepts of tension and extensity. A pressure is no more a force than an electric potential is a pressure!

IN THE THERMODYNAMICS OF IRREVERSIBLE PROCESSES, which takes the standpoint of motion, the *entropy production* $\dot{\sigma}$ is given by:

$$\dot{\sigma} = \sum J_i X_i.$$

J_i represents a *flux* or a *generalized current*, depending on whether or not the entropy production is reduced to unit of volume, whereas X_i represents a generalized force, or a *generalized affinity*. In this expression, the force X_i by which the current J_i is multiplied to create entropy may be considered as the cause of the current (a flux of heat is caused by a temperature gradient, a flux of matter by a gradient of chemical potential, etc.). The difficulties in applying this notion arise generally from the fact that, for a given process, there is not just one couple (J, X).

Gibbs, Josiah Willard (1839–1903)—American thermodynamicist. Between 1875 and 1878, he published a fundamental dissertation on 'The equilibrium of heterogeneous substances' in the *Transactions of the Connecticut Academy of Science*, a work of more than 300 pages that remained largely ignored owing to the small circulation of the journal. It was translated into German in 1897 by Ostwald and into French in 1899 by Le Chatelier.

Because of all his work, Gibbs is considered the main architect of the development of thermodynamics. Amongst his most outstanding contributions, he introduced the concept of *thermodynamic potential* and that of *chemical potential* with its notation μ, now universally adopted. He stated the *phase rule*, established theoretically the properties of *azeotropes*, later experimentally verified by Konovalov, created *phase diagrams*, and coined the expression *configurational entropy*.

Towards the end of his life, Gibbs participated in the birth of statistical thermodynamics (he also proposed the term), which issued from the works of Maxwell and Boltzmann, and introduced the concept of ensemble, now called *Gibbs ensemble*, and the *canonical* and *microcanonical ensembles*.

G

Gibbs–Duhem equation—A differential equation connecting *partial molar quantities*, obtained by differentiating the *Euler identity*:

$$Y = \sum_i n_i y_i$$

$$\left(\frac{\partial Y}{\partial T}\right)_{P,n_i} dT + \left(\frac{\partial Y}{\partial P}\right)_{T,n_i} dP + \sum_i \left(\frac{\partial Y}{\partial n_i}\right)_{T,P,n_{j \neq i}} dn_i = \sum_i n_i \, dy_i + \sum_i y_i \, dn_i.$$

On introducing the relationship defining a partial molar quantity:

$$y_i = (\partial Y/\partial n_i)_{T,P,n_{j \neq i}},$$

we obtain:

$$\left(\frac{\partial Y}{\partial T}\right)_{P,n_i} dT + \left(\frac{\partial Y}{\partial P}\right)_{T,n_i} dP = \sum_i n_i \, dy_i,$$

which is the most general form of the Gibbs–Duhem equation. In practice, at constant temperature and pressure it reduces to:

$$\sum_i n_i \, dy_i = 0.$$

Expressed in this form, the Gibbs–Duhem relation gives access to all partial molar quantities of a mixture from a knowledge of only one of them.

Applied to *chemical potential*:

$$\sum_i n_i \, d(\Delta\mu_i) = RT \sum_i n_i \, d(\ln a_i) = 0.$$

It is better to introduce $a_i = \gamma_i x_i$ and, after noting the result:

$$\sum_i n_i \, d(\ln x_i) = n \sum_i x_i \, d(\ln x_i) = n \sum_i x_i (dx_i/x_i) = n \sum_i dx_i \equiv 0,$$

to integrate in the form: $\sum n_i \, d(\ln \gamma_i) = 0$, which removes indeterminacies because, when $x_i \to 0$, $a_i \to 0$ but $\gamma_i = a_i / x_i \to$ finite, non-zero value.

For a binary system:

$$\ln \gamma_2 = \int_1^{x_2} -\frac{1-x_2}{x_2} \, d\ln \gamma_1,$$

or, in a more symmetrical and memorable form:

$$\frac{\partial \ln \gamma_1}{\partial (1-x_1)^2} = \frac{\partial \ln \gamma_2}{\partial (1-x_2)^2}.$$

Gibbs energy—The Gibbs energy G, sometimes called the Gibbs function or Gibbs *free energy*, a function introduced by Gibbs in 1878, is defined by:

$$G \equiv H - TS.$$

It may also be defined from the *Helmholtz energy*:

$$G = F + PV.$$

When a *system* evolves from a well-defined initial state to a well-defined final state, the Gibbs energy change ΔG equals the *work* exchanged by the system with its *surroundings*, less the work of the pressure forces, during a *reversible* transformation of the system from the same initial state to the same final state.

For instance, when a redox reaction occurs in a *galvanic cell* through a *reversible* path, the electromotive force $\Delta \mathscr{E}$ between the electrodes is related to ΔG, the *Gibbs energy of reaction*, by:

$$\Delta G = -n\mathscr{F}\Delta\mathscr{E}.$$

The minus sign is due to the sign convention for $\Delta \mathscr{E}$. $\mathscr{F}$ is the Faraday constant and n the number of electrons exchanged in the reaction.

GIBBS ENERGY, THERMODYNAMIC POTENTIAL: By differentiating G:

$$dG = dH - T\,dS - S\,dT$$

$$dG = -S\,dT - T\,d\sigma + V\,dP + \sum \mu_i \, dn_i + \cdots$$

At constant temperature, pressure and amount of substance (*closed system without chemical reaction*) and in the absence of any other energy exchange (magnetic, electric, etc.), the expression for dG is reduced to:

$$dG = -T\,d\sigma \leqslant 0.$$

In such conditions, the function G can only decrease and reach a minimum at equilibrium. The Gibbs energy is thus the *thermodynamic potential* of a system constrained by the condition (T, P, n_i) constant.

In practice, the $P=$const. constraint is commoner than $V=$const., which explains the success of the function G, considered the 'queen' of the great thermodynamic functions. As a consolation, the *Helmholtz function F* may boast of a more fundamental character, having been defined first, from the *second law, corollary 6*.

G

Gibbs energy of mixing—See also *Quantity of mixing, Integral molar quantity, Partial molar quantity*. The Gibbs energy of mixing $\Delta_{\text{mix}}G$ is the change in *Gibbs energy* associated with the transformation:

$$\sum n_i \langle A_i \rangle \to \langle \text{Mixture} \rangle.$$

The integral molar Gibbs energy of mixing is the integral Gibbs energy of mixing reduced to one mole:

$$\sum x_i \langle A_i \rangle \to \langle \text{Mixture} \rangle \quad (\text{with } \sum x_i = 1).$$

The partial Gibbs energy of mixing (or dissolution) of i, represented by $\Delta\mu_i$, is the difference between μ_i, the partial molar Gibbs energy (or *chemical potential*) of i in the mixture, and $\mu_1^\cup$, the molar Gibbs energy (or chemical potential) of i outside of the mixture. The change, δG, in the Gibbs energy associated with the transformation:

$$\langle \text{Mixture} \rangle + \delta n_i \langle A_i \rangle \to \langle \text{Mixture} \rangle$$

is then expressed by $\delta G - (\mu_i - \mu_i^\circ)\delta n_i = \Delta\mu_i\,\delta n_i$.

Between the partial molar Gibbs energies of dissolution and the integral Gibbs energy of mixing, there exist:

- the definition relationship: $\Delta\mu_i = (\partial\Delta_{\text{mix}}G/\partial n_i)_{T,P,n_{j\neq i}}$
- the *Euler identity*: $\Delta_{\text{mix}}G = \sum n_i\Delta\mu_i$
- the Gibbs–Duhem equation: $\sum n_i\,d(\Delta\mu_i) = 0$ (constant T and P).

EXPRESSION OF GIBBS ENERGIES OF MIXING: The chemical potential of a constituent i in a mixture of *ideal gases* is given by:

$$\mu_i(T, p_i) = \mu_i^\circ(T) + RT\ln(p_i/p_i^\circ).$$

If the mixture is formed at constant temperature and pressure, $p_i^\circ = P$ whatever the gas i considered; hence $p_i/p_i^\circ = x_i$. The partial molar Gibbs energy of dissolution of i in the mixture is then given by:

$$\Delta\mu_i = \mu_i - \mu_i^\circ = RT\ln x_i.$$

For a mixture of real gases:

$$\Delta\mu_i = \mu_i - \mu_i^\circ = RT\ln a_i,$$

where $a_i = f_i/f_i^\circ$ is the *activity* of gas i in the mixture, with the state of i outside of mixture chosen as the standard state.

It is possible to use the same expression with a condensed (solid or liquid) solution, provided that the physical meaning of a_i is kept in mind. By selecting as a standard state the constituent i taken outside the mixture (generally pure i in the condensed phase):

$$a_i \equiv \frac{f_i}{f_i^\circ} = \frac{\text{fugacity of } i \text{ in the solution}}{\text{fugacity of } i \text{ in its standard state}}.$$

Since the vapour pressures of condensed phases are generally low, fugacities can be identified with vapour pressures.

In summary, Gibbs energies of mixing may be expressed as a function of activities by means of the general relationships:

- partial molar Gibbs energy of dissolution of i: $\Delta\mu_i = RT\ln a_i$;
- integral Gibbs energy of mixing: $\Delta_{\text{mix}}G = RT\sum n_i\ln a_i$;
- integral molar Gibbs energy of mixing: $\Delta_{\text{mix}}G = RT\sum x_i\ln a_i$.

Gibbs energy of reaction—Let n_i moles of M_i be a *system* within which a *chemical reaction* $\sum v_i M_i = 0$ can proceed. When the *extent* of the reaction varies by $d\xi$, the *Gibbs energy* of the system varies by:

$$dG = \left(\sum v_i\mu_i\right)d\xi = \Delta_r G\,d\xi,$$

where μ_i is the partial molar Gibbs energy, or *chemical potential*, of i in the system:

$$\mu_i = \mu_i^\circ + RT\ln a_i.$$

$\Delta_r G = dG/d\xi$ is the Gibbs energy of the reaction:

$$\Delta_r G = \sum v_i\mu_i = \sum v_i\mu_i^\circ + RT\sum v_i\ln a_i,$$

a relation traditionally presented in the form:

$$\Delta_r G = \Delta_r G^\circ + RT\ln\prod a_i^{v_i}.$$

$\Delta_r G^\circ = \sum v_i \mu_i^\circ$, the *standard Gibbs energy* of the reaction, a function of the stoichiometric numbers v_i and of the choice of the *standard state* for each constituent M_i, is generally calculated from tables of thermodynamic data.

The *activity* a_i of the constituent M_i in the system $(a_i = f_i/f_i^\circ)$ is a function of the experimental conditions (by f_i) and of the standard states (by f_i°).

$\Delta_r G^\circ$, the Gibbs energy of the reaction, is a function of the stoichiometric numbers v_i and of the experimental conditions (by the fugacities f_i), but does not depend on the standard states selected. It is easy to verify that modification of one standard state (for instance, by changing the standard fugacity of M_i from f_i° to $f_i'^\circ$) has the effect of:

G

- modifying $\Delta_r G^\circ$ [from $\Delta_r G^\circ$ to $\Delta_r G'^\circ = \Delta_r G^\circ + v_i RT \ln(f_i'^\circ/f_i^\circ)$];
- modifying the activity of M_i [from $a_i = f_i/f_i^\circ$ to $a_i' = a_i(f_i^\circ/f_i'^\circ)$];
- leaving unchanged the Gibbs energy $\Delta_r G$ of the reaction.

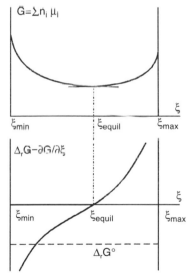

Gibbs energy of reaction: defined by $\Delta_r G = \partial G/\partial \xi$. The extent of the reaction, defined by $d\xi = dn_i/v_i$, varies between ξ_{min} (at least one constituent is lacking on the right-hand side) and ξ_{max} (at least one constituent is lacking on the left-hand side). The extent of the reaction at equilibrium, ξ_{equil}, is obtained by making the Gibbs energy of reaction $\Delta_r G$ zero or by looking for the minimum of the function G. The curve $G = f(\xi)$, a function only of the initial quantities n_i, does not depend on the choice of standard states, which is not the case for $\Delta_r G^\circ$. On the other hand, $\Delta_r G^\circ$ depends neither on ξ, nor on n_i.

Remark 1: When a chemical reaction can proceed in a mixture formed of n_i of constituents M_i, the Gibbs energy of reaction represents the *work* exchanged by the system with the surroundings without taking into account the work of pressure forces when the reaction occurs once along a *reversible* path. Hence the Gibbs energy of a redox reaction performed in a *galvanic cell* is related to the electromotive force $\Delta\mathscr{E}$ of the cell by $\Delta_r G = -n\mathscr{F}\Delta\mathscr{E}$. It is quite understandable that under these conditions, the value of $\Delta_r G$ is independent of the choice of standard states, which depends only on the personal convenience of the thermodynamicist!

Remark 2: The Gibbs energy function being the *thermodynamic potential* of a system working at constant T and P, the sign of $\Delta_r G$ indicates the direction of evolution of a system subjected to the constraints of constant temperature and pressure:

- if $\Delta_r G < 0$, the reaction as written will occur spontaneously;
- if $\Delta_r G > 0$, the reverse reaction will occur spontaneously;
- If $\Delta_r G = 0$, the system, at equilibrium, remains unchanged.

Whether or not the reaction does occur spontaneously will depend not only on the sign of $\Delta_r G$, but also on the rate of the reactions.

Remark 3: It is of the utmost importance to distinguish between a Gibbs energy change and a Gibbs energy of reaction. A Gibbs energy change defined by $\Delta G = G_f - G_i$ is expressed in joules. A Gibbs energy of reaction, defined by $\Delta_r G = \partial G / \partial \xi$, the derivative of a Gibbs energy with respect to an extent, is expressed in joules per mole.

Gibbs ensemble—A Gibbs ensemble, or Gibbs assembly, is defined by a very large number $\mathscr{N}$ of identical *systems* subject to the same *constraints*. When $\mathscr{N}$ tends towards infinity, each distinct *microscopic state* is represented by the same number of systems, such that a mean physical quantity calculated for $\mathscr{N}$ systems is interpreted as the expected value from a measurement made on a real system in thermodynamic equilibrium. This interpretation forms the *ergodic* hypothesis.

The macroscopic state of a system with only one constituent is completely determined by three physical quantities judiciously selected: its energy E, its volume V and the number of particles N. Instead of using three *extensive* quantities (E, V, N), it is possible to replace one or two by the conjugate *intensive* variable (T, P, μ). However, it is necessary to retain at least one extensive quantity to define the size of the system. When several chemical species are present, the minimum number of variables required equals the number of species plus two, for instance (E, V, N_k), N_k being the number of particles k.

- If each system is *isolated*, (N, E, V) are constant. The Gibbs ensemble is then called a *microcanonical ensemble*. Each system of this ensemble is called microcanonical.
- If each *closed* system is in thermal contact with a *reservoir*, the Gibbs ensemble is then called a *canonical ensemble*. Each system characterized by (N, T, V) constant is said to be canonical.
- If each *closed* system is in thermal and mechanical contact with a reservoir characterized by its temperature and its pressure, the Gibbs ensemble is then called an *isobaric–isothermal ensemble*. Each system characterized by (N, T, P) constant is said to be isobaric–isothermal.
- If each *open* system is in thermal and chemical contact with a reservoir characterized by its temperature and the *chemical potential* μ_k of the species k, the Gibbs ensemble is then called a *grand canonical ensemble*. Each system characterized by (μ_k, T, V) constant is said to be grand canonical.

G

Gibbs–Helmholtz equations—In the relationship defining the *Helmholtz energy*, $F = U - TS$, it is possible to express S by its value taken from the differential expression $\mathrm{d}F$:

$$\mathrm{d}F = -S\,\mathrm{d}T - P\,\mathrm{d}V, \quad \text{whence: } S = -(\partial F / \partial T)_V$$

$$F = U + T(\partial F / \partial T)_V.$$

Division of both sides by T^2 and a little manipulation yield:

$$\frac{\partial}{\partial T}\left(\frac{F}{T}\right)_V = -\frac{U}{T^2}.$$

In the same way, from the *Gibbs energy* $G = H - TS$:

$$\mathrm{d}G = -S\,\mathrm{d}T + V\,\mathrm{d}P, \quad \text{whence: } S = -(\partial G / \partial T)_P$$

$$G = H + T(\partial G / \partial T)_P$$

$$\frac{\partial}{\partial T}\left(\frac{G}{T}\right)_P = -\frac{H}{T^2}.$$

These expressions are known as Gibbs–Helmholtz equations. They yield straight away the *van't Hoff equation*:

$$\frac{\mathrm{d}}{\mathrm{d}T}\left(\frac{\Delta_r G^\circ}{T}\right) = -\frac{\Delta_r H^\circ}{T^2} \quad \text{with: } \Delta_r G^\circ(T) = -RT\ln K(T)$$

$$\frac{\mathrm{d}\ln K(T)}{\mathrm{d}T} = +\frac{\Delta_r H^\circ}{RT^2}.$$

134 Gibbs paradox

Gibbs paradox—The partial molar *entropy of mixing* and the integral entropy of mixing for ideal gases at constant pressure are given by:

$$\Delta s_i = -R \ln x_i$$

$$\Delta_{mix} S = \sum n_i \Delta s_i = -R \sum n_i \ln x_i.$$

The proof of these relations does not presuppose that the gases *i* must be different. The mixing of identical particles, at constant *T* and *P*, must then lead to *entropy* creation, which seems illogical. Such is the paradox that Gibbs raised for the first time.

This paradox is removed by statistical calculation of the entropy. The entropy of mixing that appears when mixing identical particles is linked to the assumption that the particles are distinguishable. The difficulty disappears on postulating that identical particles cannot be distinguished: the exchange of two distinguishable particles creates a new *complexion*, which does not occur in the exchange of two indistinguishable particles.

Gibbs triangle—The composition of a ternary mixture A, B, C may be represented by a point on a diagram. Indeed, between the three mole fractions x_A, x_B and x_C there is a relation: $\sum x_i = 1$. If none of the three constituents plays a special role, it is possible to use a triangle called

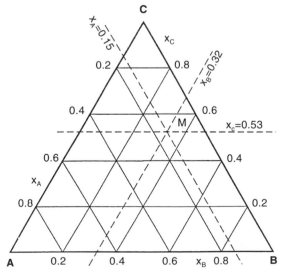

Gibbs triangle: Here the mixture containing 15% A, 32% B, and 53% C is represented by the point M. Parallels to the sides AB, BC, CA are respectively characterized by a constant value of x_C, x_A, x_B.

a Gibbs triangle. Each apex represents a pure constituent; each side represents a binary mixture.

Glass—Glass is a disordered, metastable state of matter, obtained by cooling a supercooled liquid to the point where the temperature is low enough to fix the position of the molecules. The glass transition temperature is the temperature below which the *viscosity* of the liquid is high enough (around 10^{12} Pa s) to prevent any motion of the molecules.

The glass transition is a *second-order* transition. It is revealed by a break in the slope of the molar volume–temperature or enthalpy–temperature curve, and hence by a discontinuity in the thermal expansivity or in the heat content at constant pressure. The glass transition temperature does not characterize the material: it depends on the conditions of cooling.

G

Grand canonical ensemble—The name given to a *Gibbs ensemble* in which each system is characterized by its temperature T, its volume V, and the *chemical potential* μ_i of each constituent. With such conditions, the energy and the number of particles in each system vary, but their mean value is known, which allows the grand canonical ensemble to be considered as isolated.

Let $\mathcal{N}$ be the number of systems in the isolated assembly and $\mathcal{N}_{i,j}$ be the number of systems containing $N_{i,j}$ particles of type i and being in energy state E_j. The number of states corresponding to such a distribution is given by:

$$\Omega_{i,j} - \frac{\mathcal{N}!}{\prod_i \prod_j \mathcal{N}_{i,j}}.$$

The most probable distribution is given by seeking the maximum value of $\Omega_{i,j}$, taking into account the constraints expressing the conservation relationships:

- number of systems in the assembly: $\sum_i \sum_j \mathcal{N}_{i,j}(N_{i,j}, E_j) = \mathcal{N}$;
- energy of the assembly: $\sum_i \sum_j \mathcal{N}_{i,j}(N_{i,j}, E_j) E_j = E$;
- number of particles of type i: $\sum_j \mathcal{N}_{i,j}(N_{i,j}, E_j) N_{i,j} = N_i$.

This third constraint must be written once for each kind of particle. By introducing $(2+i)$ *Lagrange multipliers*:

$$\ln \mathcal{N}_{i,j}(N_{i,j}\, E_j) + \alpha + \beta E_j + \sum_i \gamma_i N_{i,j} = 0.$$

136 Grand potential

The multiplier α is obtained by using the first constraint:

$$\mathcal{N} = \exp(-\alpha) \sum_j \exp\left(-\beta E_j - \sum_i \gamma_i N_{i,j}\right).$$

The probability of observing $N_{i,j}$ particles of type i with an energy E_j is:

$$P_{i,j} = \frac{\mathcal{N}_{i,j}}{\mathcal{N}} = \frac{\exp(-\beta E_j - \sum_i \gamma_i N_{i,j})}{\sum_j \exp(-\beta E_j - \sum_i \gamma_i N_{i,j})}.$$

$$\Xi = \sum_j \exp\left(-\beta E_j - \sum_i \gamma_i N_{i,j}\right)$$

represents the grand canonical *partition function* of a system.

Multipliers β and γ_i are obtained by comparing the expressions of the *entropy* in classical and in statistical thermodynamics:

- In classical thermodynamics: $dS = (dE + P\,dV - \sum \mu_i\,dn_i)/T$.
- In statistical thermodynamics: $dS = k\,d\ln\Omega$. By expansion:

$$dS = k\left(d\ln\Xi + \beta\,dE + \sum \gamma_i\,dn_i\right),$$

which, by identification, leads to the relationships:

$$PV = kT \ln\Xi$$

$$\beta = 1/kT$$

$$\gamma_i = -\mu_i/kT.$$

Thus we have the grand canonical partition function of the system:

$$\Xi(T, V, \mu_i) = \sum_j \exp\left(-\frac{E_j}{kT} + \sum_i \frac{\mu_i}{kT} N_{i,j}\right).$$

The probability that a system will have an energy E_j and contain $N_{i,j}$ particles of type i is given by:

$$P_{i,j} = \frac{1}{\Xi} \exp\left(-\frac{E_j}{kT} + \frac{\mu_i}{kT} N_{i,j}\right).$$

Grand potential—The function defined by:

$$\Omega \equiv F - \sum n_i \mu_i.$$

On differentiating: $d\Omega = -S\,dT - T\,d\sigma - P\,dV - \sum n_i\,d\mu_i$.

At constant temperature, volume, and *chemical potential* and without any other kind of energy exchange (electric, magnetic, etc.), this becomes:

$$d\Omega = -T\,d\sigma \leqslant 0.$$

In such conditions, the grand potential function Ω can only decrease and reach a minimum at equilibrium. The grand potential is thus the *thermodynamic potential* of a system constrained by the condition (T, V, μ_i) constant.

On introducing the *Euler identity*, which links the *Gibbs energy* to the chemical potentials: $G = \sum n_i \mu_i$, the grand potential function may be defined by:

$$\Omega = F - G = -PV.$$

G

Grüneisen parameter—From *Maxwell's equations*, a general relationship between c_P, c_V, α_P, and χ_T may be established:

$$c_P - c_V = TV\alpha_P^2 / \chi_T.$$

It is possible to simplify this relation in the case of a solid if the *Debye temperature* is expressed in a simple form:

$$\theta = CV^{-\gamma}.$$

γ is the Grüneisen parameter, a constant which characterizes the solid; independent of the temperature, its value generally lies between 1 and 2. By decomposing F, the *Helmholtz energy* of the solid is as follows:

$$F = F_\circ + F_v = F_\circ + Tf(\theta/T),$$

where $F_\circ$ is the Helmholtz energy of the solid at $0\,\mathrm{K}$, and F_v, the Helmholtz energy associated with the lattice vibrations at temperature T:

$$P = -\left(\frac{\partial F}{\partial V}\right)_T = -\left(\frac{\partial F_\circ}{\partial V}\right)_T - \left(\frac{\partial F_v}{\partial \theta}\right)_T \left(\frac{\partial \theta}{\partial V}\right)_T = -\left(\frac{\partial F_\circ}{\partial V}\right)_T - U_v \frac{\gamma}{V},$$

whence:

$$\left(\frac{\partial P}{\partial T}\right)_V = c_V \frac{\gamma}{V}.$$

Furthermore:

$$\left(\frac{\partial P}{\partial T}\right)_V = -\left(\frac{\partial P}{\partial V}\right)_T \left(\frac{\partial V}{\partial T}\right)_P = \frac{\alpha_P}{\chi_T}.$$

By comparison, we get the Grüneisen parameter:

$$\gamma = \alpha_P V / \chi_T c_V.$$

Guldberg, Cato (1836–1902)—Norwegian chemist and mathematician, who discovered, with Peter Waage, the *mass action law* (1864). In 1890, he showed that it is possible to estimate the critical temperature of a liquid by taking 1.5 times the normal boiling temperature.

H

H theorem—A famous theorem, established by Boltzmann in 1872, on which is based every discussion about the *irreversibility* of the behaviour of a large number of particles. Let $f(p, q, t)$ be a *distribution* function, or probability of finding, at a given time t a particle with position q and momentum p. The distribution evolves with time, owing to the proper motion of molecules and their collisions. Boltzmann defined a function H:

$$H = \int_p \int_q f \ln f \, dp \, dq.$$

The H theorem of Boltzmann states that this function decreases with time and tends towards a minimum which corresponds to the *Maxwell distribution*:

$$f = 4\pi N v^2 \left(\frac{m}{2\pi kT}\right)^{3/2} \exp\left(-\frac{mv^2}{2kT}\right).$$

If the distribution of the velocities is Maxwellian, the H function remains constant during time. The Boltzmann H function may be identified with *negentropy*, $-S$, which, for an isolated system, tends towards a minimum.

Heat—An interaction between two *closed systems* without exchange of *work* is a pure heat interaction when the two systems, initially *isolated* and in a stable *equilibrium* state, are placed in contact. The energy exchanged between the two systems is then called heat.

Remark 1: The above definition assigns to the system a well-defined *constraint*: that it is closed. It is clear that a heat exchange may exist without such a constraint, but then the interaction will not be a pure heat interaction: the two systems will exchange energy in the form of heat and work.

Remark 2: The concept of heat arises from the existence of exchanges which do not match the definition of work. Let A (temperature T_A) and B (temperature $T_B > T_A$) be two systems in thermal contact. At equilibrium, the assembly $(A + B)$ tends towards a final temperature T_f. The energy exchanged between A and B obviously cannot be work. Indeed, although it is possible to make the system A evolve from T_A to T_f with, as the only external effect, the lowering of a mass in a gravitational field, it is impossible to make the system B evolve from T_B to T_f with, as the only external effect, the raising of a mass in a gravitational field. The *second law* is based on such observation.

Remark 3: Heat, like work, is a transfer quantity. It is impossible to define the heat quantity enclosed in a system, but it is quite possible to define the heat quantity transferred from A to B, or the heat quantity transferred from A to the *surroundings*.

H

Remark 4: Heat may be defined as the energy which, exchanged by a system, has the effect of modifying its temperature (sensible heat) or inducing a change of state (latent heat). The statement proposed, which defines heat by contrast with work, nevertheless presents the advantage of greater generality because it does not need a previous definition of *temperature*.

Remark 5: Quantity of heat, like all energy, is expressed in *joules* and may be represented by the product of a *tension* (the temperature) by an *extensive quantity* (the entropy). According to the second law:

$$dQ = T(dS - d\sigma),$$

where dS is the entropy change of the system and $d\sigma \geqslant 0$ is the entropy change of the *universe*, or entropy created by the *irreversibility* of the transformation.

Remark 6: dQ is a *differential* which is generally neither *exact* nor even *integrable*. It is integrable only for a *reversible* transformation ($d\sigma = 0$). The *integrating factor* is then $1/T$. It is exact in particular cases, for instance during transformations at constant volume or pressure.

Heat capacity—The heat capacity, or thermal capacity, is defined by:

$$c = dQ/dT,$$

the ratio of the heat quantity exchanged by the *system* with the *surroundings* to the observed temperature change.

140 Heat capacity

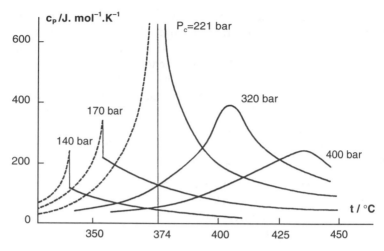

Heat capacity: *Near the critical point, the heat capacity of a fluid is strongly dependent on temperature and pressure, as shown in the above example, for water. Dashed lines represent the behaviour of the liquid; full lines that of the vapour.*

According to the conditions of the exchange, such a quotient may take every possible value between $-\infty$ and $+\infty$. It is thus necessary to specify conditions of derivation. The main heat capacities encountered are the following:

- c_V: heat capacity at constant volume: $c_V = dQ_V/dT = (\partial U/\partial T)_V$;
- c_P: heat capacity at constant pressure: $c_P = dQ_P/dT = (\partial H/\partial T)_P$;
- $c_x = dQ_x/dT$: heat capacity along the liquid–vapour equilibrium curve; the subscript x means that between T and P there exists a relation obtained by integration of the *Clapeyron equation*, in order to remain at liquid–vapour equilibrium.

Using *Maxwell's equations*, we can prove that:

$$c_P - c_V = T\left(\frac{\partial P}{\partial T}\right)_V\left(\frac{\partial V}{\partial T}\right)_P = TV\frac{\alpha_P^2}{\chi_T},$$

where α_P is the isobaric *expansivity* and χ_T the isothermal *compressibility*.

$$c_x = T\left(\frac{\partial S}{\partial T}\right)_x = c_P - T\left(\frac{\partial V}{\partial T}\right)_P\left(\frac{\partial V}{\partial T}\right)_x = c_V + T\left(\frac{\partial P}{\partial T}\right)_V\left(\frac{\partial P}{\partial T}\right)_x.$$

These expressions of c_x are valid for the liquid and for its vapour in equilibrium; for the liquid, far from the *critical* point: $c_V \approx c_x \approx c_P$, whereas for the vapour, c_x is generally negative (example: water). In some cases (example: benzene), c_x may take positive values between two temperatures called *inversion* temperatures.

When $T \rightarrow T_c$, at liquid–vapour equilibrium, $c_p \rightarrow +\infty$ for both liquid and vapour, whereas $c_V \rightarrow$ finite limit, $(c_x)_{\text{liq}} \rightarrow +\infty$, and $(c_x)_{\text{vap}} \rightarrow -\infty$.

Heat pump—A device whose object is to supply heat to a warm sink by removing it from a reservoir at a lower temperature.

Heat quantity— The unit of heat quantity is the joule (symbol: J). The joule being a quite small unit, the kilojoule is customarily used for expressing changes in *enthalpy* or *Gibbs energy*, whereas *entropies* are expressed in $J\ K^{-1}$. Heat, like *work*, is not a *state variable* but a *transfer variable*. It is possible to measure the heat quantity crossing a wall; the heat quantity enclosed in a system cannot be defined.

Helmholtz, Herman Ludwig (1821–1894)—German scientist and philosopher, universal mind, author of important contributions to physiology, optics, electrodynamics, and meteorology. In thermodynamics he recognized in 1847 that energy is conserved if all its forms are included.

Helmholtz energy—Sometimes called 'Free energy' or 'Helmholtz function', the Helmholtz (free) energy F, introduced independently in 1875 by Gibbs and in 1882 by Helmholtz, is defined by:

$$F \equiv E - TS,$$

where E represents the *energy* of the system and S its *entropy*. In the absence of any external field, the Helmholtz energy may be defined by:

$$F = U - TS.$$

Remark: *Energy*, introduced by the *first law*, is defined by $dE = (dW)_{\text{ad}}$. In the same way, the Helmholtz energy may be introduced by the *second law, corollary 6* and defined by $dF = (dW)_{\text{rev}}$. The Helmholtz energy change ΔF of a system during a real transformation equals the *work* exchanged with the surroundings during every *reversible* transformation between the same initial state and the same final state. In accordance with the second law, if the transformation is reversible, the work exchanged with the surroundings is a minimum and does not depend on the path followed. Explicitly: if $\Delta F > 0$, ΔF represents the minimum work that must be supplied to the system to carry out the transformation; if $\Delta F < 0$, $|\Delta F|$ represents the maximum work that can be extracted from the system during the transformation.

HELMHOLTZ ENERGY, THERMODYNAMIC POTENTIAL: Differentiating the function F gives:

$$dF = dE - T\,dS - S\,dT$$

$$dF = -S\,dT - T\,d\sigma - P\,dV + \sum \mu_i\,dn_i + \cdots$$

At constant temperature, volume, and amount of substance (*closed system without chemical reaction*) and in the absence of any other energy exchange (magnetic, electric, etc.), the expression of dF is reduced to:

$$dF = -T\,d\sigma \leqslant 0.$$

In such conditions, F decreases and tends towards a minimum corresponding to the equilibrium. The Helmholtz energy is thus the *thermodynamic potential* of a system subjected to the constraints constant T, V, and n_i.

Henrian activity—The *activity* of a constituent i is called 'henrian' when the *reference state* selected is the constituent i in solution at infinite dilution. Henrian activity is well determined only if the nature of the solvent and the reference composition, $c_i^\ominus$, are well specified. In order that an activity, $a_i = \gamma_i c_i / c_i^\ominus$, be Henrian, it is necessary that $\gamma_i \to 1$ when $c_i \to 0$.

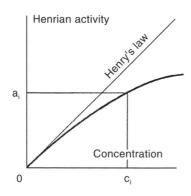

Henrian activity: Equals composition at high dilutions. Its value is thus a function of the unit selected to express compositions.

Note: Henrian activity must not be confused with *Henry's law*. A Henrian activity implies a very precise selection of the reference state, whereas Henry's law describes the general behaviour of all dilute solutions, behaviour which cannot depend on arbitrary choice by an experimenter.

Henry—Unit of inductance in the SI system (symbol: H), named from the American physicist Joseph Henry (1797–1878), who discovered self-induction:

$$1\,H = 1\,V\,A^{-1}\,s = 1\,m^2\,kg\,s^{-2}\,A^{-2}.$$

Henry's law—William Henry, English physician and chemist (1775–1836), showed in 1803 that the solubility of a gas in a solvent is proportional to its pressure, provided that no chemical reaction occurs. This statement must be applied cautiously, because it is not always satisfied (see for instance *Sievert's law*), and the more general law which now bears his name is expressed differently. Henry's law states that *activity* of a *solute i* in a dilute solution is proportional to its concentration. The use of Henry's law comes down to considering dilute solutions as ideal by replacing the activity–composition curve related to a solute *i* by its tangent at the origin.

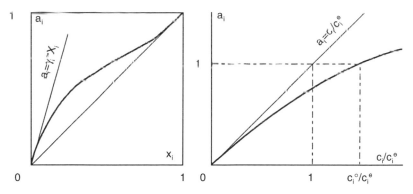

Henry's law: Amounts, at high dilutions, to identifying the activity–composition curve with its tangent at the origin. On the left, the standard state for i is the pure constituent i in the same aggregation state as the solution. On the right, the reference state for i is the constituent i in infinitely dilute solution in the pure solvent, the reference composition being $c_i^{\ominus}$.

When the *mole fraction* of i lies between 0 and 1, the standard state generally selected is the pure constituent i: $a_i = \gamma_i x_i$ with $\gamma_i \to \gamma_i^\infty = \text{const.}$ when $x_i \to 0$. For a solute, where x_i lies generally close to 0, it is more convenient to use *concentrations* rather than *titres*. The activity of i is then expressed by:

$$a_i = \gamma_i k_i c_i / c_i^\ominus,$$

where $c_i^\ominus$ is the reference concentration (generally $c_i^\ominus = 1 \, \text{mol L}^{-1}$); k_i, the coefficient of conversion from mole fractions x_i to concentrations c_i, tends towards k_i^∞ at high dilutions. The activities may be divided by $\gamma_i^\infty \cdot k_i^\infty$, which amounts to performing a change of *reference state* and hence a change of *standard state*. The new standard state ($a_i = 1$) corresponds to a solution, at best ill defined, at worst without any real existence. On the other hand, the new reference state ($\gamma_i = 1$) corresponds to an infinitely dilute solution, with a well-defined reference concentration $c_i^\ominus$.

Hildebrand, Joel H. (1881–1983)—American chemist and thermodynamicist, who introduced the concept of *regular solution*. His monograph '*Solubility of non-electrolytes*' (1924) is a standard work.

Homogeneous—Whose constitutive elements are of the same nature. Homogeneous has a meaning wider than *isotropic*. For instance, a phase is a homogeneous part of a mixture, which does not preclude within the same phase the existence of gradients of intensive quantities.

In mathematics, a function $f(x, y, z, u, v)$ is said to be homogeneous of degree k with respect to variables x, y, z when:

$$f(\lambda x, \lambda y, \lambda z, u, v) = \lambda^k \cdot f(x, y, z, u, v).$$

In thermodynamics, *extensive quantities* are homogeneous functions of the first degree with respect to amount of substance.

I

Ideal gas—A gas obeying the equation of state:

$$PV = nRT.$$

Every real gas, under low pressure (in practice, when $P \lesssim 1 \text{bar}$) behaves like an ideal gas: $PV \to nRT$ when $P \to 0$.

EXPERIMENTALLY: The behaviour laws were established by:

- Boyle (1660) and Mariotte (1676), at constant temperature: $PV = P_o V_o$;
- Charles (1787), at constant volume: $P = P_o(1 + \alpha t)$;
- Gay-Lussac (1802), at constant pressure: $V = V_o(1 + \alpha t)$.

Satisfaction of any two of these three laws implies that the third is also satisfied. The three laws may be gathered into a single equation:

$$PV = P_o V_o (1 + \alpha t).$$

Using the *Celsius* scale of temperature, $\alpha = 1/273.15$, and introducing: $T = t + 273.15$, we obtain:

$$PV = P_o V_o T / 273.15 = RT.$$

Under *normal* conditions of temperature (273.15 K) and pressure (101 325 Pa), one mole of ideal gas occupies a volume of $22.414097 \times 10^{-3} \text{ m}^3$, whence:

$$R = 101\,325 \times 22.414\,097 \times 10^{-3}/273.15 = 8.314\,510 \,\text{J}\,\text{mol}^{-1}\,\text{K}^{-1}.$$

JOULE'S LAW: 'The *internal energy* of an ideal gas is a function only of its temperature' (see *Joule's law*). Accordingly:

- The *enthalpy* of an ideal gas is a function of only the temperature. Indeed:

$$H = U + PV = U + nRT.$$

- The *heat capacities* at constant volume and pressure are functions of only the temperature, because:

$$c_V = (\partial U/\partial T)_V = dU/dT,$$

$$c_P = (\partial H/\partial T)_P = dH/dT.$$

- Mayer's relation: Differentiation with respect to temperature of both sides of the relation: $H(T) = U(T) + nRT$ gives:

$$dH/dT = dU/dT + nR;$$

thus, for one mole of an ideal gas: $c_P - c_V = R$. We must keep in mind that an expression such as dU/dT has meaning when U is a function only of T. Otherwise it is necessary to specify the conditions of differentiation!

INTERNAL ENERGY: From *Maxwell's equations*:

$$dU = c_V\,dT + \left[T\left(\frac{\partial P}{\partial T}\right)_V - P\right]dV.$$

We verify that, for an ideal gas, the square bracket equals zero, whence:

$$U(T) = U(T_\circ) + \int_{T_\circ}^{T} c_V\,dT.$$

The fact that a gas is ideal does not imply $c_V = $ const. However, it is possible to use, as a first approximation:

- For a monatomic gas: $c_V = 3R/2$. This value, resulting from the *kinetic theory of gases*, rests on the assumption that the only contribution to the heat capacity comes from the translation kinetic energy.
- For a diatomic gas: $c_V = 5R/2$. To the preceding value is added R, which represents the contribution of rotation.
- For a polyatomic gas: $c_V = 7R/2$. The value R added again represents a 'rough and ready compromise' taken to include other vibratory and electronic contributions.

In the absence of experimental data, the above values of c_V are generally acceptable; however, statistical models allow a more accurate approach.

ENTHALPY: A similar relationship may be obtained:

$$H(T) = H(T_\circ) + \int_{T_\circ}^{T} c_P\,dT.$$

$c_P = c_V + R$, from Mayer's relation, is a function only of T. It is possible to use, as a first approximation:

- $c_P = 5R/2$ for a monatomic gas;
- $c_P = 7R/2$ for a diatomic gas;
- $c_P = 9R/2$ for a polyatomic gas.

ENTROPY: From Maxwell's equations:

$$dS = c_V \frac{dT}{T} + \frac{R}{V} dV = c_P \frac{dT}{T} - \frac{R}{P} dP,$$

which on integration gives:

$$S(T,V) - S(T_o, V_o) = \int_{T_o}^{T} \frac{c_V}{T} dT + R \ln \frac{V}{V_o},$$

$$S(T,P) - S(T_o, P_o) = \int_{T_o}^{T} \frac{c_P}{T} dT - R \ln \frac{P}{P_o}.$$

With the assumption of constant heat capacity, the equations of isentropic curves are easily obtained:

$$\frac{T}{T_o} = \left(\frac{P}{P_o} \right)^{R/c_P} = \left(\frac{V_o}{V} \right)^{R/c_V},$$

or, by introducing $\gamma = c_P/c_V$:

$$PV^{\gamma} - \text{const.}$$

HELMHOLTZ ENERGY: Direct integration of the basic relationship:

$$dF = -S\,dT - P\,dV$$

gives:

$$F(T,V) - F(T_o, V_o) = \int_{T_o}^{T} -S\,dT - RT \ln \frac{V}{V_o},$$

an expression generally used in the form:

$$F(T,V) = F(T,V_o) - RT \ln \frac{V}{V_o}.$$

It is of interest to verify that at constant volume, F is a *concave* function of temperature, whereas at constant temperature, F is a *convex* function of volume.

GIBBS ENERGY: Direct integration of the basic equation:

$$dG = -S\,dT + V\,dP$$

gives:

$$G(T,P) - G(T_o, P_o) = \int_{T_o}^{T} -S\,dT + RT \ln \frac{P}{P_o},$$

an expression generally used in the form:

$$G(T, P) = G(T, P_o) + RT \ln \frac{P}{P_o}.$$

If, by convention, we introduce $P_o = P^\circ$, the *standard* pressure, and then if we apply the above relationship to one mole of an ideal gas, the following are obtained:

$$G(T, P) = \mu(T, P), \text{ chemical potential}$$

$$G(T, P^\circ) = \mu^\circ(T), \text{ standard chemical potential:}$$

$$\mu(T, P) = \mu^\circ(T) + RT \ln \frac{P}{P^\circ}.$$

It is of interest to verify that at constant temperature, G is a concave function of pressure and that at constant pressure, G is also a concave function of temperature.

Ideality—See *Ideal gas, Ideal solution, Henry's law, Raoult's law*.

Ideal solution—A solution is called ideal when the *activity* of a constituent i is proportional to its composition, that is to say, when its *activity coefficient* γ_i is constant. It is of course necessary to specify how the composition is expressed, because an ideal solution with *mole fractions* is no longer ideal with mol L^{-1}, and vice versa. In order that a solution be ideal, it is sufficient that the ideality condition be verified with only one constituent. This result, which is obvious for a binary solution owing to the *Gibbs–Duhem equation*:

$$x_1 \, d \ln \gamma_1 + x_2 \, d \ln \gamma_2 = 0,$$

$$\ln \gamma_1 = \text{const.} \Rightarrow d \ln \gamma_1 = 0 \Rightarrow d \ln \gamma_2 = 0 \Rightarrow \ln \gamma_2 = \text{const.},$$

can be generalized. Let us consider a solution with n constituents, which is ideal for constituent number 1:

$$a_1 = \gamma_1 x_1,$$

with $\gamma_1 = \text{const.}$ Let us write the relationship allowing the calculation of a *partial molar quantity* y_i from the *integral quantity* Y:

$$y_i = Y + \sum_{j=2}^{n} (\delta_{ij} - x_j) \left(\frac{\partial Y}{\partial x_j} \right)_{x_k \neq j}.$$

δ_{ij} is the *Kronecker delta* and Y the integral molar quantity, expressed as a function of $x_2, x_3, \ldots, x_n$. Applied to the excess Gibbs energies: $y_i = RT \ln \gamma_i$ and $Y = RT \sum x_i \ln \gamma_i$. The factor RT, constant, may be discarded for simplicity. For $i = 1$ (with $\ln \gamma_1 = k_1 = \text{const.}$):

$$\ln \gamma_1 = \left(\frac{\partial Y}{\partial n_1}\right)_{n_{j \neq 1}} = Y + \sum_{j=2}^{n} -x_j \left(\frac{\partial Y}{\partial x_j}\right)_{x_{k \neq j}} = k_1,$$

$$Y = k_1 + \sum_{j=2}^{n} x_j \left(\frac{\partial Y}{\partial x_j}\right)_{x_{k \neq j}}.$$

Inserting this expression of Y into the general relationship giving y_i:

$$y_i = k_1 + \left(\frac{\partial Y}{\partial x_i}\right)_{x_{k \neq i}}.$$

Taking into account the *Euler identity* ($Y = \sum x_i y_i$), we get a system of n differential equations whose solutions are of the form: $y_i = k_i = \text{const.}$ The ideality of a solution with respect to constituent number 1 ($k_1 = \ln \gamma_1 = \text{const.}$) thus involves ideality of the solution with respect to every other constituent of the solution ($k_i = \ln \gamma_i = \text{const.}$).

Remark: This result may be applied to a restricted range of compositions. If a dilute solution obeys *Henry's law* for a *solute i* ($a_i = \gamma_i^\infty x_i$ with $\gamma_i^\infty = \text{const.}$ when $x_i \to 0$), it obeys not only *Henry's law* for every solute present in the solution, but also *Raoult's law* for the *solvent*.

Imperial system—A system of units which is becoming obsolete in scientific literature, but is still in use in the USA and thus hard to ignore! The British system is 'logically' founded on the foot, the pound-mass (or pound 'avoirdupois'), the pound-force, the second and the degree Rankine:

- Foot $1\,\text{ft} = 12\,\text{in} = 0.3048\,\text{m}$
- Pound-mass $1\,\text{lb}_m = 0.453\,592\,37\,\text{kg}$
- Pound-force $1\,\text{lb}_f = 4.448\,222\,\text{N}$
- British thermal unit $1\,\text{Btu} = 1055.056\,\text{J}$
- Pound-force per square inch $1\,\text{psi} = 6894.757\,\text{Pa}$
- Degree Rankine $1°\text{R} = (5/9)\text{K}$

The Rankine and Fahrenheit degrees are related by: $°\text{R} = °\text{F} + 459.67$. The temperature of the triple point of water equals $491.69°\text{R}$.

150 Implicit

Implicit—From Latin *implicitus*, shrouded. It is said of what is virtually included in a statement, but which must nevertheless remain clear! For instance, if I express the *standard Gibbs energy* of a reaction by:

$$\Delta_r G^\circ(T) = \Delta_r H^\circ(298) - T\Delta_r S^\circ(298),$$

I implicitly make two simplifications. First, I assume $\Delta_r H^\circ$ and $\Delta_r S^\circ$ are independent of the temperature; then I neglect the possible phase transitions of constituents between 298 and T. Finally, I tacitly assume that my reader knows how to write the balance equation and is aware of the *standard states* I have selected when I omit to give these specifications!

Incongruent—A transformation i → f undergone by a phase i is called incongruent when the composition of the final product f differs from that of the phase i. The material balance of course involves the formation of a second phase during the transformation.

Index—The ratio of two physical quantities having the same dimension, used to characterize a property. For instance, the basicity index of a slag represents the mass ratio of basic oxides (such as CaO, MgO) to acid oxides (such as SiO_2, P_2O_5).

Inductance—Electrical induction is a *coupling* phenomenon between magnetic flux and electric current. Inductance is the ratio of the induction flux to the inducting current. It is known as self-inductance when the induction flux Φ_1 through a circuit is generated by a current I_1 which flows in the circuit: $\Phi_1 = L_{11} I_1$; it is called mutual when the induction flux Φ_2 through a circuit is generated by an inducting current I_1 in the first circuit: $\Phi_2 = L_{12} I_1$. Conversely, if the second circuit is traversed by a current I_2, it induces in the first circuit a flux Φ_1 given by: $\Phi_1 = L_{21} I_2$. *Onsager reciprocity* prescribes $L_{12} = L_{21}$.

The electromotive force generated by induction is expressed by: $\mathcal{E} = -d\Phi/dt$. In the SI system, the unit of inductance is the Henry:

$$1\,H = 1\,Wb\,A^{-1} = 1\,V\,s\,A^{-1} = 1\,m^2\,kg\,s^{-2}\,A^{-2}.$$

Information—A message comprising N symbols, each of which is subject to a probability $P(j_i)$, conveys an information content $\mathscr{I}$, or entropy, of the message, expressed by a function proposed in 1948 by the American mathematician Claude Elwood Shannon:

$$\mathcal{I} = -kN\sum_i P(j_i)\ln P(j_i).$$

In information theory, the entropy of a message is sometimes expressed in bits per symbol, in the form: $\mathcal{H} = -\sum_i P(j_i)\log_2 P(j_i)$.

It equals zero when the probability of each symbol is 0 or 1: the message does not convey any information. It is a maximum when the probability of each symbol is the same. There is an analogy with the entropy of an assembly of N particles distributed between states so that N_i particles are in state i:

$$S = k\ln\left(\frac{N!}{\prod N_i!}\right) = -kN\sum x_i\ln x_i.$$

The entropy equals zero when all the particles are in the same state; it reaches its maximum when they are evenly distributed between the different states. The entropy measures the lack of information about the detailed structure of a system: the entropy of a solid is lower than that of a gas, because our information about the positions of the atoms is greater.

We obtain information by taking measurements, which consumes *negentropy*. *Maxwell's demon*, for instance, uses negentropy to obtain information about the position and the velocity of the particles, then uses this information to generate negentropy, that is, to establish order.

Insulating—The salient feature of a material which cannot be traversed by electricity (electrical insulator) or by heat (thermal insulator). An *adiabatic* wall is composed of thermal insulators.

Integrable differential—Sometimes called a 'holonomic differential', an erudite term from the Greek ολος, whole, and νομος, rule. A *differential* dQ that is not *exact* is said to be integrable when there is a function $1/\tau$ such that the new differential dQ/τ is exact. The function $1/\tau$ is called the integrating factor, τ being the integrating denominator.

DIFFERENTIALS THAT ARE FUNCTIONS OF TWO VARIABLES may be exact or not, but they are always integrable. Indeed, the solution of the equation:

$$dQ = A(x, y)\,dx + B(x, y)\,dy = 0$$

defines on the plane (x, y) a family of curves with one parameter. The tangent to the curve going through a point (x, y) of the plane has a slope $dy/dx = -A/B$. Let $\sigma(x, y) = $ const. be one of these curves.

The differentiation:

$$\left(\frac{\partial \sigma}{\partial x}\right)_y + \left(\frac{\partial \sigma}{\partial y}\right)_x \frac{dy}{dx} = 0,$$

followed by the substitution

$$\frac{dy}{dx} = -\frac{A}{B}, \quad \text{gives: } B\left(\frac{\partial \sigma}{\partial x}\right)_y = A\left(\frac{\partial \sigma}{\partial y}\right)_x = \frac{AB}{\tau(x, y)}.$$

Hence $A = \tau(\partial\sigma/\partial x)_y$ and $B = \tau\,(\partial\sigma/\partial y)_x$. By insertion into dQ:

$$dQ = \tau\left[\left(\frac{\partial \sigma}{\partial x}\right)_y dx + \left(\frac{\partial \sigma}{\partial y}\right)_x dy\right] = \tau\,d\sigma.$$

The division of the differential dQ by τ gives an exact differential. The replacement of σ by any other function $S(\sigma)$ gives another integrating factor:

$$dS = \frac{dS}{d\sigma} d\sigma = \frac{dS}{d\sigma} \frac{dQ}{\tau} = \frac{dQ}{T(x, y)},$$

with $T(x, y) = \tau(x, y)\,d\sigma/dS$. In other words, if a differential admits one integrating factor, it admits an infinity. However, to know the existence of an integrating factor does not mean that it is always easy to find!

DIFFERENTIALS THAT ARE FUNCTIONS OF THREE VARIABLES: These may be neither exact nor integrable, because the equation $dQ = A\,dx + B\,dy + C\,dz = 0$ does not always define, in the space (x, y, z), a family of surfaces. In order that the differential be integrable, the functions A, B and C must satisfy:

$$A\left[\left(\frac{\partial B}{\partial z}\right)_{x,y} - \left(\frac{\partial C}{\partial y}\right)_{x,z}\right] + B\left[\left(\frac{\partial C}{\partial x}\right)_{y,z} - \left(\frac{\partial A}{\partial z}\right)_{x,y}\right]$$

$$+ C\left[\left(\frac{\partial A}{\partial y}\right)_{x,z} - \left(\frac{\partial B}{\partial x}\right)_{y,z}\right] = 0.$$

Let λ be the integrating factor. $\lambda\,dQ$ is an exact differential when:

$$\lambda\left(\frac{\partial A}{\partial y}\right)_{x,z} + A\left(\frac{\partial \lambda}{\partial y}\right)_{x,z} = \lambda\left(\frac{\partial B}{\partial x}\right)_{y,z} + B\left(\frac{\partial \lambda}{\partial x}\right)_{y,z}$$

$$\lambda\left(\frac{\partial B}{\partial z}\right)_{x,y} + B\left(\frac{\partial \lambda}{\partial z}\right)_{x,y} = \lambda\left(\frac{\partial C}{\partial y}\right)_{x,z} + C\left(\frac{\partial \lambda}{\partial y}\right)_{x,z}$$

$$\lambda\left(\frac{\partial C}{\partial x}\right)_{y,z} + C\left(\frac{\partial \lambda}{\partial x}\right)_{y,z} = \lambda\left(\frac{\partial A}{\partial z}\right)_{x,y} + A\left(\frac{\partial \lambda}{\partial z}\right)_{x,y}.$$

On adding the three relations after multiplication of the first by C, the second by A, and the third by B, the terms carrying λ cancel out and the above condition remains which functions A, B and C must satisfy in order that the differential dQ be integrable. Differentials that are functions of two variables $(C=0)$ and exact differentials are always integrable.

Integral molar quantity—An integral molar quantity Y_m is an integral quantity Y reduced to one mole:

$$Y_m = \frac{Y}{\sum n_i}.$$

Consider for instance the transformation:

$$\sum x_i A_i \rightarrow \langle \text{Mixture} \rangle \qquad (\text{with } \sum x_i = 1)$$

$$\Delta_{mix}Y_m = Y_f - Y_i = Y_m - \sum x_i y_i^\circ = Y_m - Y_m^\circ.$$

$Y_f = Y_m$ is the integral molar quantity of the mixture whereas $\Delta_{mix}Y_m = Y_m - Y_m^\circ$ is the integral molar quantity of mixing.
y_i° is the value of Y related to one mole of i outside the mixture.
$Y_i = Y_m^\circ = \sum x_i y_i^\circ$ is the value of Y_m before mixing.
An integral quantity is an *extensity* whereas an integral molar quantity is a *reduced extensity*. Clearly this means that an integral molar quantity is not an extensity but an *intensity*!

Integrating factor—The name given to any function Φ which, multiplied by a differential dF, gives an exact differential ΦdF. A differential is said to be integrable if it admits an integrating factor, which is not always satisfied. Nevertheless, differentials which are not exact but are functions of two variables: $dF = A(x, y) dx + B(x, y) dy$, always admit an integrating factor (See *Integrable differential*).

In thermodynamics, during a *reversible* transformation, the differential dQ is not exact, but admits $1/T$ as an integrating factor:

$$dS(x, y, z) = dQ(x, y, z)/T(x, y, z).$$

The attentive reader will recognize in Q, S, and T the quantity of heat, entropy, and temperature (See *Caratheodory*). For an *irreversible* transformation, dQ is a differential which is neither exact nor even integrable.

Intensity—An intensive quantity, or intensity, is the partial derivative of an *extensive* function with respect to an extensive state variable. *Integral molar quantities, partial molar quantities*, or *reduced extensities* that are derivatives of an extensive function with respect to amount of substance are intensities, but any confusion between them must be avoided. If Y is an integral quantity (with $n = \sum n_i$), it is possible to define:

- the partial molar quantity: $y_i = (\partial Y / \partial n_i)_{T,P,n_{j \neq i}}$;
- the integral molar quantity: $Y_m = (\partial Y / \partial n)_{T,P,x_i} = Y/n$.

Intensities are defined at one point of the space, and of degree zero with respect to amount of substance. To multiply all n_i by the same factor k does not modify intensities.

When the extensive function is *internal energy*, the corresponding intensive quantities are then called *tensions*. A tension takes the same value on both sides of a boundary separating two systems at equilibrium, a characteristic not shared by any other intensive quantity.

Interaction parameter—A concept introduced by Carl Wagner in 1950 with the object of expressing straightforwardly the solubility of a gas in a molten alloy, but it is of more general interest; interaction parameters may be used for calculating the solubility of a *solute i* in a *solvent* that has dissolved other solutes $j, k, l, \ldots$ If we select as *reference state* the solute i in infinitely dilute solution in the pure solvent, the reference titre being $x_i^{\ominus} = 1$, the *activity* of i may be expressed by: $a_i = \gamma_i x_i$. The *activity coefficient* γ_i of the solute is a more or less complex function of T, P, and x_i, but $\gamma_i \to 1$ when $x_i \to 0$ according to *Henry's law* and our selection of reference state.

Suppose now that the solute i, for which the same reference state is taken, is in solution in a solvent containing other solutes: $j, k, l, \ldots$ The activity coefficient γ_i no longer tends towards 1 when $x_i \to 0$, owing to the interaction of the other solutes $j, k, l, \ldots$ If we compare, at an imposed activity of i (by fixing for instance the partial pressure of gaseous i above the solution, or by saturating the solution with solid i), x_i°, the solubility of i in the pure solvent, with x_i, the solubility of i in the solution, the activity coefficient γ_i is given experimentally by:

$$\gamma_i = x_i^{\circ} / x_i.$$

It is then possible to define the interaction parameters of first and second order between the solutes:

$$\varepsilon_i^{(j)} \equiv \lim_{x_j \to 0} \left(\frac{\partial \ln \gamma_i}{\partial x_j} \right)_{T,P} ;$$

$$\varepsilon_i^{(j,j)} \equiv \lim_{x_j \to 0} \left(\frac{\partial^2 \ln \gamma_i}{\partial x_j^2} \right)_{T,P} ; \qquad \varepsilon_i^{(j,k)} \equiv \lim_{x_j, x_k \to 0} \left(\frac{\partial^2 \ln \gamma_i}{\partial x_j \partial x_k} \right)_{T,P} .$$

A knowledge of interaction parameters gives access to the solubility of a constituent i in a solvent that has already dissolved solutes $j, k, l, \ldots$:

$$\ln \gamma_i = \ln \frac{x_i^\circ}{x_i} = x_i \varepsilon_i^{(i)} + \sum_j x_j \varepsilon_i^{(j)} + \sum_j \frac{1}{2} x_j^2 \varepsilon_i^{(j,j)} + \sum_j \sum_{k \neq j} x_j x_k \varepsilon_i^{(j,k)} + \cdots$$

The self-interaction parameter $\varepsilon_i^{(i)}$ equals zero as long as the solute follows *Henry's law*. Parameters $\varepsilon_i^{(j)}$ and $\varepsilon_j^{(i)}$ are linked by the *Wagner reciprocity*:

$$\varepsilon_i^{(j)} = \varepsilon_j^{(i)}.$$

It is often convenient to express solubilities in mass per cent [%i] and to select as reference state for the solute i, the solute i in an infinitely dilute solution in the pure solvent, the reference composition being $\lfloor \%i \rfloor^\triangleright = 1$ mass%. The activity coefficient of i in a solvent that has dissolved $j, k, l, \ldots$ is then experimentally given by:

$$f_i = [\%i]^\circ / [\%i],$$

where $[\%i]^\circ$ and $[\%i]$ represent respectively the solubility of i in the pure solvent and in the solution for a given activity of i. Interaction parameters of first and second order between i and j are then:

$$e_i^{(j)} \equiv \lim_{[\%j] \to 0} \left(\frac{\partial \log_{10} f_i}{\partial [\%j]} \right)_{T,P}$$

$$e_i^{(j,j)} \equiv \lim_{[\%j] \to 0} \left(\frac{\partial^2 \log_{10} f_i}{\partial [\%j]^2} \right)_{T,P}$$

$$e_i^{(j,k)} \equiv \lim_{[\%j],[\%k] \to 0} \left(\frac{\partial^2 \log_{10} f_i}{\partial [\%j] \partial [\%k]} \right)_{T,P}.$$

Hence the solubility of i in a solvent in the presence of $j, k, l, \ldots$ is given by:

$$\log_{10} f_i = \log_{10} \frac{[\%i]^\circ}{[\%i]} = [\%i] e_i^{(i)} + \sum_j [\%i] e_i^{(j)} + \sum_j \frac{1}{2} [\%j]^2 e_i^{(j,j)}$$

$$+ \sum_j \sum_{k \neq j} [\%j][\%k] e_i^{(j,k)} + \cdots$$

The self-interaction parameter $e_i^{(i)}$ equals zero as long as the solute i follows Henry's law. It is easy to find the relation between e and ε:

$$\varepsilon_i^{(j)} = 230 \frac{M_j}{M_1} e_i^{(j)} + \frac{M_1 - M_j}{M_1},$$

where M_j and M_1 are the molar masses of the element j and the solvent 1.

In practice, expansions used to evaluate a solubility are limited to the first- or second-order interaction parameters.

A negative interaction parameter $\varepsilon_i^{(j)} < 0$ implies:

$$\left[\frac{\partial \ln \gamma_i}{\partial x_j} \right]_{a_i} < 0, \quad \text{whence:} \quad \left[\frac{\partial \ln x_i}{\partial x_j} \right]_{a_i} > 0.$$

The presence of j in the solvent has the effect of increasing the solubility of i; conversely, if $\varepsilon_i^{(j)} > 0$, the presence of j in the solvent decreases the solubility of i.

Internal energy—The part of the *energy* of a system which depends only on its microscopic parameters. External contributions most frequently encountered being the *kinetic energy* $E_k = mv^2/2$ and the *potential energy* $E_p = mgz$, the internal energy U is then defined by:

$$U = E - \frac{1}{2} mv^2 - mgz.$$

The kinetic and potential energies being *state functions* of the system, the internal energy is also a state function. It is possible to assimilate the energy of a system with its internal energy when neither E_k nor E_p varies.

EXPRESSION OF THE INTERNAL ENERGY: The change in internal energy undergone by a system during a transformation is the sum of energies exchanged by the system with the surroundings. It is expressed in a general way by a sum of various contributions: $dU = \sum X_j dx_j$.

Explicitly:

$$dU = T(dS - d\sigma) + \Gamma\, d\mathscr{S} - P\, dV + \sum \mu_i\, dn_i + \cdots$$

$T(dS - d\sigma)$: heat quantity dQ (see *Second law*)

$\Gamma\, d\mathscr{S}$: work of surface tension

$-P\, dV$: work of pressure forces

$\mu_i\, dn_i$: work related to the change in amount of i.

Not all contributions are always present. However, it must be kept in mind that an expression such as $dU = T\, dS - P\, dV$ implicitly supposes a *reversible* transformation in a *thermoelastic* system.

INTERNAL ENERGY, THERMODYNAMIC POTENTIAL: Let us write the general expression of dU by emphasizing only $d\sigma$, the *entropy* created by the *irreversibility* of the transformation:

$$dU = -T\, d\sigma + \sum X_j\, dx_j.$$

For a system subject to the *constraints* $x_j = $const. (*closed* system without chemical reaction), dU is reduced to $dU = -T\, d\sigma < 0$. In such conditions, the function U decreases, reaching a minimum which represents the equilibrium state of the system. Internal energy is thus the *thermodynamic potential* of a system subject to the constraints $x_j = $const. For a thermoelastic system, these constraints are reduced to constant S and V.

Remark 1: Such constraints are seldom encountered in practice, and internal energy used only as thermodynamic potential would be uninteresting (see, however, *Fountain effect*). On the other hand, at constant volume and amount of substance, there remains $dU = dQ_V$. The change in internal energy of a system that evolves in such conditions then equals the heat quantity exchanged with the surroundings, the subscript V signifying that the volume remains constant.

Remark 2: A number of authors directly infer the concept of internal energy from the *first law*, that is to say, they do not distinguish it from *external energy*. There is nevertheless a difference in nature between these two kinds of energy. Kinetic and potential energies are state functions related to the presence of an external field and to the macroscopic properties of the system, such as mass; internal energy is also a state function, but related to the microscopic properties of the system such as temperature, surface tension, pressure, electric potential, chemical potential, etc. Finally, although it is possible to express

158 Internal pressure

dE_k and dE_p in the form: $dE_k = v\,d(mv), dE_p = mg\,dz$, it is clear that neither velocity nor weight matches the definition of a tension.

Internal pressure—The measurement of pressure requires the presence of a wall which transmits forces exerted by impacts of molecules. Now, within a fluid, interaction forces exerted between the molecules contribute to the pressure and are not taken into account by the measurement. It is thus necessary to add to the measured pressure P a pressure Π, called internal pressure, whose influence increases with the proximity of the molecules. In the *van der Waals equation*, internal pressure is proportional to the number of pairs of molecules, and thus to the square of the fluid density, whence its expression: $\Pi = a/V^2$. The parameter a characterizes the amplitude of attraction.

Invariant—Whose *variance* equals zero. An equilibrium is said to be invariant when it is impossible to modify an *intensive* parameter without simultaneously changing the number of phases. The equilibrium of a pure substance in three phases is invariant. More generally, the equilibrium of n pure substances in $(n+2)$ phases is invariant.

Inversable—Antonym of *renversable*.

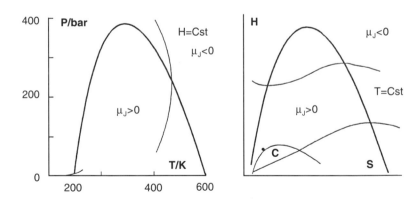

Inversion: The inversion curve of the Joule–Thomson effect for nitrogen, N_2, is shown in bold. On the pressure–temperature diagram on the left, it is obtained by taking, on the isenthalpic curves, the locus of points with a vertical tangent. On the enthalpy–entropy diagram, on the right, it is obtained by taking, on the isothermal curves, the locus of points with a horizontal tangent. C is the critical point.

Inversion—Inversion of a phenomenon occurs when one of the quantities which characterize it changes in sign. For instance:

- The *Joule–Thomson effect* is characterized by an inversion temperature, depending on the pressure, a temperature at which the Joule–Thomson coefficient equals zero: $\mu_j = (\partial T/\partial P)_H = 0$.
- The inversion temperature of a chemical reaction corresponds to the change of sign of the *standard Gibbs energy* $\Delta G°$. At the inversion temperature, the equilibrium constant equals unity.
- A saturated vapour may show two inversion temperatures, between which the *heat capacity* along the saturated vapour curve $(c_x)_{vap} > 0$. Without any inversion, $(c_x)_{vap} < 0$.

Ionic strength—A concept introduced by Lewis and Randall in 1921, the ionic strength of a solution is defined by: $I = \frac{1}{2}\sum c_i z_i^2$, where c_i and z_i are respectively the *molarity* and the charge of the ion i. Such a concept is of interest because in dilute solutions the *activity coefficient* γ_i depends not on the nature of the ions present in the solution, but only on the ionic strength of the solution (see *Debye–Hückel theory*).

Irreversibility—A transformation is said to be irreversible when it is impossible to cancel out its effects on the system and on the surroundings. In order to demonstrate the irreversible character of a transformation, it is sufficient to verify that at least one of the three criteria of *reversibility* is not obeyed:

$$dS = (dQ/T) + d\sigma.$$

The entropy $d\sigma$ created by the irreversibility of the transformation is positive in accordance with the *second law*, and tends towards zero when the transformation becomes reversible.

Isenergetic—At constant *internal energy*. From $dU = T\,dS - P\,dV$ combined with the expression of dS as a function of dT and dV, we get the differential equation governing isenergetic curves:

$$dU = c_V\,dT + \left[T\left(\frac{\partial P}{\partial T}\right)_V - P\right]dV = 0.$$

For an ideal gas, the bracket equals zero: the internal energy of an ideal gas depends only on the temperature.

160 Isenthalpic

An isenergetic expansion is one that is irreversible with no *work* exchanged between the system and its surroundings. It is achieved approximately during *adiabatic* expansion of a gas into a vacuum. In this case:

$$dU = dQ + dW = 0 + 0 = 0(!).$$

Isenthalpic—At constant *enthalpy*. From $dH = T\,dS + V\,dP$ combined with the expression of dS as a function of dT and dP, we get the differential equation governing isenthalpic curve:

$$dH = c_P\,dT + \left[V - T\left(\frac{\partial V}{\partial T}\right)_P \right] dP = 0.$$

For an ideal gas, the bracket equals zero: the enthalpy of an ideal gas depends only on the temperature, and isenthalpic curves identify with isotherms.

An isenthalpic expansion is one that is irreversible with no *work* exchanged between the system and its surroundings. It is achieved approximately during *adiabatic* expansion of a gas through a reducing or needle valve. Let us consider a gas undergoing an adiabatic expansion ($dQ = 0$) through a reducing valve which makes it change from the state (P_1, V_1, T_1) to the state (P_2, V_2, T_2):

$$\Delta U = U_2 - U_1 = \Delta W = \int_{V_1}^{0} -P_1\,dV + \int_{0}^{V_2} -P_2\,dV = P_1 V_1 - P_2 V_2,$$

whence: $U_2 + P_2 V_2 = U_1 + P_1 V_1$, or $H_2 = H_1$. Such an expansion is thus an isenthalpic one.

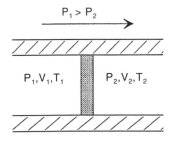

P₁ > P₂

P₁,V₁,T₁   P₂,V₂,T₂

Isenthalpic: The adiabatic and irreversible expansion of a gas through a reducing or needle valve is an isenthalpic one. With an ideal gas, it occurs without any temperature change.

Isentropic—At constant *entropy*. An *adiabatic* ($\mathrm{d}Q = 0$) and *reversible* ($\mathrm{d}\sigma = 0$) transformation is indeed an isentropic one, because:

$$\mathrm{d}S = \frac{\mathrm{d}Q}{T} + \mathrm{d}\sigma = 0 + 0 = 0!,$$

but the converse is not always true. On a pressure–temperature diagram, the differential equation governing isentropic curves is obtained as:

$$\mathrm{d}S = \left(\frac{\partial S}{\partial T}\right)_P \mathrm{d}T + \left(\frac{\partial S}{\partial P}\right)_T \mathrm{d}P = \frac{c_P}{T}\mathrm{d}T - \left(\frac{\partial V}{\partial T}\right)_P \mathrm{d}P = 0.$$

It is possible to derive other expressions from $\mathrm{d}S$ as a function of $\mathrm{d}T$ and $\mathrm{d}V$ or of $\mathrm{d}P$ and $\mathrm{d}V$ (see *Maxwell's equations*). Applied to an ideal gas and integrated on the assumption of constant *heat capacities*, these expressions yield:

$$\frac{T}{T_\circ} = \left(\frac{P}{P_\circ}\right)^{R/C_P} = \left(\frac{V_\circ}{V}\right)^{R/c_V},$$

which may be expressed more simply with $\gamma = c_P/c_V$:

$$PV^\gamma = \text{const.}$$

Iso The Greek root ισος, from which has been formed·

- isobaric: at constant pressure (1877), from βαρος, gravity;
- isochoric: at constant volume (1881), from χωρα, space;
- isomorphic: with the same crystal structure (1840), from μορφος, shape;
- isopleth: with the same quantity (1952), from πλεθος, number;
- isothermal: at constant temperature (1816) from θερμον, heat;
- isotropic: with the same properties in every direction (1840), from τροπος, turn.

Isobar, isochor, isotherm are corresponding nouns. Isobars also represent nuclides with the same atomic mass but different atomic numbers, e.g. ${}^3_1\mathrm{H}$ and ${}^3_2\mathrm{He}$.

The term isopleth is now widely used, but without good reason, to denote a vertical section of constant composition through a ternary diagram.

Isobaric–isothermal ensemble—The name given to the *Gibbs ensemble* in which each system is characterized by its temperature T, its pressure P, and the number of particles N. Such an isolated ensemble of $\mathcal{N}$ systems, sometimes called, for lack of imagination, 'grand canonical' (with the specification T, P, N) is dealt with under *grand*

162 Isobaric–isothermal ensemble

canonical ensemble (T, V, μ). The energy and the volume of each system are not constant, but their mean values are known. Let $\mathcal{N}_{i,j}(V_i, E_j)$ be the number of systems whose volume is V_i and energy E_j. The number of states for such a distribution is:

$$\Omega_{i,j} = \frac{\mathcal{N}!}{\prod_i \prod_j \mathcal{N}_{i,j}}.$$

The most probable distribution is obtained by seeking the maximum of $\Omega_{i,j}$ taking into account the three balance equations:

- number of systems in the ensemble: $\sum_i \sum_j \mathcal{N}_{ij}(V_i, E_j) = \mathcal{N}$;
- energy of the ensemble: $\sum_i \sum_j \mathcal{N}_{ij}(V_i, E_j) \cdot E_j = E$;
- volume of the ensemble: $\sum_i \sum_j \mathcal{N}_{ij}(V_i, E_j) \cdot V_j = V$.

The introduction of three *Lagrange multipliers* gives:

$$\ln \mathcal{N}_{i,j}(V_i, E_j) + \alpha + \beta E_j + \delta V_i = 0.$$

We obtain the multiplier α by using the first constraint:

$$\mathcal{N} = \exp(-\alpha) \sum_i \sum_j \exp(-\beta E_j - \delta V_i),$$

whence the probability of a volume V_i and an energy E_j:

$$P_{i,j} = \frac{\mathcal{N}_{i,j}}{\mathcal{N}} = \frac{\exp(-\beta E_j - \delta V_i)}{\sum_i \sum_j \exp(-\beta E_j - \delta V_i)}$$

$$Y = \sum_i \sum_j \exp(-\beta E_j - \delta V_i)$$

represents the isobaric–isothermal *partition function* of the system.

The multipliers β and δ are obtained by comparison of the expressions for the *entropy* in classical and statistical thermodynamics:

- In classical thermodynamics: $dS = (dE + P\,dV - \sum \mu_k \, dn_k)/T$.
- In statistical thermodynamics: $dS = k\,d \ln \Omega$. By expanding:

$$dS = k(d \ln Y + \beta\,dE + \delta\,dV),$$

which, by identification, yields the relations:

$$G = \sum n_k \mu_k = -kT \ln Y$$

$$\beta = 1/kT$$

$$\delta = P/kT,$$

whence the expression of the isobaric–isothermal partition function:

$$Y(T, P, N) = \sum_i \sum_j \exp\left(-\frac{E_j + PV_i}{kT}\right).$$

The probability of an energy E_j and a volume V_i is then given by:

$$P_{i,j} = \frac{1}{Y} \exp\left(-\frac{E_j + PV_i}{kT}\right).$$

Isolated—A *system* is said to be isolated (from the Italian *isola*, island) when it cannot have any interaction with the surroundings. From the *first law*, the *internal energy* of an isolated system is constant ($dU = 0$); from the *second law*, the *entropy* of an isolated system may only increase.

J

Jacobian—From Carl Jacobi (1804–1851), German mathematician. In the system of two equations:

$$f(u, v, x, y) = 0, \qquad g(u, v, x, y) = 0,$$

u and v may be considered as two implicit functions of the variables x and y. To calculate a partial derivative, for instance $(\partial u/\partial x)_y$, it is sufficient to write $df = 0$, $dg = 0$, to substitute $dy = 0$ into the expression of the *differentials*, and then to solve the system. Thus:

$$\left(\frac{\partial u}{\partial x}\right)_y = -\begin{vmatrix} f'_x & f'_v \\ g'_x & g'_v \end{vmatrix} \bigg/ \begin{vmatrix} f'_u & f'_v \\ g'_u & g'_v \end{vmatrix},$$

with $f'_x = (\partial f/\partial x)_{u,v,y}$, ... The determinant

$$\frac{\partial(f, g)}{\partial(u, v)} = \begin{vmatrix} f'_u & f'_v \\ g'_u & g'_v \end{vmatrix}$$

is the Jacobi determinant, or Jacobian, of the two functions f and g. We can verify easily that:

$$\frac{\partial(f, g)}{\partial(u, v)} = \frac{\partial(g, f)}{\partial(v, u)}.$$

Partial derivatives may be expressed as a function of Jacobians:

$$\left(\frac{\partial u}{\partial x}\right)_y = -\frac{\partial(f,g)}{\partial(x,v)} \Big/ \frac{\partial(f,g)}{\partial(u,v)} \qquad \left(\frac{\partial v}{\partial x}\right)_y = -\frac{\partial(f,g)}{\partial(u,x)} \Big/ \frac{\partial(f,g)}{\partial(u,v)}.$$

Generalization is possible. For instance, with three functions:

$$f(x,y,z,u,v,w)=0, \quad g(x,y,z,u,v,w)=0, \quad h(x,y,z,u,v,w)=0$$

$$\left(\frac{\partial u}{\partial x}\right)_{y,z} = - \begin{bmatrix} f'_x & f'_v & f'_w \\ g'_x & g'_v & g'_w \\ h'_x & h'_v & h'_w \end{bmatrix} \Big/ \begin{bmatrix} f'_u & f'_v & f'_w \\ g'_u & g'_v & g'_w \\ h'_u & h'_v & h'_w \end{bmatrix} = -\frac{\partial(f,g,h)/\partial(x,v,w)}{\partial(f,g,h)/\partial(u,v,w)}.$$

The use of Jacobians in thermodynamics rests on two important relationships which are easily verified:

$$\left(\frac{\partial y}{\partial z}\right)_x = \frac{\partial(y,x)}{\partial(z,x)} = \frac{\partial(y,x)}{\partial(u,v)} \times \frac{\partial(u,v)}{\partial(z,x)}.$$

Example:

$$\left(\frac{\partial U}{\partial T}\right)_S = \frac{\partial(U,S)}{\partial(T,S)} = \frac{\partial(U,S)}{\partial(U,V)} \times \frac{\partial(U,V)}{\partial(T,V)} \times \frac{\partial(T,V)}{\partial(T,S)} = \left(\frac{\partial S}{\partial V}\right)_U \left(\frac{\partial U}{\partial T}\right)_V \left(\frac{\partial V}{\partial S}\right)_T,$$

$$\left(\frac{\partial U}{\partial T}\right)_S = \frac{P}{T} c_V \left(\frac{\partial T}{\partial P}\right)_V = \frac{c_V}{T \cdot \beta_V},$$

where β_V is the isochoric *coefficient of bulk expansion*. Another example:

$$\left(\frac{\partial P}{\partial V}\right)_S = \frac{\partial(P,S)}{\partial(V,S)} = \frac{\partial(P,S)}{\partial(P,T)} \times \frac{\partial(P,T)}{\partial(V,T)} \times \frac{\partial(V,T)}{\partial(V,S)} = \left(\frac{\partial S}{\partial T}\right)_P \left(\frac{\partial P}{\partial V}\right)_T \left(\frac{\partial T}{\partial S}\right)_V,$$

$$\left(\frac{\partial P}{\partial V}\right)_S = \frac{c_P}{T} \left(\frac{\partial P}{\partial V}\right)_T \frac{T}{c_V} = -\gamma V \chi_T,$$

where χ_T is the isothermal *compressibility* and $\gamma = c_P/c_V$. In conclusion, manipulation of Jacobians is an entertainment which gives access to many relationships between differentials, but it is not a mandatory procedure!

Joule—The unit of energy in the SI system:

$$1\,J = 1\,N\,m = 1\,Pa\,m^3 = 1\,C\,V = 1\,W\,s = 1\,kg\,m^2\,s^{-2}.$$

Joule, James Prescott (1818–1889)—English physicist, who established that mechanical work, electrical work, and heat represent various kinds of energy, described the *Joule effect* (1840), stated *Joule's law* (1845), determined the mechanical equivalent of heat (1845), and, with William Thomson, later Lord Kelvin, demonstrated the *Joule–Thomson effect* (1852). He introduced the word 'thermodynamics' (1858).

Joule cycle—See *Brayton cycle.*

Joule effect—The transformation of electrical energy into heat, discovered by Joule in 1840. The corresponding law was expressed the following year: an electric current I flowing through a conductor of resistance R during a time t generates a heat quantity $Q = RI^2 t$.

Joule's law—The *internal energy* of an ideal gas is a function only of its temperature. The *kinetic theory of gases* together with statistical considerations (see *Partition function*) give the contribution of translational motion to internal energy: $U = 3nRT/2$ ($u = 3kT/2$ for one molecule).

Immediate consequence: the *enthalpy* of an ideal gas is also dependent only on its temperature. Indeed: $H = U + PV = U + nRT$.

Joule's historic experiment allows it to be verified that isenergetic expansion of an ideal gas proceeds without temperature change. In a calorimeter, two compartments can be interconnected by means of a tap. Initially, the gas stays inside one compartment, the other being empty. Opening the tap allows the gas to expand into the two compartments but does not produce any observable temperature change in the calorimeter. The system having exchanged neither heat nor work with the surroundings, its internal energy remains constant. As the volume and the pressure of the gas have been modified during the expansion, its internal energy depends neither on its volume nor on its pressure; it can only be a function of the temperature, the only parameter remaining unmodified.

A moment's critical thought shows that Joule's experiment is of low accuracy, because the heat capacity of the calorimeter is much higher than that of the gas. Joule's law may be verified better experimentally by showing that the *isenthalpic* expansion of an ideal gas occurs without temperature change.

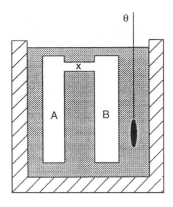

Joule's law: Adiabatic expansion of a gas into a vacuum is isenergetic. For an ideal gas, it occurs without temperature change.

Joule–Thomson effect—Discovered in 1852, the Joule–Thomson, sometimes called Joule–Kelvin, effect is the thermal effect observed during *adiabatic* and *isenthalpic* expansion of a real gas. The Joule–Thomson coefficient:

$$\mu_J = \left(\frac{\partial T}{\partial P}\right)_H = \frac{1}{c_P} \cdot \left[T\left(\frac{\partial V}{\partial T}\right)_P - V \right]$$

equals zero for an ideal gas. For a real gas, the curve of the equation $\mu_J = 0$ on a state diagram (P vs V, PV vs P, P vs T) appears roughly parabolic and separates the plane into two fields. In the field which includes the liquid–vapour equilibrium curve, $\mu_J > 0$, which expresses cooling of the gas upon isenthalpic expansion; in the field furthest from the critical point, $\mu_J < 0$, which expresses heating of the gas upon isenthalpic expansion.

K

Kamerlingh Onnes, Heikke (1853–1926)—Dutch physicist, who was the first to succeed in liquefying gaseous helium in 1908; he discovered superconductivity in 1911. Nobel prize for physics in 1913.

Kelvin—Named from Lord Kelvin (see *Thomson*), one of the seven SI base units in the SI system, the kelvin (symbol K; the notation °K, dropped in 1967, must be proscribed) was defined in 1968 by the 13th Conférence Générale des Poids et Mesures as the 1/273.16 fraction of the *thermodynamic temperature* of the triple point of water.

Kilogram—One of the seven SI base units, the kilogram (symbol kg) was defined in 1889 by the 1st Conférence Générale des Poids et Mesures as the mass of the prototype in platinum–iridium alloy deposited at the International Bureau of Weights and Measures at Sèvres, near Paris.

In contrast to the definitions of the metre and the second, which have been subjected to several changes, that of the kilogram has never been modified.

Kinetic energy—In order to impart a velocity v to a mass m, initially at rest in a given frame of reference, it is necessary to provide a quantity of work $mv^2/2$. The mass m has then gained, in this frame of reference, a kinetic energy of:

$$E_k = mv^2/2.$$

Remark 1: From the *first law*, the *energy* of an *isolated* system is constant. When a mass m is placed in a terrestrial gravitational field, its *internal energy* also remains constant, whence $E_k + E_p = \text{const}$. It is wrong to say that the earth imparts energy to the mass m. A more rigorous description of the process consists in saying that kinetic energy of the system increases at the expense of its *potential energy*.

Remark 2: The kinetic energy of a system of mass m, propelled at a velocity v, contributes to its *external energy* because it depends on its macroscopic parameters (mass and velocity), whereas the kinetic energy of individual particles which form the system contributes to its internal energy because it depends on its microscopic parameters. The macroscopic properties measured for the whole system (pressure, temperature) express only the average motion of individual particles.

Kinetic theory of gases—The kinetic theory of gases is a model which links the temperature of a gas to the velocity of its molecules. In this model, the molecules, supposed spherical, are not subjected to interactions other than perfectly elastic collisions.

Let us consider N molecules, each with a mass m, in a volume V, each molecule moving with a velocity v. The velocity vector has components v_x, v_y, and v_z. The number of molecules whose velocity lies

K

between v and $v+dv$ equals $n_v dv$. The number of these particles colliding with an area $\mathscr{S}$ of the wall perpendicular to the x axis during a time t equals $n_v v_x t \mathscr{S} dv$. The momentum transferred to the surface $\mathscr{S}$ during time t equals:

$$2mv_x^2 n_v t \mathscr{S} \, dv.$$

The pressure exerted on the wall is obtained by integrating all the velocities per unit surface:

$$P = \int_0^\infty 2mv_x^2 n_v \, dv = \int_{-\infty}^{+\infty} mv_x^2 n_v \, dv$$

$$P = m\frac{N}{V} \langle v_x^2 \rangle,$$

where $\langle v_x^2 \rangle$ represents the mean of the component along the x axis of the squared particle velocities. On the assumption of identical molecules moving in random directions:

$$\langle v_x^2 \rangle = \langle v_y^2 \rangle = \langle v_z^2 \rangle = \frac{1}{3}\langle v^2 \rangle,$$

whence the pressure $P = (2/3) \cdot (N/V) \cdot \langle (1/2)mv^2 \rangle$, an expression which can be compared with that of the *internal energy* of the system on the assumption that only the kinetic energy due to translational motion of the molecules contributes to the internal energy of the gas:

$$U = \frac{1}{2}Nm\langle v^2 \rangle \Rightarrow PV = \frac{2}{3}U.$$

If the temperature scale is defined by the ideal gas thermometer, k being the *Boltzmann constant*:

$$PV = NkT = \frac{2}{3}U$$

$$\left\langle \frac{1}{2}mv^2 \right\rangle = \frac{3}{2}kT.$$

The mean kinetic energy of a particle without interaction thus equals $kT/2$ for each degree of freedom.

Kirchhoff, Gustav Robert (1824–1887)—German physicist, who expressed the laws of radiation, introduced the concept of black body, contributed to the development of spectroscopy and electricity, and applied spectrum analysis to the study of the composition of the sun.

Kirchhoff's equations—These allow calculation of $\Delta_r H(T)$ or $\Delta_r S(T)$, from a knowledge of $\Delta_r H(T_o)$ or $\Delta_r S(T_o)$.

At constant pressure: $dH = c_P \, dT$ and $dS = c_P \, dT/T$, whence for a *chemical reaction* $\sum v_i M_i = 0$:

$$d(\Delta_r H) = \Delta c_P \, dT \quad \text{and} \quad d(\Delta_r S) = \Delta c_P \, dT/T,$$

with $\Delta c_P = \sum v_i c_P(M_i)$. Integration yields:

$$\Delta_r H(T) = \Delta_r H(T_o) + \int_{T_o}^{T} \Delta c_P \, dT \text{ and } \Delta_r S(T) = \Delta_r S(T_o) + \int_{T_o}^{T} \frac{\Delta c_P}{T} \, dT.$$

Thermodynamic data are generally given for a *standard state* at a temperature $T_o = 298.15 \, K$. These equations assume implicitly that, between T_o and T, no one constituent M_i undergoes a phase transition. If it does so:

$$\Delta_r H(T) - \Delta_r H(T_o) + \int_{T_o}^{T} \Delta c_P \, dT + \sum v_i \Delta_{tr} H(M_i)$$

$$\Delta_r S(T) - \Delta_r S(T_o) + \int_{T_o}^{T} \frac{\Delta c_P}{T} \, dT + \sum v_i \frac{\Delta_{tr} H(M_i)}{T_{tr}}.$$

These expressions represent curves with as many branches plus one as there are phase transitions between the temperatures T_o and T. Any phase change is characterized by a discontinuity in the curves $\Delta_r H = f(T)$ and $\Delta_r S = f(T)$, but only by a break in the slope of the curve $\Delta_r G = f(T)$.

Knudsen effect—The name given to the *thermomolecular effect* in the particular case where the system is a gas shared between two compartments separated by capillaries or small apertures. Here, 'small' means smaller than the *mean free path* of the molecules; thus every molecule which presents itself in front of the aperture will cross it freely. It follows from the *kinetic theory of gases* that the number of molecules that cross an aperture is proportional to the pressure of the compartment vacated and inversely proportional to the mean velocity of the molecules, i.e. proportional to $P/\sqrt{T}$. In a steady-state regime:

$$P_A / P_B = \sqrt{T_A / T_B}.$$

This is the Knudsen relationship giving the difference in thermomolecular pressure, an equation which may be expressed in a differential form:

$$P = K\sqrt{T}, \quad \frac{\mathrm{d}P}{\mathrm{d}T} = \frac{1}{2} K \frac{\sqrt{T}}{T} = \frac{1}{2}\frac{P}{T} = \frac{R}{2V}.$$

Kronecker delta—Symbolized by δ_{ij}, this index represents 1 when $i = j$ and 0 when $i \neq j$. It must not be confused with the *Dirac delta function*.

Kuhn, Thomas S. (1923–1996)—American philosopher, famous for his essay *Structure of the scientific revolution* (1962), in which he discusses the concept of *paradigm* and rejects the notion of falsifiability developed by *Karl Popper*. According to Thomas Kuhn, a statement is true or false only in the context of a particular paradigm. Thermodynamicists interested in the history of science will appreciate his *Black body theory and the quantum discontinuity (1894–1912)*, published in 1978.

L

Lagrange, Joseph-Louis, Comte de (1736–1813)—French mathematician. Of his considerable work, thermodynamicists will recall the development of differential calculus and the method of *Lagrange multipliers*, together with the introduction of the notation $f'(x)$, $f''(x)$, still in regular use.

Lagrange multipliers—The method of Lagrange multipliers allows calculation of the extrema of a function $F(x_1, x_2, \ldots, x_n)$ taking into account k constraints $g_j(x_1, x_2, \ldots, x_n) = 0$ (with $j = 1, 2, \ldots, k$ and $k < n$). Without any constraint, extrema of the function F are obtained by equating to zero the *differential* $\mathrm{d}F$. Since the differentials $\mathrm{d}x_i$ are independent, their coefficients must be zero, which comes down to solving a system of n equations with n unknowns: $\partial F/\partial x_i - 0$ (with $i = 1, 2, \ldots, n$).

In the presence of k constraints, the k conditions $\mathrm{d}g_j = 0$ are added to the condition $\mathrm{d}F = 0$. The usual method consists in expressing k differentials as a function of the remaining $(n - k)$ taken as independent variables, thus giving $(n-k)$ equations to which we must add the k constraints in order to obtain a system with n equations and n unknowns.

The use of multipliers provides the result more elegantly. The extrema of a function F subject to k constraints $g_j = 0$ are also those of the function $F - \sum \lambda_j g_j$, whence $dF - \sum \lambda_j dg_j = 0$. The coefficients λ_j introduced here are the Lagrange multipliers. On expansion:

$$\sum_i \left(\frac{\partial F}{\partial x_i} - \sum_j \lambda_j \frac{\partial g_j}{\partial x_i} \right) dx_i = 0,$$

an expression which can be satisfied only if every bracket equals zero, which yields a system of n equations of first degree in λ_j. It is thus possible to use k equations to obtain the multipliers λ_j as a function of x_i, then to insert the results into the remaining $(n-k)$ equations. These $(n-k)$ equations, together with the k constraints $g_j = 0$, give an ensemble of n equations with n unknowns. The solution gives the extrema of the function F, taking into account the constraints imposed.

Latent—From the Latin *latere*, to lie hidden. A latent heat is the heat quantity absorbed or released by a substance during a change of state. Since a phase transition proceeds at constant temperature and pressure, it is more convenient to speak of the enthalpy of the transformation rather than to use the old, poetic but obsolete term of latent heat.

L

Lattice energy—The enthalpy of formation of a crystal lattice from ions taken in the gaseous state:

$$(Na^+)_{(gas)} + (Cl^-)_{(gas)} \rightarrow \langle NaCl \rangle_{(crystal)}.$$

Lavoisier, Antoine-Laurent de (1743–1794)—French chemist, who, with Pierre-Simon de Laplace in 1780, made the first calorimetric measurements. He is considered the father of modern chemistry by his introduction of the systematic use of the balance and expressing the laws of conservation of mass.

Law of stable equilibrium—According to Hatsopoulos and Keenan (1972), it would be possible to base thermodynamics on a single principle, called the law of stable equilibrium:

A constrained system, evolving in such a way that the surroundings remain unmodified, can reach one and only one state of stable equilibrium.

172 Law of stable equilibrium

Corollary 1: The *work* exchanged by a *system* with its *surroundings* during a transformation from a stable state S_1 to a stable state S_2 is the same for every *adiabatic process*.

Let a and b be two adiabatic processes bringing the same system S from a state S_1 to a state S_2 and having the effect respectively of moving a mass m_a and a mass m_b in a gravitational field from height z_1 to height z_2. Now consider the *isolated* system E, constituted by the system S and the displaced masses. Then it would be possible from the same stable state E_1 (masses m_a and m_b at height z_1) to reach two stable states E_{2a} (mass m_a transferred from z_1 to z_2) and E_{2b} (mass m_b transferred from z_1 to z_2), a situation which violates the law of stable equilibrium. Every adiabatic process subjecting a system to the same transformation exchanges the same work with the surroundings.

Corollary 2: The *energy* of a system in a state of stable equilibrium is a physical quantity, the change in which equals the work exchanged by the system with its surroundings during an adiabatic transformation which makes it pass from a stable state 1 to a stable state 2: $\Delta E = E_2 - E_1 = \Delta W_{ad}$. It follows directly from corollary 1 that energy so defined is a state function of the system in a state of stable equilibrium.

Corollary 3: The energy of a system out of equilibrium may be defined as the energy of the system returned to its state of stable equilibrium after having been isolated from its surroundings, the constraints being maintained. Indeed, from the law of stable equilibrium, this state is the only one.

Corollary 4: The stable state of a constrained system is completely identified when its energy is known. Indeed, the law of stable equilibrium forbids the existence of several stable states with the same energy, arising from one and the same state out of equilibrium. This result was established by Kline and Koenig in 1957 under the name of *state principle*.

Corollary 5: The energy of a system is a state function whose change equals the work exchanged by the system with its surroundings during an adiabatic transformation which makes it pass from a state 1 to a state 2: $\Delta E = E_2 - E_1 = \Delta W_{ad}$. This result is a generalization of corollary 2 to states out of equilibrium, a generalization that is possible owing to corollary 3. This statement is that of the *first law* proposed by Clausius, which can also be presented in a differential form: $dE = dW_{ad}$.

Corollary 6: It is impossible to bring a constrained system from a stable state S_1 to an available state S_2 with the only external effect that of raising a mass in a gravitational field. This amounts to stating that it is impossible for a system in a stable equilibrium state to evolve with the only external effect that of supplying work to the surroundings ($\Delta W < 0$). Indeed, if that were possible, one could conceive of using the work produced to bring the system into a state different from the initial equilibrium state, which contradicts the law of stable equilibrium. For instance, by bringing the mass to its initial level it would be possible to transform the recovered work (mgz) into kinetic energy imparted to the system.

Remark: From the law of stable equilibrium alone, we have again found a possible statement of the first law (corollary 5) and of the *second law* (corollary 6). Nevertheless, such a seductive presentation is still far from being universally accepted.

Lebedev, Pyotr Nikolayevitch (1866–1912)—Russian physicist, who in 1901 demonstrated by very incisive experiments the existence of radiation pressure foreseen by Maxwell in 1874.

Le Chatelier's laws These laws were expressed in 1888 by Henri Le Chatelier, French chemist and metallurgist (1850–1936), translator in 1899 of the works of Gibbs and inventor of thermal analysis. They express in non-mathematical language the laws of equilibrium displacement. They often cause unjustified panic, but may be applied correctly by a 12-year-old once the message conveyed is understood.

In a word: an equilibrium is expressed by an equality between two terms. If I have an equality, I have an equilibrium! If external intervention results in modifying one of the two terms, there is no longer equality, hence there is no longer equilibrium! The system will evolve towards a new equilibrium state, so it will react in the direction which restores the equality between the two terms. It is possible to gather Le Chatelier's laws into a general statement:

> *Given a system at equilibrium, every modification of a constraint results in making the system evolve in such a way as to neutralize the disturbance and restore equilibrium.*

This statement may be applied for instance:

- to pressure: an increase in pressure displaces the equilibrium in the direction which tends to reduce the volume of the system;

174 Length

- to *dilution*: an addition of solvent displaces the equilibrium in the direction which tends to increase the dissociation of the dissolved species;
- to temperature: an increase in temperature displaces the equilibrium in the *endothermic* direction, that is, the direction which absorbs heat;
- to *chemical potential*: an increase in the chemical potential of only one constituent displaces the equilibrium in the direction which consumes it.

Note: When applying Le Chatelier's laws, it must be kept in mind that intervention in a system may modify more than one parameter. For instance, with a chemical equilibrium, the addition of a gas increases only the partial pressure of the added gas when the operation is performed at constant volume, whereas it alters the partial pressure of every gas when the operation takes place at constant pressure. In such a situation, the system does not inevitably react in the direction which consumes the added constituent. It is thus necessary to examine very closely the influence of the modification on the *mass action product* of the system.

Length—The unit of length in the SI system, the metre (symbol m) is one of the seven base units. In thermodynamics, length is an extensive quantity whose conjugate tension is force.

Lenz's law—Lenz's law in electromagnetism corresponds to *Le Chatelier's law* in thermodynamics. It was established in 1833 by the Russian physicist Heinrich Lenz (1804–1865): 'The electric current induced by a change in magnetic flux flows in a direction such that its effect is to oppose the change in the inductive flux'.

If the inductive flux increases, the induced current generates a flux in the direction opposite to that of the inductive flux; if the inductive flux decreases, the flux of the induced current has the same direction as the inductive flux.

Lewis, Gilbert Newton (1875–1946)—American chemist, who introduced the concepts of *fugacity* and *activity*, proposed a general definition of acids, explained covalent bonding, and isolated deuterium. In 1923, with Merle Randall, he published *Thermodynamics and free energy of chemical substances*, the first thermodynamics textbook really accessible to students.

Liquid—The state of matter in which molecules are free to change their position but obliged to occupy a finite volume. This definition is satisfactory well below the critical point, a region in which it is hard to confuse liquid and vapour: there is the same volume ratio between liquid water and its vapour in equilibrium at 25°C as between a rabbit and an elephant! The liquid–vapour transition, of the first *order*, is characterized by a non-zero enthalpy of vaporization and volume change.

At high pressures, above the critical pressure, there is no marked boundary separating the liquid region from the vapour region. Generally speaking, it is customary to consider a fluid as gaseous when the kinetic energy of a particle is greater than the interaction potential energy between two particles; a fluid is considered as a liquid when both energies are of the same magnitude.

A metastable liquid under a pressure lower than its saturated vapour pressure is said to be superheated. A metastable liquid at a temperature lower than its melting temperature is said to be super-cooled: the presence of 'seeds' in the liquid is generally sufficient to trigger crystallization.

Liquidus—On an isobaric binary diagram, the liquidus is a curve, or an ensemble of curves, separating the plane into two domains. On the high-temperature side, no solid phase may be stable; on the low-temperature side, a liquid phase may be stable only in equilibrium with a solid.

On an isobaric ternary diagram, where the composition is defined by two parameters, it is possible to use the third dimension for the temperature. The liquidus appears then in the form of a surface, or an ensemble of surfaces, above which no stable solid phase may be stable.

LIQUID–PURE SOLID EQUILIBRIUM: When the liquid is in equilibrium with a pure solid constituent A, the equation of the liquidus is easily obtained by expressing the equality of *chemical potentials* of A in the solid s and liquid l phases: $\mu_{A,s} = \mu_{A,l}$, whence:

$$\mu_{A,s}^{\circ} = \mu_{A,l}^{\circ} + RT \ln a_{A,l}.$$

By selecting as *standard state* for A in the solid and liquid phases respectively the solid and liquid pure constituent A:

$$\mu_{A,l}^{\circ} - \mu_{A,s}^{\circ} = \Delta_{fus} G^{\circ}(A),$$

$\Delta_{fus}G°(A)$ being the *standard Gibbs energy* of fusion of pure A, whence:

$$a_{A,l} = \exp[-\Delta_{fus}G°(A)/RT].$$

If the liquid solution is *ideal*: $a_{A,l} = x_{A,l}$. The shape of the liquidus depends only on the properties of A. Now, owing to *Raoult's law*, the behaviour of actual solutions tends towards ideality when the composition of the solution tends towards the pure solvent: if $x_{A,l} \to 1$, $a_{A,l} \to 1$ and $da_{A,l}/dx_{A,l} \to 1$. It follows that for every real solution, the slope of a liquidus arising from a pure constituent A is the same, whatever the nature of solutes B, C, etc. present in A:

$$\lim_{x_{A,l} \to 1}\left(\frac{dx_{A,l}}{dT}\right) = \frac{\Delta_{fus}H°(A)}{RT^2_{fus}(A)}.$$

LIQUID–SOLID SOLUTION EQUILIBRIUM: If the liquid is not in equilibrium with pure solid A, but with A having dissolved B, C, etc. in solid solution, the equality of chemical potentials gives:

$$\mu°_{A,s} + RT \ln a_{A,s} = \mu°_{A,l} + RT \ln a_{A,l}$$

$$a_{A,l}/a_{A,s} = \exp[-\Delta_{fus}G°(A)/RT].$$

In the case of a binary system, the replacement of A by B gives another expression. These two equations with two unknowns yield the liquidus and *solidus* curves, provided that the activities of A and B in the two phases are known. If the solid and liquid solutions are ideal, the two equations are of the first degree as a function of the composition, and the diagram obtained is cigar-shaped. For a ternary system, the calculation of the liquidus and solidus surfaces needs three equations expressing the equality of chemical potentials of A, B, and C in the two phases.

LIQUID–SOLID EQUILIBRIUM IN THE NEIGHBOURHOOD OF A PURE CONSTITUENT: In the vicinity of a pure constituent A, $x_{A,l}$ and $x_{A,s}$ tend towards 1. Raoult's law may be applied to both phases. It follows that for any real solution, the difference between the slopes of the liquidus and the solidus arising from a pure constituent A is independent of the nature of the solutes B, C, etc. present in A:

$$\lim_{x_A \to 1}\left(\frac{dx_{A,l}}{dT} - \frac{dx_{A,s}}{dT}\right) = \frac{\Delta_{fus}H°(A)}{RT^2_{fus}(A)}.$$

If the difference between the slopes arising from A is dependent only on A, each slope taken alone depends on the nature of the solutes B, C, etc. present in solid or liquid solution in A.

LIQUID–DEFINED COMPOUND EQUILIBRIUM: If the liquid is in equilibrium with a defined compound $\gamma = A_m B_n$, it is always possible to express the equality of the chemical potentials of A and B between the phases l and γ. However, there is a difference in nature between a pure constituent A and a defined compound γ. Indeed, at a given T and P, the chemical potential of A in pure A remains constant whereas the chemical potential of A in γ varies as a function of the potential of A in the vapour. The chemical potentials of A and B in the γ phase are not independent, but are related by the *Euler identity*:

$$G_\gamma = m\mu_A + n\mu_B.$$

If γ is treated as a stoichiometric compound and the pure solids are selected as standard state for A, B, and γ:

$$G_\gamma - G_\gamma^\circ = m\mu_{A,\gamma}^\circ + n\mu_{B,\gamma}^\circ + RT(m\ln a_{A,\gamma} + n\ln a_{B,\gamma})$$

$$\mu_{A,\gamma}^\circ = \mu_{A,s}^\circ; \quad \mu_{B,\gamma}^\circ = \mu_{B,s}^\circ; \quad G_\gamma^\circ - m\mu_{A,\gamma}^\circ - n\mu_{B,\gamma}^\circ = \Delta_f G^\circ(\gamma).$$

$\Delta_f G^\circ(\gamma)$ is the *standard Gibbs energy* of formation of the compound γ; between the activities of A and B in the γ phase exists the relation:

$$RT(m\ln a_{A,\gamma} + n\ln a_{B,\gamma}) = \Delta_f G^\circ(\gamma),$$

which may be presented in the form of a *solubility product*:

$$(a_{A,\gamma})^m (a_{B,\gamma})^n = \exp[\Delta_f G^\circ(\gamma)/RT].$$

The equilibrium of A and B between the γ and l phases is expressed by:

$$\mu_{A,\gamma}^\circ + RT\ln a_{A,\gamma} = \mu_{A,l}^\circ + RT\ln a_{A,l}$$

$$\mu_{B,\gamma}^\circ + RT\ln a_{B,\gamma} = \mu_{B,l}^\circ + RT\ln a_{B,l}.$$

Addition of the above equations, after multiplying both sides of the first by m and both sides of the second by n, yields:

$$m\mu_{A,\gamma}^\circ + n\mu_{B,\gamma}^\circ + \Delta_f G^\circ(\gamma) = m\mu_{A,l}^\circ + n\mu_{B,l}^\circ + RT(m\ln a_{A,l} + n\ln a_{B,l})$$

$$RT(m\ln a_{A,l} + n\ln a_{B,l}) = \Delta_f G^\circ(\gamma) - m\Delta_{fus}G^\circ(A) - n\Delta_{fus}G^\circ(B).$$

This last equation is that of the liquidus arising from $\gamma = A_m B_n$. The activities of A and B in the liquid must be known, because the mere existence of a compound γ makes it risky to assume ideality for the liquid.

CONSTRUCTION OF THE LIQUIDUS FROM CURVES $\Delta_{mix}G = f(x,T)$: It is of interest to examine the geometric meaning of the equation of the

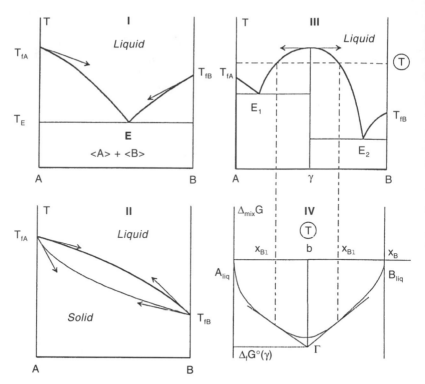

Liquidus: On each diagram, the liquidus is represented in bold.
I – Diagram with only one eutectic. Miscibility between A and B is complete in the liquid phase, non-existent in the solid phases. The slope of the liquidus arising from A (or B) depends only on the properties of the constituent A (or B).

II – Cigar-shape diagram. Solubility between A and B is complete in the solid and liquid phases. The difference between the slopes of the solidus and liquidus curves arising from A (or B) depends only on the properties of the constituent A (or B). For large deviations from ideality, the diagram may show a congruent point corresponding to an extremum on the liquidus and on the solidus.

III – Diagram characterized by the formation of a defined compound γ with congruent melting. It cannot be considered as the superposition of two diagrams with one eutectic, because the slope of the liquidus arising from γ is zero.

IV – Construction of the above diagram from the Gibbs energy of the liquid mixture and the standard Gibbs energy of formation of the defined compound γ. Here, the standard state selected for A and B is the liquid state.

liquidus arising from γ. Let $(m+n)=1$, which amounts to dividing both members by $(m+n)$; then draw, as a function of $x_{B,l}$ (with $x_{A,l}=1-x_{B,l}$), the Gibbs energy of the transformation:

$$x_{A,l}(A)_l + x_{B,l}(B)_l \rightarrow (\text{liquid solution})$$

$$\Delta_{\text{mix}}G = RT(x_{A,l} \ln a_{A,l} + x_{B,l} \ln a_{B,l}).$$

The left-hand side of the equation of the liquidus represents the ordinate at point $x_B = n$ of the tangent from the abscissa point $x_{B,l}$ of the curve $\Delta_{\text{mix}}G = f(x_B, T)$. The right-hand side of the equation of the liquidus represents merely the standard Gibbs energy of formation of the solid compound $\gamma - A_m B_n$ from the pure liquids A and B:

$$m(A)_l + n(B)_l \rightarrow \langle AB \rangle_s.$$

The abscissa point $x_{B,l}$ on the liquidus from γ at the temperature T is thus easily obtained by a geometric construction, by drawing on a diagram of G vs x_B the tangent from the point $\Gamma[x_B = n, G = \Delta_f G^\circ(\gamma)]$ on the curve $\Delta_{\text{mix}}G = f(x_B, T)$. This construction has been made by selecting as standard state for A and B the pure liquid constituents A and B, but the construction must remain the same and the value of x obtained must not depend on the choice of a standard state. The equation of the liquidus from γ has generally two solutions, because there are two branches arising from γ. When the point Γ on the diagram (G, x_B) lies above the curve $\Delta_{\text{mix}}G = f(x_B, T)$, the equation of the liquidus has no solution, because the compound γ is less stable than the liquid solution.

LIQUID–SOLID EQUILIBRIUM IN THE NEIGHBOURHOOD OF A DEFINED COMPOUND: The difference in nature between a pure constituent A and a defined compound γ is again encountered when calculating the slope of the tangent to the liquidus at the point of composition $\gamma = A_m B_n$. The equation of the liquidus is of the type:

$$m \ln a_{A,l} + n \ln a_{B,l} = \varphi(T).$$

Let us differentiate, then write the *Gibbs–Duhem equation*:

$$m\,\mathrm{d}(\ln a_{A,l}) + n\,\mathrm{d}(\ln a_{B,l}) = \varphi'(T)\mathrm{d}T$$

$$x_{A,l}\,\mathrm{d}(\ln a_{A,l}) + x_{B,l}\,\mathrm{d}(\ln a_{B,l}) = 0.$$

The elimination of $\mathrm{d}(\ln a_{A,l})$ between the two equations yields:

$$[n - m(x_{B,l}/x_{A,l})]\mathrm{d}(\ln a_{B,l}) = \varphi'(T)\mathrm{d}T,$$

where $a_{B,l}$ is a function of $x_{B,l}$ and T:

$$d(\ln a_{B,l}) = \Psi_x dT + \Psi_T dx,$$

with $\Psi_x = [d(\ln a_{B,l})/dT]_{x_{B,l}}$ and $\Psi_T = [d(\ln a_{B,l})/dx_{B,l}]_T$.
By using the developed expression of $d(\ln a_{B,l})$, we obtain:

$$[n - m(x_{B,l}/x_{A,l})]\Psi_T dx_{B,l} = \{\varphi'(T) - [n - m(x_{B,l}/x_{A,l})]\Psi_x\}dT.$$

In the neighbourhood of the γ compound: $x_{B,l} \to n$, $x_{A,l} \to m$, $T \to T_{fus}(\gamma)$, the bracket $[n - m(x_{B,l}/x_{A,l})] \to 0$, whence $(dT/dx_{B,l}) \to 0$; the liquidus arising from a defined compound shows a horizontal tangent at the melting temperature of this compound.

Litre—The term 'litre' was adopted in 1964 as a specific name for the dm^3 by the 12th Conférence Générale des Poids et Mesures. $1\,L = 10^{-3}\,m^3$.

Local equilibrium—In the thermodynamics of irreversible processes, the postulate of local equilibrium admits that space–time relationships between thermal and mechanical properties of a system out of equilibrium are the same as for a system at equilibrium.

Such a statement supposes the system to be divided into cells large enough for each cell to be considered as a macroscopic sub-system, but small enough for the assumption of equilibrium to be a realistic one. The intensive quantities such as temperature, pressure, and *partial molar quantities* remain uniform in each cell but vary from one cell to another. Hence all relationships established at equilibrium between state variables which can be defined at any moment in each cell remain valid.

The postulate of local equilibrium applied to *diffusion* allows one to write the equality of *chemical potentials* of a species i on the two sides of the interface separating two phases. Prigogine showed that this postulate remains valid as long as relationships between *flux* and *forces* retain a linear character.

Long-range order—In a solid solution AB whose structure presents two types of site α and β, long-range order appears when constituent A prefers the α sites or, what amounts to the same thing, when B prefers the β sites. Let $p_{A,\alpha}$ be the probability of finding A at α; long-range order is expressed by:

$$p_{A,\alpha} > x_A, \quad p_{B,\alpha} < x_B, \quad p_{A,\beta} < x_A, \quad p_{B,\beta} > x_B.$$

Long-range order is all the more easily evidenced when there is a good match between the composition of the solution and the number of available sites. For instance, in a body-centred cubic CuZn alloy, the zinc tends at low temperature to occupy the centre of the cube and the copper to occupy the vertices. Such order is also encountered in a face-centred cubic alloy of the type Cu_3Au, copper occupying preferentially the centre of the faces and gold the vertices of the cube. Experimentally, the formation of long-range order is detected in X-ray diffraction patterns by the appearance of superstructure peaks, an ordered lattice having a crystal parameter that is a multiple of that of a disordered one.

Let us consider the simple case of an equiatomic alloy AB in a structure comprising as many α sites as β sites. The order parameter $\mathcal{P}$ is defined by:

$$\mathcal{P} = \frac{(p_{A,\alpha})_{obs} - (p_{A,\alpha})_{sta}}{(p_{A,\alpha})_{max} - (p_{A,\alpha})_{sta}} = 2\left(p_{A,\alpha} - \frac{1}{2}\right).$$

Indeed, the order parameter is zero for complete disorder, when $(p_{A,\alpha}) = (p_{A,\alpha})_{sta} = 1/2$; it equals unity for complete order, when $(p_{A,\alpha}) = (p_{A,\alpha})_{max} = 1$. A negative order parameter would mean that A prefers β sites; by inverting the α and β designations, $\mathcal{P}$ again becomes positive.

The order parameter must take an equilibrium value, owing to the opposing effects of the enthalpic and entropic terms on the function $G = H - TS$. Indeed, if the A–B interactions are attractive, an increase in order decreases the enthalpic term (the number of A–B bonds increases), but at the same time decreases the entropic term

The numbers of atoms A and B at the sites α and β are given by:

$$N_{A,\alpha} = N_{B,\beta} = N(1 + \mathcal{P})/4$$

$$N_{A,\beta} = N_{B,\alpha} = N(1 - \mathcal{P})/4.$$

If z is the coordination number of the crystal, the total number of pairs is $zN/2$; the number of A–A pairs equals the total number of pairs multiplied by the probability of finding an A–A pair:

$$P_{AA} = P_{BB} = N(1 - \mathcal{P}^2)z/8, \quad P_{AB} = N(1 + \mathcal{P}^2)z/4.$$

If ε_{AA}, ε_{BB}, and ε_{AB} are the enthalpies of formation of the A–A, B–B, and A–B bonds, the enthalpy of the solution is:

$$H = P_{AA}\varepsilon_{AA} + P_{BB}\varepsilon_{BB} + P_{AB}\varepsilon_{AB}$$

$$H = \frac{1}{8}zN[(1 - \mathcal{P}^2)(\varepsilon_{AA} + \varepsilon_{BB}) + 2(1 + \mathcal{P}^2)\varepsilon_{AB}].$$

The configurational entropy is calculated by the relation $S = k \ln \Omega$, taking into account the two types of sites:

$$S = k \ln \left[\frac{(N/2)!}{(N_{A,\alpha})!(N_{B,\alpha})!} \cdot \frac{(N/2)!}{(N_{A,\beta})!(N_{B,\beta})!} \right].$$

The replacement of $N_{A,\alpha}$, $N_{B,\alpha}$, ... by their expression using the *Stirling formula* ($\ln N! \approx N \ln N - N$) yields:

$$\frac{S}{R} = \ln 2 - \frac{1}{2}[(1 + \mathscr{P}) \ln(1 + \mathscr{P}) + (1 - \mathscr{P}) \ln(1 - \mathscr{P})].$$

In the absence of order ($\mathscr{P} = 0$), we find $S = R \ln 2$.

Complete order ($\mathscr{P} = 1$) gives $S = 0$.

The order parameter at equilibrium is calculated by seeking the minimum of the function G. Without taking into account the non-configurational entropy (mainly the entropy of vibration), we easily obtain:

$$\frac{dG}{d\mathscr{P}} = \mathscr{P}(2\varepsilon_{AB} - \varepsilon_{AA} - \varepsilon_{BB}) + \frac{RT}{4} \ln \frac{1 + \mathscr{P}}{1 - \mathscr{P}} = 0.$$

When $(2\varepsilon_{AB} - \varepsilon_{AA} - \varepsilon_{BB}) > 0$ (repulsive A–B interactions), the only solution of this equation is $\mathscr{P} = 0$. Long-range order is then impossible, whatever the temperature.

When $(2\varepsilon_{AB} - \varepsilon_{AA} - \varepsilon_{BB}) < 0$ (attractive A–B interactions), the solution $\mathscr{P} = 0$ corresponds to a minimum on the curve $G = f(\mathscr{P})$ above a critical temperature T_c. There is then no long-range order. When $T < T_c$, the solution $\mathscr{P} = 0$ corresponds to a maximum on the curve $G = f(\mathscr{P})$: the complete disorder is unstable and there is a second solution of the equation $dG/d\mathscr{P} = 0$ corresponding to a minimum of the function $G = f(\mathscr{P})$, and hence to a value of the order parameter at equilibrium between 0 and 1. The order parameter $\mathscr{P}$ remains close to 1 for a long time and tends quickly towards 0 when T tends towards T_c.

An order–disorder transition is generally of the second order (see *Order of a transition*), but it may be of the first order. In this case, at $T = T_c$ we observe a sharp transition of the order parameter $\mathscr{P}$ from a non-zero value (for $T < T_c$) to a zero value (for $T > T_c$).

M

Macroscopic state—The state of a system defined by macroscopic quantities: temperature, pressure, volume, composition, etc. The statistical *probability* of a macroscopic state is given by the number of *microscopic states* allowing its realization.

Magnetic energy—A change in magnetic *flux* $d\Phi$ induces in an external circuit an electromotive force of induction $-d\Phi/dt$, and hence an induced current I developing a power $I\,d\Phi/dt$, which corresponds to an element of *work*:

$$dW = I\,d\Phi.$$

There is a more general relationship:

$$\Delta W = \iiint_V \int_{\mathscr{B}} \mathscr{H}\,dV\,d\mathscr{B}.$$

$\mathscr{H}$ is the *magnetic excitation* vector and $d\mathscr{B}$ the increment of the *magnetic field* vector in the volume dV. If the magnetic field is independent of the coordinates:

$$dW = V\,\mathscr{H}\,d\mathscr{B}.$$

The magnetic field may be expressed as a function of the magnetic excitation:

$$\mathscr{B} = \mu_\circ\,(\mathscr{H} + \mathscr{M}) - \mu\mathscr{H}.$$

In vacuum ($\mathscr{H} = \mu_\circ^{-1}\mathscr{B}$, $\mu_\circ$ being a scalar), the work needed to raise the magnetic field from 0 to $\mathscr{B}$ is given by:

$$\frac{dW}{dV} = \int_0^{\mathscr{B}} \mu_\circ^{-1}\mathscr{B}\,d\mathscr{B} = \frac{1}{2}\mu_\circ^{-1}\mathscr{B}^2.$$

$\mathscr{B}^2/2\mu_\circ$, in $J\,m^{-3}$, is the energy density of vacuum in a magnetic field $\mathscr{B}$. In the presence of a homogeneous magnetic material, the energy density may be expressed by $dW/dV = \mathscr{B}^2/2\mu$. From $\mathscr{B} = \mu_\circ\mathscr{H}$, we obtain $\mathscr{B}^2/2\mu_\circ = \mu_\circ\mathscr{H}^2/2$, a relation valid in a vacuum but not in a material medium. In the presence of matter, the energy density of a vacuum is always $\mu_\circ\mathscr{H}^2/2$, but is no longer equal to $\mathscr{B}^2/2\mu_\circ$.

184 Magnetic excitation

A material introduced into a pre-existing magnetic field gains work:

$$dW = V \mathcal{H} \, d\mathcal{B} - V \mu_\circ \, \mathcal{H} \, d\mathcal{H} = V \mu_\circ \mathcal{H} \, d\mathcal{M}.$$

In summary, the magnetic energy has two contributions:

- an external energy which is the magnetic energy of vacuum:

$$\Delta W = \iiint_V \int_{\mathcal{H}} \mu_\circ \mathcal{H} \, dV \, d\mathcal{H};$$

- an internal energy which is the magnetic energy given to a material introduced into a magnetic field:

$$\Delta W = \iiint_V \int_{\mathcal{M}} \mu_\circ \mathcal{H} \, dV \, d\mathcal{M}.$$

Magnetic excitation—The *magnetic field* in a solenoid whose length is l, comprising n turns, and traversed by a current I is given by: $\mathcal{B} = \mu_\circ n I / l$ ($\mu_\circ$: *permeability* of vacuum). Expression in terms of the volume V of the solenoid and the surface $\mathcal{S}$ of one coil turn gives:

$$\mu_\circ^{-1} \mathcal{B} = n I \mathcal{S} / V,$$

which represents a *magnetic moment* per unit volume. In the presence of a magnetic material, $(\mu_\circ^{-1} \times \mathcal{B})$ is the result of two contributions, that of the current and that of the *magnetization* $\mathcal{M}$ of the material:

$$\mu_\circ^{-1} \mathcal{B} = (n I \mathcal{S} / V) + \mathcal{M}.$$

$(\mu_\circ^{-1} \mathcal{B} - \mathcal{M})$ is related to the current I in the solenoid in the presence of a magnetic material, in the same way that $(\mu_\circ^{-1} \times \mathcal{B})$ is related to the current I in the empty solenoid. Generally, $\mathcal{B}$ and $\mathcal{M}$ are vectors, not always collinear. By definition:

$$\mathcal{H} \equiv \mu_\circ^{-1} \mathcal{B} - \mathcal{M}.$$

The vector $\mathcal{H}$ is called 'magnetic excitation'. This rigorous designation, due to Sommerfeld, demonstrates the analogy between magnetic excitation and *electric excitation*. Alas, it is still encountered under various names such as: 'magnetic field, magnetic field intensity, derived magnetic vector, auxiliary magnetic field, etc. In the SI system, magnetic excitation is expressed in $A \, m^{-1}$.

Magnetic field—A maximum of confusion has always been the rule in the designation of magnetic units. The quantity represented by the vector $\mathcal{B}$ and introduced historically under the name 'magnetic induction' is now called, following Sommerfeld, 'magnetic field strength' or, for simplicity 'magnetic field'. Still designated 'magnetic flux density', it is defined by starting from the force exerted by a circuit A, in which flows a current I_A, on a circuit B, in which flows a current I_B:

$$F_{AB} = \frac{\mu_o}{4\pi} I_A I_B \oint_A \oint_B \frac{d l_B \times (d l_A \times r_u)}{r^2}.$$

F_{AB} is the force exerted by I_A on I_B. It is easy to verify that: $F_{AB} = -F_{BA}$.

$\mu_o = 4\pi \times 10^{-7}\,\mathrm{H\,m^{-1}}$ is the *permeability* of vacuum.

r_u is the unit vector oriented from A towards B.

$d l_A$ and $d l_B$ are elements of length $d l_A$ and $d l_B$, taken on the circuits A and B, and oriented in the direction of the currents I_A and I_B.

To resume:

$$F_{AB} = I_B \oint_B d l_B \times \left[\frac{\mu_o}{4\pi} I_A \oint_A \frac{d l_A \times r_u}{r^2} \right]$$

$$F_{AB} = I_B \oint_B d l_B \times \mathscr{B}_A$$

M

$\mathscr{B}_A$ is the magnetic field, or magnetic field strength, created by the circuit A at the point where the element $d l_B$ of circuit B is located:

$$\mathscr{B}_A = \frac{\mu_o}{4\pi} I_A \oint_A \frac{d l_A \times r_u}{r^2}.$$

The unit vector r_u is oriented from the source $d l_A$ towards the point where $\mathscr{B}_A$ is calculated. No magnetic material is assumed to be present at this point.

If a current I flows through a solenoid having n/l turns per unit length, $\mathscr{B}$ is oriented in the direction of the axis and, in the absence of any magnetic material, the above relationship becomes:

$$\mathscr{B} = \mu_o n I / l.$$

The unit of magnetic field in the SI system is the tesla (symbol T):

$$1\,\mathrm{T} = 1\,\mathrm{kg\,s^{-2}A^{-1}} = 1\,\mathrm{Wb\,m^{-2}} = 1\,\mathrm{V\,s\,m^{-2}}.$$

The gauss ($1\,\mathrm{G} = 10^{-4}\,\mathrm{T}$) is an old cgs unit of magnetic field whose use must be proscribed. Very weak magnetic fields are sometimes expressed in 'gammas', or, better, in nanoteslas ($1\,\gamma = 10^{-9}\,\mathrm{T}$).

186 Magnetic moment

'Auxiliary magnetic field' is the name sometimes given to the vector *magnetic excitation* $\mathcal{H}$ defined by:

$$\mathcal{H} = \mathcal{B}/\mu_\circ - \mathcal{M},$$

where $\mathcal{M}$ is the *magnetization*, expressed, like $\mathcal{H}$, in $\mathrm{A\,m^{-1}}$.

Magnetic moment—When an electric current I flows in a loop whose surface is $\mathcal{S}$, the magnetic moment of the loop is given by: $M = I\mathcal{S}$. It is a vector perpendicular to the plane of the circuit, directed from the feet to the head of a dwarf standing at the centre of the circuit and looking so that the current flows counterclockwise. The magnetic moment has the dimensions of current times surface, or of energy divided by *magnetic field*. The unit of magnetic moment in the SI system is the $\mathrm{A \times m^2}(1\,\mathrm{A\,m^2} = 1\,\mathrm{J\,T^{-1}})$.

The magnetic moment (or sometimes 'dipolar magnetic moment') measures the tendency of a magnetic dipole to line up in the direction of a magnetic field. The resulting magnetic moment of a material is the sum of all the elementary magnetic moment vectors. The electronic and nuclear magnetic moments are quantized, which means that they can take only some directions with respect to that of the imposed magnetic field. They are expressed in *magnetons*.

Magnetic susceptibility—Between *magnetic field* $\mathcal{B}$, *magnetic excitation* $\mathcal{H}$ and *magnetization* $\mathcal{M}$ there exist basic relationships:

- in vacuum: $\mathcal{H} = \mu_\circ^{-1}\mathcal{B}$,
- in presence of matter: $\mathcal{H} = \mu_\circ^{-1}\mathcal{B} - \mathcal{M} = \mu^{-1}\mathcal{B}$,

$\mu_\circ$ and μ being respectively the *permeability* of vacuum and that of the material considered; $\mu_r = \mu/\mu_\circ$ is the relative permeability.

The magnetic susceptibility is defined by $\kappa = \mu_r - 1$. It is slightly negative (of the order of -10^{-6}) for a diamagnetic compound (whose presence in a magnetic field has the effect of decreasing the field); it is slightly positive (of the order of $+10^{-4}$) for a paramagnetic compound (whose presence in a magnetic field has the effect of increasing the field). It is very high and field-dependent for a ferromagnetic compound.

Magnetism—Magnetic systems are described by:

- the *magnetic field* $\mathcal{B}$, expressed in tesla ($1\,\mathrm{T} = 1\,\mathrm{kg\,s^{-2}A^{-1}}$);
- the *magnetic excitation* $\mathcal{H}$, expressed in $\mathrm{A\,m^{-1}}$;
- the *magnetization* $\mathcal{M}$, expressed in $\mathrm{A\,m^{-1}}$.

There are two relationships between these three quantities:

- an equation of state: $\varphi(\mathcal{M}, \mathcal{H}, T, \ldots) = 0$;
- a basic relationship: $\mathcal{H} = \mu_\circ^{-1}\mathcal{B} - \mathcal{M}$.

$\mu_\circ = 4\pi \times 10^{-7}$ H m^{-1}(or kg m s^{-2}A^{-2}) is the *permeability* of vacuum.

It is convenient to introduce some auxiliary quantities:

- the *magnetic moment*: $M = \mathcal{M}V$, expressed in A m^2;
- the permeability: $\mu = \mathcal{B}/\mathcal{H}$, expressed in H m^{-1}(or kg m s^{-2}A^{-2}) ;
- the relative permeability: $\mu_r = \mu/\mu_\circ$ (dimensionless);
- the *magnetic susceptibility*: $\kappa = \mu_r - 1$ (dimensionless);
- the differential susceptibility: $\kappa = (\partial\mathcal{M}/\partial\mathcal{H})_{x,y}$ (dimensionless).

Thermodynamic treatment of magnetic systems rests on the expression of the element of work (see *Magnetic energy*):

$$dW = V\mu_\circ \, \mathcal{H} \, d\mathcal{M}.$$

At constant volume, the magnetic moment M is naturally introduced:

$$dW = \mu_\circ \mathcal{H} \, dM.$$

The main thermodynamic functions may then be defined:

- *Internal energy*: $dU = T(dS - d\sigma) + \mu_\circ\mathcal{H}\,dM + \cdots$
- *Enthalpy*: $dH^* = d(U - \mu_\circ\mathcal{H}M) = T(dS - d\sigma) - \mu_\circ M\,d\mathcal{H} + \cdots$
- *Helmholtz energy*: $dF = d(U - TS) = -S\,dT - T\,d\sigma + \mu_\circ\mathcal{H}\,dM + \cdots$
- *Gibbs energy*: $dG^* = d(H^* - TS) = -S\,dT - T\,d\sigma - \mu_\circ M\,d\mathcal{H} + \cdots$

The functions H^* and G^* bear asterisks to distinguish them from the traditional functions of *enthalpy* and *Gibbs energy*, because new names have never been proposed for these new functions.

Functions U, H^*, F, and G^* are *thermodynamic potentials* of systems subjected to the conditions: S and M constant (for the internal energy U); S and $\mathcal{H}$ constant (for the enthalpy H^*); T and M constant (for the Helmholtz energy F); and T and $\mathcal{H}$ constant (for the Gibbs energy G^*). These functions being state functions, their differentials are exact:

$$dU = T\,dS + \mu_\circ\mathcal{H}\,dM \ \Rightarrow \ \mu_\circ\left(\frac{\partial\mathcal{H}}{\partial S}\right)_M = \left(\frac{\partial T}{\partial M}\right)_S,$$

$$dH^* = T\,dS - \mu_\circ M\,d\mathcal{H} \ \Rightarrow \ \mu_\circ\left(\frac{\partial M}{\partial S}\right)_{\mathcal{H}} = -\left(\frac{\partial T}{\partial\mathcal{H}}\right)_S,$$

M

$$dF = -S\,dT + \mu_\circ \mathcal{H}\,dM \;\Rightarrow\; \mu_\circ \left(\frac{\partial \mathcal{H}}{\partial T}\right)_M = -\left(\frac{\partial S}{\partial M}\right)_T,$$

$$dG^* = -S\,dT - \mu_\circ M\,d\mathcal{H} \;\Rightarrow\; \mu_\circ \left(\frac{\partial M}{\partial T}\right)_{\mathcal{H}} = \left(\frac{\partial S}{\partial \mathcal{H}}\right)_T.$$

Between the eight differentials dU, dH^*, dS, dF, dG^*, dM, $d\mathcal{H}$, and dT, there are six relations because U, H^*, S, F, and G^* are state functions of the system, and there is one equation of state, φ $(M, \mathcal{H}, T) = 0$. The above Maxwell equations allow any differential to be easily expressed as a function of any two others. For instance:

$$dS = \left(\frac{\partial S}{\partial T}\right)_M dT + \left(\frac{\partial S}{\partial M}\right)_T dM = \frac{c_M}{T}\,dT - \mu_\circ \left(\frac{\partial \mathcal{H}}{\partial T}\right)_M dM,$$

$$dS = \left(\frac{\partial S}{\partial T}\right)_{\mathcal{H}} dT + \left(\frac{\partial S}{\partial \mathcal{H}}\right)_T d\mathcal{H} = \frac{c_{\mathcal{H}}}{T}\,dT + \mu_\circ \left(\frac{\partial M}{\partial T}\right)_{\mathcal{H}} d\mathcal{H}.$$

The identification of these two expressions yields relationships between the heat capacities c_M and $c_{\mathcal{H}}$. For instance, at constant M:

$$c_M - c_{\mathcal{H}} = T\mu_\circ \left(\frac{\partial M}{\partial T}\right)_{\mathcal{H}} \left(\frac{\partial \mathcal{H}}{\partial T}\right)_M.$$

At constant $\mathcal{H}$, we get the same expression. On the other hand, elimination of $(d\mathcal{H}/dT)_M$ yields another relation:

from:
$$\left(\frac{\partial \mathcal{H}}{\partial T}\right)_M \left(\frac{\partial T}{\partial M}\right)_{\mathcal{H}} \left(\frac{\partial M}{\partial \mathcal{H}}\right)_T = -1,$$

we obtain
$$c_M - c_{\mathcal{H}} = -T\mu_\circ \left(\frac{\partial M}{\partial T}\right)_{\mathcal{H}}^2 \left(\frac{\partial \mathcal{H}}{\partial M}\right)_T.$$

We recognize $(d\mathcal{H}/dM)_T = 1/\kappa_T V$ (κ_T: isothermal differential suscepibility). Heat capacities c_M and $c_{\mathcal{H}}$ can be calculated separately. For instance, $c_{\mathcal{H}}$ is obtained by extracting dT from the expression $dS = f(dT, d\mathcal{H})$ and by identifying with the general expression for dT:

$$dT = \frac{T}{c_{\mathcal{H}}}\,dS - \frac{T}{c_{\mathcal{H}}}\mu_\circ \left(\frac{\partial M}{\partial T}\right)_{\mathcal{H}} d\mathcal{H} = \left(\frac{\partial T}{\partial S}\right)_{\mathcal{H}} dS + \left(\frac{\partial T}{\partial \mathcal{H}}\right)_S d\mathcal{H}.$$

By comparing the coefficients of $d\mathcal{H}$:

$$c_{\mathcal{H}} = -T\mu_\circ \left(\frac{\partial M}{\partial T}\right)_{\mathcal{H}} \left(\frac{\partial \mathcal{H}}{\partial T}\right)_S.$$

We verify easily that comparison of the coefficients of dS yields the same relationship. c_M is obtained from a similar calculation, by extracting dT from the differential $dS = f(dT, dM)$, resulting in:

$$c_M = T\mu_\circ \left(\frac{\partial \mathcal{H}}{\partial T}\right)_M \left(\frac{\partial M}{\partial T}\right)_S.$$

The ratio $c_{\mathcal{H}}/c_M$ equals the ratio κ_T/κ_S of the differential susceptibilities, an interesting result:

$$\frac{c_{\mathcal{H}}}{c_M} = -\frac{(\partial M/\partial T)_{\mathcal{H}}(\partial \mathcal{H}/\partial T)_S}{(\partial \mathcal{H}/\partial T)_M(\partial M/\partial T)_S} = -\frac{(\partial T/\partial M)_S(\partial \mathcal{H}/\partial T)_S}{(\partial \mathcal{H}/\partial T)_M(\partial T/\partial M)_{\mathcal{H}}}$$

$$= \frac{(\partial \mathcal{H}/\partial M)_S}{(\partial \mathcal{H}/\partial M)_T} = \frac{\kappa_T}{\kappa_S}.$$

Magnetization—The name given to the *magnetic moment* per unit volume. In the SI system, magnetization is expressed in $A\,m^{-1}$.

The presence of matter in a *magnetic field* has the effect of modifying the field, because the elementary magnetic moments carried by electrons create a magnetic field which is added to the pre-existing field.

Let for instance I be a current flowing in a solenoid whose length is l, and having n turns. The magnetic field created at its axis without any material present is given by the well-known relationship:

$$\mathcal{B} = \mu_\circ nI/l$$

($\mu_\circ$ being the *permeability* of vacuum), which may be expressed by introducing the surface $\mathcal{S}$ of a turn and the volume V of the solenoid:

$$\mu_\circ^{-1}\mathcal{B} = nI\mathcal{S}/V.$$

The above expression represents the magnetic moment of the solenoid per unit volume. If the solenoid is not empty, the magnetization is the sum of two contributions, that of the current and that of the material which fills it:

$$\mu_\circ^{-1}\mathcal{B} = (nI\mathcal{S}/V) + \mathcal{M}.$$

$(\mu_\circ^{-1}\mathcal{B} - \mathcal{M})$, which defines the *magnetic excitation* $\mathcal{H}$, is related to the current I in the filled solenoid as $\mu_\circ^{-1}\mathcal{B}$ is related to the current I in the empty solenoid.

The magnetization $\mathcal{M}$ of a magnetic material is analogous to the *polarization* P of a dielectric substance. In the same way, to the *magnetic excitation* $\mathscr{H} \equiv \mu_\circ^{-1}\mathscr{B} - \mathscr{M}$ corresponds the *electric excitation* $\mathbf{D} \equiv \varepsilon_\circ \mathbf{E} + \mathbf{P}$.

Magneton—Elementary magnetic moment:

- Bohr magneton: $\mu_B = eh/4\pi m_e = 9.2740154 \times 10^{-24}\,\text{A m}^2$,
- Nuclear magneton: $\mu_N = eh/4\pi m_p = 5.0507866 \times 10^{-27}\,\text{A m}^2$.

Actually, measured magnetic moments differ from the above values:

- Magnetic moment of the electron: $\mu_e = 9.2847701 \times 10^{-24}\,\text{A m}^2$,
- Magnetic moment of the proton: $\mu_p = 1.41060761 \times 10^{-26}\,\text{A m}^2$.

The magneton must not be confused with the *permeability*, expressed in H m^{-1} (or $\text{m kg s}^{-2}\text{A}^{-2}$), which is also symbolized by μ!

Mass action law—A law formulated in 1864 by Guldberg and Waage, stating that a *reaction rate* is proportional to the product of the 'active masses' elevated to a power corresponding to the number of moles taking part in the reaction. Applied to a *chemical reaction* $\sum v_i M_i = 0$, this law may be expressed as: the product of active masses a_i at equilibrium $\prod (a_i)^{v_i}$ is constant (see *Equilibrium constant*). Nobody now uses the indefinite concept of active mass; instead, the more accurate concept of *activity* is preferred.

Mass action product—Consider a mixture of n_i moles of M_i able to react in accordance with the *chemical reaction* $\sum v_i M_i = 0$. The product $\Pi = \prod (a_i)^{v_i}$, where a_i represents the *activity* of component M_i, is called the mass action product. It is sometimes encountered under the name of 'mass action ratio'. It can take any value from zero to infinity and must not be confused with the *equilibrium constant K*. The equality $\Pi = K$ is satisfied only by a system at equilibrium.

Massieu function—The name given to the function $J = -F/T$. Introduced in 1869 by François Massieu (1832–1896) together with the *Planck function* $(Y = -G/T)$, it is now of only historical interest.

Matthew effect—The name given by economists to the growth of large structures at the expense of small ones, from Matthew 25:29. In thermodynamics, this effect describes the same phenomenon: large crystals develop at the expense of small ones because their solubility in a solvent is lower; large drops develop at the expense of small ones owing to their lower vapour pressure (see *Droplet*). Is it necessary to

point out that a critical faculty must never be absent when using such comparisons?

Maxwell, James Clerk (1831–1879)—Scottish physicist, considered, like Newton, as one of the greatest physicists of his time. In 1859, on the basis of probabilistic arguments, he gave the distribution of velocities among the molecules of a gas. He showed in 1864 that light is an electromagnetic wave and created the theory of electromagnetic fields. He provided the relation between information and energetics in 1871 and established theoretically the existence of radiation pressure in 1874. His main contributions may be found in his books *Theory of heat* (1872) and *Treatise on electricity and magnetism* (1873). The maxwell is also an old cgs unit of magnetic flux, whose use is not recommended ($1\,\text{Mx} = 10^{-8}\,\text{Wb}$).

Maxwell–Boltzmann statistics—The name given to the common *distribution* towards which *quantum statistics* tend when $g_i \gg N_i$. The expansion of Ω_{BE} and Ω_{FD} gives:

$$\Omega_{\text{BE}} = \prod_i \frac{(N_i + g_i - 1)!}{N_i!(g_i - 1)!} = \prod_i \frac{g_i(g_i + 1)\cdots(g_i + N_i - 1)}{N_i!},$$

$$\Omega_{\text{FD}} = \prod_i \frac{g_i!}{N_i!(g_i - N_i)!} = \prod_i \frac{g_i(g_i - 1)\cdots(g_i - N_i + 1)}{N_i!}.$$

When $g_i \gg N_i$, it is possible to neglect before g_i the terms lower than N_i. The two distributions tend towards the same limit, which is called the Maxwell–Boltzmann distribution:

$$\Omega_{\text{MD,ind}} = \prod_i \frac{g_i^{N_i}}{N_i!}.$$

The subscript 'ind' signifies that the above expression applies only to indistinguishable particles: the exchange of two particles does not produce a new microscopic state. If we abandon the assumption of indistinguishability, a characteristic of quantum statistics, the above relationship becomes:

$$\Omega_{\text{MB,dis}} = N! \prod_i \frac{g_i^{N_i}}{N_i!}.$$

$\Omega_{\text{MB,dis}}$ then represents the number of *complexions* and the exchange of two particles gives a new complexion.

M

Remark 1: This last expression, known before the birth of quantum mechanics, may be established by calculating the number of complexions corresponding to N_i particles at the energy level ε_i whose *degeneracy* is g_i: They are $C_{N_1}^N$ ways of placing N_1 particles at the energy level ε_1. This level having a degeneracy g_1, there are $g_1^{N_1}$ possibilities of allotting N_1 particles amongst g_1 quantum states, whence in total there are $C_{N_1}^N \times g_1^{N_1}$ possibilities. In the same way, there are $C_{N_2}^{N-N_1} \times g_2^{N_2}$ ways to allot N_2 particles, taken from the $(N-N_1)$ remaining, at the energy level ε_2 whose degeneracy is g_2. Hence the total number of complexions is:

$$\Omega_{\text{MB,dis}} = C_{N_1}^N \times C_{N_2}^{N-N_1} \times C_{N_3}^{N-N_1-N_2} \times \cdots \times C_{N_k}^{N-N_1-\cdots-N_{k-1}} \times \prod_i g_i^{N_i}.$$

On expanding:

$$\Omega_{\text{MB,dis}} = \left[\frac{N!}{N_1!(N-N_1)!} \right]\left[\frac{(N-N_1)!}{N_2!(N-N_1-N_2)!} \right] \cdots \prod_i g_i^{N_i}$$

$$\Omega_{\text{MB,dis}} = \frac{N!}{N_1! N_2! \cdots N_k!} \prod_i g_i^{N_i} = N! \prod_i \frac{g_i^{N_i}}{N_i!}.$$

If the particles are indistinguishable, it is sufficient to divide $W_{\text{MB,dis}}$ by the number of permutations of N particles:

$$\Omega_{\text{MB,ind}} = \frac{\Omega_{\text{MB,dis}}}{N!} = \prod_i \frac{g_i^{N_i}}{N_i!}.$$

Remark 2: The two expressions of Ω_{MB} give two expressions of the *entropy* $S = k \ln \Omega$. The *Stirling formula* ($\ln N! \approx N \ln N - N$), used for large numbers, yields:

$$S_{\text{ind}} = k \ln \Omega_{\text{MB,ind}} = k \sum_i N_i \left[\ln\left(\frac{g_i}{N_i}\right) + 1 \right],$$

$$S_{\text{dis}} = k \ln \Omega_{\text{MB,dis}} = k \sum_i N_i \ln\left(\frac{g_i}{N_i}\right) + k N \ln N.$$

The last expression, resulting from the assumption of discernible particles, has a severe drawback: it does not confer extensive properties on the entropy function. The entropy of a system containing $2N$ distinguishable particles is no longer twice the entropy of a system containing N distinguishable particles, because we find that:

$$S_{\text{dis}}(2N) - 2S_{\text{dis}}(N) = 2kN \ln 2.$$

On the other hand, dropping the discernibility assumption eliminates the term $2kN\ln 2$, which represents an *entropy of mixing*:

$$S_{ind}(2N) - 2S_{ind}(N) = 0.$$

Maxwellian distribution—The name given to the *distribution* of velocities in a gas. Calculations are based on the hypothesis of equivalent distributions for the three components v_x, v_y, v_z of the vector velocity v. Moreover, the distribution must depend only on v^2, because the same result must be obtained by transformation of v into $-v$. These two conditions impose:

$$f(v^2) = f(v_x^2 + v_y^2 + v_z^2) = \varphi(v_x^2) \times \varphi(v_y^2) \times \varphi(v_z^2).$$

We have to find the nature of two functions f and φ such that:

$$f(x+y) = \varphi(x) \times \varphi(y).$$

Differentiation of both sides with respect to x yields:

$$f_x'(x+y) = \varphi'(x) \times \varphi(y).$$

The ratio of the last two equations gives:

$$\frac{f_x'(x+y)}{f(x+y)} = \frac{\varphi'(x)}{\varphi(x)}, \text{ independent of } y.$$

For the same reason, the ratio $\varphi'(y)/\varphi(y)$ must be independent of x, whence:

$$\frac{\varphi'(x)}{\varphi(x)} = \frac{\varphi'(y)}{\varphi(y)} = \text{const.} = -\lambda.$$

By integration: $\varphi(x) = A\exp(-\lambda x), \varphi(y) = A\exp(-\lambda y)$, whence:

$$f(x+y) = A^2 \exp[-\lambda(x+y)].$$

Generalization with three variables v_x, v_y, v_z yields:

$$\varphi(v_x^2) = A\exp(-\lambda v_x^2)$$
$$f(v^2) = A^3 \exp(-\lambda v^2).$$

Integration over all velocities gives the number of molecules N:

$$N = \int_0^\infty 4\pi v^2 A^3 \exp(-\lambda v^2)\, dv = \left(\frac{\pi}{\lambda}\right)^{3/2} A^3,$$

whence $A^3 = N(\lambda/\pi)^{3/2}$ and $N_v = 4\pi v^2 N(\lambda/\pi)^{3/2}\exp(-\lambda v^2)$.

M

194 Maxwell's demon

The *kinetic theory of gases* gives a relation between λ and temperature:

$$\frac{3}{2}kT = \left\langle \frac{1}{2}mv^2 \right\rangle = \frac{1}{2}m\left(\frac{\lambda}{\pi}\right)^{3/2}\int_0^\infty 4\pi v^4 \exp(-\lambda v^2)\,dv = \frac{3m}{4\lambda},$$

whence $\lambda = m/2kT$, m being the mass of a molecule. The Maxwellian distribution of the velocities is thus given by:

$$N_v = 4\pi v^2 N \left(\frac{m}{2\pi kT}\right)^{3/2} \exp\left(-\frac{mv^2}{2kT}\right).$$

Remember that $N_v\,dv$ represents the number of molecules whose velocity lies between v and $(v+dv)$. From this distribution, we calculate:

- the most probable velocity (N_v maximum): $v_{mp} = \sqrt{\dfrac{2kT}{m}}$,

- the mean velocity: $\langle v \rangle = \dfrac{1}{N}\int_0^\infty v\,N_v\,dv = \sqrt{\dfrac{8kT}{\pi m}}$,

- the mean square velocity: $\langle v^2 \rangle = \dfrac{1}{N}\int_0^\infty v^2 N_v\,dv = \dfrac{3kT}{m}$.

With respect to the most probable velocity::

$$\langle v \rangle = 1.1284\,v_{mp}, \quad \sqrt{\langle v^2 \rangle} = 1.2248\,v_{mp}.$$

For helium at $0°C$, we calculate $\langle v \rangle = 1201\,\mathrm{m\,s^{-1}}$. The velocity of the particles is independent of pressure, whereas the *mean free path* depends on the pressure.

Maxwell's demon—This little devil represents one of the most famous paradoxes in thermodynamics. Maxwell in 1871 imagined a closed box filled with a gas and separated by a wall into two compartments. A microscopic valve operated by a little demon allows communication between the two compartments. If the demon allows passage of the fastest molecules in one direction and the slowest in the other, it will establish a temperature difference from which it will be possible to get *work*! The result will be the same if the demon allows the molecules to pass in only one direction. It will establish a pressure difference from which it will also be possible to get work!

It seems clear that the existence of such a devil contradicts the *second law*, which forbids obtaining work from only one heat source. In 1951, Brillouin exorcized Maxwell's demon by showing that, to operate properly, it needs information about the positions and the velocities of the molecules. Now, in a closed system with uniform temperature, it can see only black body radiation and therefore cannot operate its valve. The situation may be altered by letting external radiation penetrate the system, allowing the demon to see the molecules. To get the information needed, according to the second law the demon will consume more *negentropy* than it will create by working the valve.

Maxwell's equations— The name given to the equations expressing the fact that thermodynamic functions U, H, F, and G are state functions; their differentials are thus *exact differentials*. For instance, with transformations at equilibrium for a thermoelastic system:

$$\mathrm{d}U = T\,\mathrm{d}S - P\,\mathrm{d}V \;\Rightarrow\; \left(\frac{\partial P}{\partial S}\right)_V = -\left(\frac{\partial T}{\partial V}\right)_S$$

$$\mathrm{d}H = T\,\mathrm{d}S + V\,\mathrm{d}P \;\Rightarrow\; \left(\frac{\partial V}{\partial S}\right)_P = \left(\frac{\partial T}{\partial P}\right)_S$$

$$\mathrm{d}F = -S\,\mathrm{d}T - P\,\mathrm{d}V \;\Rightarrow\; \left(\frac{\partial P}{\partial T}\right)_V = \left(\frac{\partial S}{\partial V}\right)_T$$

$$\mathrm{d}G = -S\,\mathrm{d}T + V\,\mathrm{d}P \;\Rightarrow\; \left(\frac{\partial V}{\partial T}\right)_P = -\left(\frac{\partial S}{\partial P}\right)_T.$$

Other relationships may also be obtained by making the differentials explicit. However, the above equations are by far the most commonly used. Recall that it is important to pay attention to the differentiation conditions! It is true that $(\partial P/\partial S)_V = 1/(\partial S/\partial P)_V$; nevertheless, we must not confuse $(\partial S/\partial P)_V$ and $(\partial S/\partial P)_T$!

The great interest of Maxwell's equations arises from the fact that they lead to the partial derivatives of the *entropy* as a function of physical quantities directly available by experiment. Thus, $(\partial S/\partial V)_T$ has no evident physical meaning for those who do not use thermodynamics as their daily bread, but $(\partial P/\partial T)_V$ is directly related to β_V, the isochoric *bulk expansion coefficient*.

Between the eight differentials $\mathrm{d}U$, $\mathrm{d}H$, $\mathrm{d}S$, $\mathrm{d}F$, $\mathrm{d}G$, $\mathrm{d}P$, $\mathrm{d}V$, and $\mathrm{d}T$ there are six relations, because U, H, S, F, and G are *state functions* and $f(P,V,T)=0$ is an equation of state. Maxwell's equations allow any differential to be expressed as a function of any two others. For instance:

$$dS=\left(\frac{\partial S}{\partial T}\right)_P dT+\left(\frac{\partial S}{\partial P}\right)_T dP=\frac{c_P}{T}dT-\left(\frac{\partial V}{\partial T}\right)_P dP,$$

$$dS=\left(\frac{\partial S}{\partial T}\right)_V dT+\left(\frac{\partial S}{\partial V}\right)_T dV=\frac{c_V}{T}dT+\left(\frac{\partial P}{\partial T}\right)_V dV.$$

Identification of both expressions yields relationships between c_P and c_V. For instance, with P or V constant:

$$c_P-c_V=T\left(\frac{\partial V}{\partial T}\right)_P\left(\frac{\partial P}{\partial T}\right)_V =TVP\alpha_P\beta_V=\frac{TV\alpha_P^2}{\chi_T}.$$

dS may also be expressed as a function of dP and dV:

$$dS=\left(\frac{\partial S}{\partial P}\right)_V dP+\left(\frac{\partial S}{\partial V}\right)_P dV= -\left(\frac{\partial V}{\partial T}\right)_S dP+\left(\frac{\partial P}{\partial T}\right)_S dV.$$

Those who find the derivatives $(\partial V/\partial T)_S$ and $(\partial P/\partial T)_S$ abstruse may always make them illuminating by writing '$=0$' at the end of the expressions $dS=\varphi(dT,dV)$ and $dS=\psi(dT,dP)$, which yields:

$$dS=\frac{1}{T}c_V\left(\frac{\partial T}{\partial P}\right)_V dP+\frac{1}{T}c_P\left(\frac{\partial T}{\partial V}\right)_P dV.$$

Heat capacities c_P and c_V may be calculated separately. For instance, c_P is obtained by extracting dT from the expression $dS=f(dT,dP)$, then by identifying with the general expression for dT:

$$dT=\frac{T}{c_P}dS+\frac{T}{c_P}\left(\frac{\partial V}{\partial T}\right)_P dP=\left(\frac{\partial T}{\partial S}\right)_P dS+\left(\frac{\partial T}{\partial P}\right)_S dP.$$

By comparison of the coefficients of dP:

$$c_P=T\left(\frac{\partial V}{\partial T}\right)_P\left(\frac{\partial P}{\partial T}\right)_S.$$

c_V is obtained by extracting dT from the expression $dS=f(dT,dV)$:

$$c_V=-T\left(\frac{\partial P}{\partial T}\right)_V\left(\frac{\partial V}{\partial T}\right)_S.$$

The ratio $\gamma = c_P/c_V$ equals the ratio χ_T/χ_S of the isothermal and isentropic *compressibilities*, a well-known result:

$$\frac{c_P}{c_V} = -\frac{(\partial V/\partial T)_P(\partial P/\partial T)_S}{(\partial P/\partial T)_V(\partial V/\partial T)_S} = -\frac{(\partial T/\partial V)_S(\partial P/\partial T)_S}{(\partial P/\partial T)_V(\partial T/\partial V)_P} = +\frac{(\partial P/\partial V)_S}{(\partial P/\partial V)_T} = \frac{\chi_T}{\chi_S}.$$

Mean—The name given to the order-1 *moment* of a stochastic variable X. If $f(x_i) > 0$ is the probability for the variable X to take the value x_i [with $\sum f(x_i) = 1$], or if, the variable X being continuous, $f(x)\,dx$ [with $\int f(x)\,dx = 1$] is the probability for X to take a value included between x and $(x + dx)$, the mean value of X is defined by:

$$\langle x \rangle = \sum x_i f(x_i) \quad \text{or} \quad \langle x \rangle = \int x f(x)\,dx.$$

Mean free path—The mean free path of a molecule represents the distance covered by the molecule between two collisions. Its determination rests on the assumption that the probability of collision is independent of the path covered since the last collision. Let N be the number of molecules whose mean free path is greater than x and dN the number of molecules whose mean free path is included between x and $x + dx$; it is natural to imagine a proportionality: $dN = -KNdx$. The constant K having the dimensions of reciprocal length l, the expression becomes: $dN/N = -dx/l$, a differential equation whose integration gives:

$$N = N_o \exp(-x/l).$$

The mean free path may be defined from the mean value of x:

$$\langle x \rangle = \frac{1}{N_o} \int_0^\infty x\,dN = \int_0^\infty \frac{x}{l} \exp\left(-\frac{x}{l}\right) dx = l.$$

It represents physically the length l introduced when seeking an expression for N.

EVALUATION OF l: For this purpose, we shall imagine each molecule with a sphere of radius r moving with a velocity v. Each molecule scans during unit time, a volume $\pi r^2 v$. If unit volume contains $\mathcal{N}$ molecules, each sphere encounters $v = \pi r^2 v \mathcal{N}$ centres of other molecules, which represents the number of collisions undergone during unit time. Since

M

the distance covered during unit time equals v, the mean free path is finally given by the relation $l = v/\nu = 1/\pi r^2 \mathcal{N}$.

In fact the above result assumes other molecules to be motionless. To take into account their movement, the velocity v of the molecule must be replaced by the mean relative velocity of two molecules. From a knowledge of velocity distribution in gases (*Maxwellian distribution*), the correct expression is:

$$l = 1/\sqrt{2}\pi r^2 \mathcal{N}.$$

With a mean velocity of molecules $\langle v \rangle = \sqrt{8kT/\pi m}$, a molecule undergoes during unit time an average number of collisions given by:

$$\nu = \frac{v}{l} = 4\sqrt{\frac{\pi kT}{m}} \cdot r^2 \mathcal{N}.$$

NUMERICAL APPLICATION: For a gas under atmospheric pressure, near ambient temperature, and whose molecular diameter is around 0.5 nm, the calculation of l gives about 50 nm with a mean value of 10^{10} collisions per second. Under 1 Pa pressure, the mean free path calculated is around a few centimetres, with about 10^4 collisions per second. Under pressures lower than 1 Pa, the mean free path becomes greater than the dimensions of the enclosure and the number of collisions between molecules becomes negligible compared with the number of collisions against the walls of the container. The gas then enters the Knudsen state, or molecular state.

Membrane—A name given to a *semipermeable* boundary.

Membrane equilibrium—When a *membrane* is permeable to a species i, but impermeable to other species, the equilibrium of the system obtained by equating the *chemical potentials* of i between the two compartments does not imply equality of chemical potentials for other constituents. Such an equilibrium is called membrane equilibrium. In the same way, thermal equilibrium (equality of temperature on both sides of a *diathermanous* boundary) does not imply equality of pressures or that of chemical potentials.

Mesomorphous—A term proposed by G. Friedel in 1922, from the Greek $\mu\varepsilon\sigma o\varsigma$, middle, and $\mu o\rho\varphi o\varsigma$, form, to designate a state of matter half-way between the solid and liquid states. A mesomorphous phase shows a more ordered character than a liquid phase, but differs from a solid in the mobility of its molecules. The melting of a molecular crystal is generally characterized by the simultaneous appearance of disorder in both position and orientation of the molecules. In a mesomorphous crystal, the orientation disorder appears after the position disorder. The main mesomorphous phases are the *smectic* and *nematic* phases.

Metacritical temperature—See *Boyle temperature.*

Metastable equilibrium—A system is in a metastable state of equilibrium if it is possible to modify its state permanently without changing either the *constraints* exerted on it or the state of the *surroundings.* The presence of a catalyst in the system triggers evolution of the system from a metastable state towards another state, which may be stable or metastable.

Many substances may exist in several states, only one of which is stable under fixed conditions of temperature and pressure. Other states are metastable. For instance, at 25°C under 1 bar, sulfur-β is metastable. On immersion in carbon disulfide, the spontaneous transformation $\beta S \rightarrow \alpha S$ is observed. In fact, the solution, saturated with respect to the metastable sulfur-β, is supersaturated with respect to the stable sulfur-α. The carbon disulfide, which plays the role of catalyst, is found intact at the end of the operation.

It is not always easy to find a good catalyst allowing it to be decided which is the stable phase. This is sometimes a happy circumstance, because it would be a pity if metastable diamond spontaneously transformed itself into stable graphite! On the other hand, life on our earth exists and develops because organic and biological molecules are metastable with respect to their products of combustion or decomposition.

Metatectic—See *Catatectic.*

Metre—One of the seven base units of the SI system, the metre (symbol m) was defined in 1983 by the 17th Conférence Générale des Poids et Mesures as the length of path covered by light in vacuum during 1/299 792 458 second. As a consequence, the velocity of light in vacuum is exactly 299 792 458 m s^{-1}.

Microcanonical ensemble—The name given to the *Gibbs ensemble* in which each system is characterized by its temperature T, volume V, and energy E. In such conditions, every system is isolated; the energy

M

of a particle and the number of particles in each energy level may be defined.

Let us consider an isolated ensemble with N particles and let N_i be the number of particles at the energy level ε_i, whose *degeneracy* is g_i. The number of states corresponding to such a distribution is given by:

$$\Omega = \prod \frac{g_i^{N_i}}{N_i!}.$$

This expression of Ω supposes implicitly that the particles are indistinguishable (see *Maxwell–Boltzmann statistics*). However, the multiplication of Ω by a constant ($N!$) does not modify the position of its extremum. The most probable distribution is obtained by seeking the maximum of Ω, taking into account the two *constraints* expressing the conservation relations:

- number of particles: $\phi = \sum N_i - N = 0$,
- energy of the system: $\psi = \sum N_i \varepsilon_i - E = 0$.

After introducing two *Lagrange multipliers* α and β, the extremum of the function $(\ln \Omega - \alpha \phi - \beta \psi)$ is given by:

$$\sum [\ln(g_i/N_i) - \alpha - \beta \varepsilon_i] \, dN_i = 0.$$

The dN_i being independent, the bracket equals zero:

$$\ln(g_i/N_i) - \alpha - \beta \varepsilon_i = 0.$$

The multiplier α is obtained by using the first constraint:

$$N = \exp(-\alpha) \sum_i g_i \exp(-\beta \varepsilon_i),$$

whence the number N_i of particles having an energy ε_i:

$$\frac{N_i}{N} = \frac{g_i \exp(-\beta \varepsilon_i)}{\sum g_i \exp(-\beta \varepsilon_i)}.$$

$$Z = \sum g_i \exp(-\beta \varepsilon_i)$$

represents the *partition function* for one particle.

The multiplier β is obtained by comparison of *entropy* expressed in classical and statistical thermodynamics:

- In classical thermodynamics: $dS = dE/T$ (V and n_i constant).

• In statistical thermodynamics: $dS = k\, d\ln\Omega$. On expanding:

$$dS = k \sum_i [\ln(g_i/N_i)]\, dN_i.$$

Now:
$$\ln(g_i/N_i) = +\beta\varepsilon_i + \ln(Z/N),$$

whence: $dS = k\beta\, dE + k\sum_i[\ln(Z/N)]\, dN_i = k\beta\, dE$, because Z/N is constant and $\sum dN_i = 0$. By identification:

$$\beta = 1/kT,$$

whence the partition function for one particle:

$$Z(T,V) = \sum g_i \exp(-\varepsilon_i/kT).$$

Microscopic reversibility—An expression used to show the invariance of equations describing the movement of particles by transformation of t to $-t$.

Consider a system described by r parameters $\xi_1, \ldots, \xi_r$ (for instance: temperature, pressure, extent of reaction) and let α_i be the deviation of a parameter ζ_i with respect to its equilibrium value. The mean value for the product of a fluctuation of ξ_i at time t by a fluctuation of ξ_j at time $(t+\tau)$ is given by:

$$\langle \alpha_i(t)\cdot\alpha_j(t+\tau)\rangle = \lim_{t\to\infty}\frac{1}{t}\int_0^t \alpha_i(t)\cdot\alpha_j(t+\tau)\, dt$$

$$= \lim_{t\to\infty}\frac{1}{t}\int_0^t \alpha_j(t)\cdot\alpha_i(t+\tau)\, dt = \langle \alpha_j(t)\cdot\alpha_i(t+\tau)\rangle.$$

Microscopic reversibility expresses the fact that to reverse the order of fluctuations, that is, to transform t to $-t$, does not modify the mean value of the product of the fluctuations. It is possible to express this result in a differential form, by making $\tau \to 0$:

$$\langle \alpha_i(t)\cdot\dot{\alpha}_j(t)\rangle = \langle \alpha_j(t)\cdot\dot{\alpha}_i(t)\rangle,$$

with $\dot{\alpha} = d\alpha/dt$ a derivative which must be considered as the ratio of two differences because the time dt must be greater than the time separating two collisions. Close to equilibrium, microscopic reversibility appears as macroscopic reversibility, which is no longer true far from equilibrium. The hypothesis of microscopic reversibility leads to the *Onsager reciprocity* relationships: $L_{ij} = L_{ji}$.

Microscopic state—A microscopic state is defined when the number of particles in each cell of the *phase space* is known. When the coordinates of the phase space are not quantified, it is necessary to assign a size to the cells of the phase space. For that purpose, we rely on the Heisenberg uncertainty principle, which classifies pairs of physical quantities as compatible or incompatible, depending on whether their accurate and simultaneous measurement is possible or impossible.

For instance, the *momentum* and position of a particle cannot be simultaneously known with accuracy. The product of their uncertainties is:

$$\Delta p_x \times \Delta x \approx h.$$

Also, energy and time are incompatible:

$$\Delta E \times \Delta t \approx h.$$

In a phase space with $2n$ dimensions, two-by-two incompatible, the region of indistinguishability of coordinates has an extent $\approx h^n$. The regions of indistinguishability are naturally identified with the cells of the phase space.

Minimum entropy production—Two theorems, due to Prigogine, are related to the *entropy production* during an irreversible process.

Theorem 1: For any irreversible process, the minimum entropy production corresponds to the *stationary state*.

Let us consider the expression of the entropy production during a process in which only two coupled phenomena occur:

$$\dot{\sigma} = J_1 X_1 + J_2 X_2 > 0.$$

In the linear domain: $J_i = \sum_j L_{ij} X_j$ with $L_{ij} = L_{ji}$:

$$\dot{\sigma} = L_{11} X_1^2 + 2 L_{12} X_1 X_2 + L_{22} X_2^2.$$

The extremum of the entropy production can be only a minimum. Using the constraint $X_2 = \text{const.}$:

$$\left(\frac{\partial \dot{\sigma}}{\partial X_1} \right)_{X_2} = 2(L_{11} X_1 + L_{12} X_2) = 2 J_1 = 0.$$

If the *generalized affinity* X_2 is kept constant, the minimum entropy production corresponds to a flux J_1 equal to zero, which is the characteristic of a stationary state. This result may be extended to n generalized affinities $X_1, \ldots, X_n$, the first k of which remain constant. For a stationary state, the $(n-k)$ remaining fluxes equal zero: $J_{k+1} = \cdots = J_n = 0$. Hence:

$$\dot\sigma = \sum_{i=1}^{n} J_i X_i = \sum_{i=1}^{n} \sum_{j=1}^{n} L_{ij} X_i X_j \quad \text{(with } L_{ij} = L_{ji}).$$

With the constraints $X_1, \ldots, X_k$ constant, the minimum entropy production is obtained by writing $(m = k+1, \cdots, n)$:

$$\left(\frac{\partial \dot\sigma}{\partial X_m}\right)_{X_1,\ldots,X_k} = \sum_{j=1}^{n} L_{mj} X_j + \sum_{i=1}^{n} L_{im} X_i = 2J_m = 0.$$

Each sum in fact represents J_m. Since the fluxes $J_{k+1}, \ldots, J_n$ equal zero, the state is stationary.

Theorem 2: Irreversible processes originating inside a system always make the entropy production decrease in the course of time.

Let us take the expression for entropy production limited to the coupling between two phenomena:

$$\dot\sigma = L_{11} X_1^2 + 2 L_{12} X_1 X_2 + L_{22} X_2^2$$

$$\frac{1}{2}\left(\frac{\mathrm{d}\sigma}{\mathrm{d}t}\right) = (L_{11} X_1 + L_{12} X_2)\left(\frac{\mathrm{d}X_1}{\mathrm{d}t}\right) + (L_{12} X_1 + L_{22} X_2)\left(\frac{\mathrm{d}X_2}{\mathrm{d}t}\right),$$

$$\frac{1}{2}\left(\frac{\mathrm{d}\dot\sigma}{\mathrm{d}t}\right) = J_1\left(\frac{\mathrm{d}X_1}{\mathrm{d}t}\right) + J_2\left(\frac{\mathrm{d}X_2}{\mathrm{d}t}\right).$$

By introducing ξ_i, the extent of the process i:

$$\frac{1}{2}\left(\frac{\mathrm{d}\dot\sigma}{\mathrm{d}t}\right) = J_1\left(\frac{\partial X_1}{\partial \xi_1}\frac{\partial \xi_1}{\partial t} + \frac{\partial X_1}{\partial \xi_2}\frac{\partial \xi_2}{\partial t}\right) + J_2\left(\frac{\partial X_2}{\partial \xi_1}\frac{\partial \xi_1}{\partial t} + \frac{\partial X_2}{\partial \xi_2}\frac{\partial \xi_2}{\partial t}\right).$$

Now the generalized affinity has the form:

$$X = -T(\partial G/\partial \xi)_{P,T},$$

whence: $(\partial X_1/\partial \xi_2)_{P,T} = (\partial X_2/\partial \xi_1)_{P,T} = -T(\partial^2 G/\partial \xi_1 \partial \xi_2)_{P,T}$.
Likewise, $J_i = \partial \xi_i/\partial t$. Hence:

$$\frac{1}{2}\left(\frac{\mathrm{d}\dot\sigma}{\mathrm{d}t}\right) = \frac{\partial X_1}{\partial \xi_1} J_1^2 + 2\frac{\partial X_1}{\partial \xi_2} J_1 J_2 + \frac{\partial X_2}{\partial \xi_2} J_2^2 < 0.$$

The entropy production decreases in the course of time and the system reaches a stationary state, at which the entropy production is a minimum. This very general result has direct consequences for the stability of stationary states. If a fluctuation makes a system deviate slightly from its stationary state, it will react so that it will return to its initial state, corresponding to the minimum entropy production. Outside the linear domain, fluxes are no longer proportional to affinities. Prigogine and Glansdorff in 1954 proved the following inequality:

$$\iiint_V \left[\sum J_i \left(\frac{\partial X_i}{\partial t} \right) \right] dV < 0.$$

The linear domain, near equilibrium, is a deformation of the equilibrium state. When constraints are progressively applied, the various stationary states form, with the equilibrium state a continuous trace, known as a thermodynamic branch. Sufficiently far from equilibrium, fluctuations may cease to regress and then impose a new macroscopic organization in space and time, characterized by the appearance of what Prigogine called *dissipative structures*.

Mixing—See *Enthalpy of mixing, Entropy of mixing, Excess Gibbs energy of mixing, Excess quantity, Gas mixture, Gibbs energy of mixing, Ideal solution, Integral molar quantity, Partial molar quantity, Partial molar quantity of dissolution, Partial molar volume, Partial volume, Quantity of mixing, Regular solution, Solid solution, Solute, Solution, Solvent, Volume of mixing, Wagner reciprocity.*

Molality—Or 'mass molar concentration', symbolized by m_i, is expressed in $mol \, kg^{-1}$ and represents the amount of *solute i* per unit mass unit of *solvent*. Knowledge of molalities gives access to *mole fractions* or to *mass fractions*, but does not give volumetric masses. Moreover, the molality of a solution, in comparison with *molarity*, has the advantage of remaining independent of temperature.

Molar—Reduced to one *mole*. The adjective molar may be applied to a pure substance or a mixture. For instance, if Y is an integral quantity, the *integral molar quantity* will be given by $(n = \sum n_i)$:

$$Y_m = Y/n = \sum x_i y_i = (\partial Y / \partial n)_{T,P,x_i}.$$

A *partial molar quantity* is defined by: $y_i = (\partial Y / \partial n_i)_{T,P,x_{j \neq i}}.$

Molarity—Or 'volumetric molar concentration', symbolized by c_i, is expressed either in $mol \, m^{-3}$ or in $mol \, L^{-1}$ and represents the quantity of *solute i* per unit volume of the solution. Volumetric masses, mole fractions, or mass fractions cannot be obtained from only a knowledge of molarities. Besides, molarities have the drawback of depending on temperature. Nevertheless, for the choice of a *standard* concentration, the most popular is $c_i^\circ = 1\,\text{mol}\,\text{L}^{-1}$, which raises other difficulties, because such a solution is generally not dilute enough to obey *Henry's law*!

Mole—One of the seven base units of the SI system (symbol mol), defined in 1971 by the 14th Conférence Générale des Poids et Mesures as the amount of substance in a system containing as many elementary entities as there are atoms in 0.012 kg of carbon-12. As a consequence, the molar mass of isotope 12 of carbon is exactly $12.0000\,\text{g}\,\text{mol}^{-1}$. Care is needed as sometimes the kilomole (symbol kmol) is used. The molar mass of the isotope carbon-12 is exactly $12.0000\,\text{kg}\,\text{kmol}^{-1}$.

Mollier diagram—A name given to the *enthalpy–entropy diagram*, from Richard Mollier (1863–1935), German engineer.

Moment When a stochastic variable X can take several values $\{x_1, x_2, \ldots, x_i, \ldots\}$, each value x_i having a probability $f(x_i) > 0$ [with $\sum f(x_i) = 1$], the moment of order n of X is defined by:

$$\langle x^n \rangle = \sum x_i^n f(x_i).$$

If the variable X is continuous, it is possible to define a probability density $f(x)$. The probability for the variable X to be comprised between x and $x + \mathrm{d}x$ is $f(x)\,\mathrm{d}x$ with $\int f(x)\,\mathrm{d}x = 1$. The moment of order n of X is then:

$$\langle x^n \rangle = \int x^n f(x)\,\mathrm{d}x.$$

The interest of moments lies in the fact that they are more easily available than the function $f(x)$. The moment of order 1, $\langle x \rangle$, is the *mean*.

$\sigma_x^2 = \langle x^2 \rangle - \langle x \rangle^2$ is the *variance*; σ_x is the *standard deviation*. A *normal distribution* is entirely defined by a knowledge of the mean and the standard deviation; generally, a distribution is defined only when all the moments $\langle x^n \rangle$ are known.

Momentum—The name given to the product mv of mass by velocity. If a force F is exerted during a time $\mathrm{d}t$ on a mass m moving with a velocity v, the variation of momentum is given by $F = d\,(mv)/dt$, a result established by Newton as early as 1687.

Monotectic—From the Greek $\mu o\nu o\varsigma$, one, and $\tau\eta\kappa\tau o\varsigma$, fusible. The name given to the following transformation observed on heating:

$$(\text{Liquid 1}) + \langle\text{Solid}\rangle \to (\text{Liquid 2}).$$

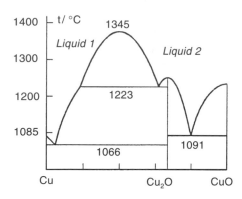

Monotectic: Metal–oxygen or metal–sulfur systems often show monotectic transitions owing to immiscibility between metal and oxide or between metal and sulfide in the liquid phase. The figure shows an example of a monotectic transition at 1223°C on the copper–oxygen diagram.

Monotropic—Two phases are called monotropic when they cannot coexist in a stable equilibrium. Monotropic is the antonym of *enantiomorphic*.

Monovariant—Whose *variance* equals unity. An *equilibrium* is called monovariant if it is impossible to modify independently more than one intensive parameter without modifying the number of phases. The equilibrium of a pure substance in two phases is monovariant because it is impossible to modify the temperature and pressure independently.

Mutually stable equilibrium—Two systems A and B are said to be in a mutually stable equilibrium state if the assembly (A + B) is in a stable equilibrium state. It is clear that two systems taken separately may be in stable equilibrium without being in mutually stable equilibrium (for instance, oxygen and hydrogen!). On the other hand, if two systems are in mutually stable equilibrium, each system taken separately is in a stable equilibrium state (see *Second law, corollary 4*).

N

Negative temperature—The *Boltzmann distribution* gives at equilibrium the number N_i of particles at the energy level ε_i whose *degeneracy* is g_i:

$$N_i = \frac{N}{Z} g_i \exp\left(-\frac{\varepsilon_i}{kT}\right).$$

The ratio between the populations of two levels is then given by:

$$\frac{N_i}{N_j} = \frac{g_i}{g_j} \exp\left(\frac{\varepsilon_j - \varepsilon_i}{kT}\right).$$

$\varepsilon_i > \varepsilon_j \Rightarrow N_i/N_j < g_i/g_j$; at equilibrium, the quantum states corresponding to the higher energies are the least populated. When $T \to \infty$, $N_i/N_j \to g_i/g_j$, the quantum states tend towards the same population. If we now imagine that the temperature could be negative, the following conclusion holds: $\varepsilon_i > \varepsilon_j \Rightarrow N_i/N_j > g_i/g_j$. The most populated quantum states correspond to the highest energies: there is population inversion.

Negative temperature states would have a higher energy than positive temperature states. From this point of view, a negative temperature would be 'higher' than a positive one, because it would be situated beyond infinity. The basic relationships may apply to negative temperature states:

$$dU = T\,dS - P\,dV \Rightarrow T = (\partial U/\partial S)_V.$$

At equilibrium, if $T > 0$, the entropy increases with internal energy, whereas if $T < 0$, it varies inversely. Consider for instance an assembly of moment-carriers in a magnetic field. If every moment is oriented in the reverse direction to the field, the energy of the system is a maximum and its entropy a minimum. This state corresponds to the temperature $T = -0\,\mathrm{K}$. If all the moments are oriented in the direction of the field, the energy of the system is then a minimum and its entropy is also a minimum. This state corresponds to a temperature $T = +0\,\mathrm{K}$. Between these two states, the entropy of the system goes through a maximum (directions of moments haphazard), a state corresponding to an infinite temperature.

208 Negentropy

Negentropy—The name given to the function $(-S)$. Negentropy is the *thermodynamic potential* of an *isolated* system. When one links *entropy* and *information* theory, it is easier, following Brillouin, to use negentropy.

Nematic—From the Greek νεματος, wire. A mesomorphic state in which molecules, generally of elongated shape, are parallel but without any order in the position of the centres of gravity.

Nernst, Walter (1864–1941)—German physicist and chemist, who established the theory of electrochemical cells in 1906. For his formulation of the third law in 1918, he earned the Nobel prize for chemistry in 1920.

Nernst heat theorem—Still called the 'Nernst principle' or *third law*, it may be stated as: 'The *entropy* change of a *system* during a *reversible* isothermal process tends towards zero when the *thermodynamic temperature* of the system tends towards zero'. This postulate, confirmed by experience, was proposed by Nernst in 1906. In 1911, Planck suggested that for every system in a stable equilibrium state, $S=0$ for $T=0$, which allows an absolute value to be assigned to the entropy function and the Nernst–Planck postulate to be stated:

The entropy of a system at equilibrium is 0 at 0 K.

In 1912, Nernst put forward the principle of the inaccessibility of absolute zero, and in 1918 he showed (see *Third law*) that:

- the inaccessibility of absolute zero may be deduced from the *second law*;
- the Nernst heat theorem proceeds from the inaccessibility of absolute zero.

The demonstration of the latter proposition is universally accepted; that of the former has been criticized by Einstein. According to him, the possibility of extracting or giving up heat at 0 K is not proved.

Nernst relation—Proposed by Nernst in 1889, it gives the *electrode potential*:

$$\mathscr{E} = \mathscr{E}^\circ + \frac{RT}{n\mathscr{F}} \ln \frac{[\text{Ox}]}{[\text{Red}]}.$$

$\mathcal{E}°$: *standard electrode potential*
R: gas constant
$\mathcal{F}$: Faraday constant
n: number of electrons exchanged during the reaction $Ox + ne^- \to Red$
$[Ox]/[Red]$: *mass action product* of the reaction at the electrode.

For instance:

$$Cr_2O_7^{2-} + 6e^- + 14H^+ \to 2Cr^{3+} + 7H_2O$$

$$\mathcal{E} = \mathcal{E}° + \frac{RT}{6\mathcal{F}} \ln \frac{a_{Cr_2O_7^{2-}} \cdot a_{H^+}^{14}}{a_{Cr^{3+}}^2 \cdot a_{H_2O}^7}.$$

Consider a galvanic cell in which the following reactions proceed:

$$\left.\begin{array}{l} -\text{at electrode 1: } Ox_1 + ne^- \to Red_1 \\ -\text{at electrode 2: } Ox_2 + ne^- \to Red_2 \end{array}\right\} \Rightarrow Ox_2 + Red_1 \to Ox_1 + Red_2$$

Its electromotive force is given by:

$$\Delta\mathcal{E} = \mathcal{E}_2 - \mathcal{E}_1 = \Delta\mathcal{E}° + \frac{RT}{n\mathcal{F}} \ln \frac{[Ox_2][Red_1]}{[Red_2][Ox_1]}.$$

Remark: For a redox reaction, $\Sigma v_i M_i = 0$, the Nernst relation is deduced from the *Gibbs energy of reaction*:

$$\Delta_r G = \Delta_r G° + RT \ln \Pi(a_i)^{v_i}.$$

Since the electromotive force of the galvanic cell is related to the Gibbs energy of the reaction by $\Delta G = -n\mathcal{F}\Delta\mathcal{E}$, it follows that, for the overall reaction:

$$Ox_2 + Red_1 \to Ox_1 + Red_2$$

$$\Delta_r G = -n\mathcal{F}\Delta\mathcal{E} = \Delta_r G° + RT \ln \frac{[Ox_1][Red_2]}{[Red_1][Ox_2]}.$$

Newton, Sir Isaac (1642–1727)

Newton, Sir Isaac (1642–1727)—English author of one of the most important works in science. As a mathematician, he invented the infinitesimal calculus; as a physicist, he developed the theory of light; as an astronomer, he stated the three fundamental laws of mechanics, leading to the law of gravitation. His two main works are *Principia mathematica* (1687) and *Optics* (1704). The newton, symbol N, is the unit of force in the SI system: $1 N = 1 kg\,m\,s^{-2}$.

Normal—This adjective, when qualifying a physical quantity, gives it a particular value, determined by convention. For instance:

- normal acceleration: $g = 9.80665 \, \mathrm{m\,s^{-2}}$
- normal concentration: $c = 1 \, \mathrm{mol\,L^{-1}}$
- *normal conditions*: $T = 273.15 \, \mathrm{K}$ and $P = 101\,325 \, \mathrm{Pa}$
- *normal pressure*: $P = 101\,325 \, \mathrm{Pa}$
- *normal distribution*
- normal solution: one whose *normality* equals unity.

The conventions may sometimes evolve with time (see for instance *Normal pressure*), which is quite abnormal! Nevertheless, excessive standardization must not conceal improvements brought about by the scientific community.

Normal conditions—Normal conditions of temperature and pressure are 0°C (273.15 K) and 1 atm (101 325 Pa). In such conditions, 1 mole of an ideal gas occupies a volume of 22.414 097 litres. There is an unfortunate tendency to call 'normal' a standard pressure of 1 bar. If such a definition were legal, 1 mole of an ideal gas would occupy in normal conditions a volume of 22.711 084 litres.

Normal distribution—A normal, or Gaussian, distribution is the name given to the *binomial distribution* when n_A and n_B are large, n_A and n_B being respectively the number of occurrences of events A and B in N experiments ($N = n_A + n_B$). Let us write the expression of the probability of event A in the case of a binomial distribution:

$$P_N(n_A) = \frac{N!}{n_A! n_B!} p^{n_A} q^{n_B};$$

then, by applying the *Stirling formula* to large numbers:

$$N! \approx \sqrt{2\pi N}(N/e)^N$$

we obtain:

$$P_N(n_A) = \frac{1}{\sqrt{2\pi N}} \exp\left(n_A \ln \frac{pN}{n_A} + n_B \ln \frac{qN}{n_B}\right).$$

It is easily shown that the maximum of this function, which gives the most probable value of n_A, is obtained for $n_A = Np$. By expanding in a Taylor series around $\langle n_A \rangle$:

$$P_N(n_A) = P_N(\langle n_A \rangle) \exp\left(\frac{1}{2} B_2 \varepsilon^2 + \frac{1}{6} B_3 \varepsilon^3 + \cdots \right),$$

with:

$$\varepsilon = n_A - \langle n_A \rangle \text{ and } B_k = \left\{ \frac{d^k \ln[\sqrt{2\pi N} P_N(n_A)]}{dn_A^k} \right\}_{n_A = \langle n_A \rangle}.$$

We calculate:

$$B_2 = -1/Npq, \quad B_3 = (q^2 - p^2)/N^2 p^2 q^2, \ldots$$

More generally:

$$|B_k| < 1/(Npq)^{k-1}.$$

By stopping the expansion at the first term, we have a 'Gaussian curve':

$$P_N(n_A) \approx P_N(\langle n_A \rangle) \exp(-\varepsilon^2/2Npq).$$

When ε is much larger than the *standard deviation* $\sqrt{Npq}$, that is, when n_A is far from the mean value, $P_N(n_A)$ tends very quickly towards zero. For high values of N, $Npq \gg 1$; the Gaussian then gives a very good approach to the binomial distribution. The use of the normalization condition:

$$\int P_N(n_A) dn_A = 1$$

yields:

$$P_N(\langle n_A \rangle) = 1/\sqrt{2\pi Npq}.$$

By introducing the standard deviation $\sigma = \sqrt{Npq}$, the Gaussian distribution may be expressed, with $x = n_A$ and $x_0 = \langle n_A \rangle$, by the relation:

$$P_N(x) = \frac{1}{\sqrt{2\pi} \cdot \sigma} \cdot \exp\left[-\frac{(x - x_0)^2}{2\sigma^2} \right],$$

where x_0, the mean value of the quantity x, corresponds also to its most probable value. The Gaussian distribution is entirely determined by its first two moments $\langle x \rangle$ and $\langle x^2 \rangle$.

- The probability that $(x-\sigma) < x < (x+\sigma)$ equals 0.68.
- The probability that $(x-2\sigma) < x < (x+2\sigma)$ equals 0.95.
- The probability that $(x-3\sigma) < x < (x+3\sigma)$ equals 0.9975.
- The probability that $(x-4\sigma) < x < (x+4\sigma)$ equals 0.99993.

Normal pressure—The normal pressure, defined in 1954 by the 10th Conference Générale des Poids et Mesures, is, by convention, 1 atm, that is, 101 325 Pa. In 1983, in legal texts, the standard pressure shifted, from 1 atm (101 325 Pa) to 1 bar (100 000 Pa). So far, the definition of normal pressure has not followed the same course. When that will be done, the normal boiling point of water, that is, its boiling point under normal pressure, will shift from 100°C to 99.2°C, an interesting revolution in prospect!

Normality—The ratio of the amount of active substance in a solution to the volume of the solution. By 'active substance', we must understand H_3O^+ ions for an acid solution, OH^- ions for a basic solution, electrons capable of being accepted or provided for an oxidizing or reducing solution. Thus, a normal acid solution holds 1 mole of an acid AH per litre of solution, but 0.5 mole of an acid AH_2. Also, a normal solution of MnO_4^- holds 1/5 or 1/3 mol L^{-1} depending on whether the MnO_4^- ion is to be reduced to Mn^{2+} or to MnO_2. When quoting a normality, it is necessary to specify which reaction is involved, thus making this concept ambiguous. There are some who would like to drop it. Its interest in analytical chemistry arises from the assumption that the reactions used are often complete often prevents this.

O

Oersted, Christian (1777–1851)—Danish physicist, author of studies on the compressibility of condensed matter. He demonstrated the existence of the magnetic field created by a current. His name has been given to a former cgs unit of *magnetic excitation*, whose use is now to be avoided:

$$1\,\text{Oe} = (1000/4\pi)\,\text{A}\,\text{m}^{-1} = 79.577\,47\,\text{A}\,\text{m}^{-1}.$$

Ohm, Georg (1787–1854)—German physicist, who in 1827 demonstrated the proportionality between potential and electric current

$(V = RI)$. The ohm (symbol Ω) is the unit of electrical resistance in the SI system:

$$1\,\Omega = 1\,\mathrm{V\,A}^{-1} = 1\,\mathrm{m^2\,kg\,s^{-3}\,A^{-2}}.$$

Onsager, Lars (1911–1976)—Norwegian physicist and chemist, who in 1931 established, with his reciprocity relations, the basis of the thermodynamics of irreversible processes. He earned the Nobel prize for chemistry in 1968.

Onsager reciprocity—During an irreversible process, the *entropy production* is a sum of products of *generalized forces* by the corresponding *fluxes*:

$$\dot{\sigma} \equiv \frac{\mathrm{d}\sigma}{\mathrm{d}t} = \sum_i J_i X_i > 0.$$

In linear thermodynamics of irreversible processes: $J_i - \sum_k L_{ik} X_k$.

The *phenomenological coefficients* L_{ik} characterize the response of the system to external stimuli. The cross-coefficients L_{ik} $(i \neq k)$ describe the *coupling* between two irreversible processes i and k. Without any magnetic interaction, statistical mechanics, together with *microscopic reversibility*, lead to the Onsager reciprocity relations (1931):

$$L_{ij} = I_{ji}.$$

If α_i represents the fluctuation of a parameter ξ_i around its equilibrium value, microscopic reversibility implies:

$$\langle \alpha_i(t) \cdot \dot{\alpha}_j(t) \rangle = \langle \alpha_j(t) \cdot \dot{\alpha}_i(t) \rangle.$$

By assuming that the regression of a fluctuation follows the laws of linear thermodynamics of irreversible processes:

$$\dot{\alpha}_i(t) = J_i = \sum_k L_{ik} X_k.$$

Let us introduce the last relation into the preceding one:

$$\sum_k L_{jk} \langle \alpha_i X_k \rangle = \sum_k L_{ik} \langle \alpha_j X_k \rangle.$$

By taking into account the independence of the fluctuations:

$$\langle \alpha_i X_k \rangle = \lambda \delta_{ik},$$

where δ_{ik} is the *Kronecker delta*, it follows that $L_{ij} = L_{ji}$.

The Onsager reciprocity relations, proved here only for small fluctuations around equilibrium, remain valid as long as relationships between fluxes and forces remain linear.

Note: The reciprocity relations do not apply to coefficients relating gradients to currents of any kind: they must be forces and fluxes in the sense of the thermodynamics of irreversible processes. If one of the processes i or j implies a *magnetic field*, the reciprocity relations take another form proposed by Casimir in 1945:

$$L_{ij} = -L_{ji}.$$

The reciprocity relations in fact follow from the reversibility of the laws of mechanics upon time inversion. In a magnetic field, not only the velocities but also the direction of the field must be inverted in order that the particles follow the same path in the reverse direction.

Open—Said of a *system* which may exchange *work, heat,* and matter with its surroundings.

Order of a transition—The equilibrium of a substance i between two phases α and β is expressed by equating *chemical potentials*: $\mu_{i,\alpha} = \mu_{i,\beta}$. During the transition $\alpha \to \beta$, the chemical potential does not undergo any discontinuity, but its derivatives may show breaks.

A transition $\alpha \to \beta$ is said to be of the first order, or of the 'first kind' if the first and subsequent derivatives of the chemical potential are discontinuous. The first derivatives of the Gibbs energy are:

- volume: $v_i = (\partial \mu_i / \partial P)_T$;

- entropy: $s_i = -(\partial \mu_i / \partial T)_P$;

- enthalpy: $h_i = -T^2 [\partial(\mu_i/T)/\partial T]_P$.

The transition proceeds with non-zero enthalpy and volume changes. The equilibrium curve on a (P,T) diagram separating the two domains α and β is obtained by integration of the *Clapeyron equation*:

$$(dP/dT)_{eq} = \Delta_{tr}S/\Delta_{tr}V = \Delta_{tr}H/T\Delta_{tr}V.$$

The transition $\alpha \to \beta$ is said to be of the second order, or of the 'second kind', if, the first derivatives of the Gibbs energy being continuous, the second and subsequent derivatives are discontinuous. A second-order transformation is thus characterized, for a pure substance, by continuity of the functions V, S, and H; the Clapeyron equation leads to an indeterminacy ($\Delta_{tr}S = 0$ and $\Delta_{tr}V = 0$), which is removed by using the *Ehrenfest equations*. The discontinuity appears in the first derivatives of the volume (*expansivities, compressibilities*), of the enthalpy (*heat capacities*), and of the entropy (*bulk expansion coefficients*).

In a solid, a first-order transition implies considerable atomic displacement and, most often, the α and β structures show no relationship. On the other hand, for a second-order transition, the α and β structures are closely related and most often identical. Amongst second-order transitions are:

• ferromagnetic–paramagnetic transitions

• ferroelectric–paraelectric transitions

• the He II (superfluid)–He I (normal fluid) transition

• superconductor–normal conductor transitions.

The above transitions are orientation order–disorder transitions. The disordered phase (paramagnetic, paraelectric, normal fluid helium, normal conductor) is stable at high temperature. The order in the low-temperature modification concerns the orientation of magnetic or electric moments (ferromagnetic or ferroelectric phases), or the orientation of angular momentum of He atoms or electrons (superfluid helium or superconductors).

The order–disorder transitions linked to the positions of atoms (see *Long-range order, Short-range order*) may be of the first or second order. The same holds for ferroelectric–paraelectric transitions, which involve a slight shift of the atoms. However, a clear distinction between $\Delta_{tr}H = 0$ and $\Delta_{tr}H \approx 0$ is not always evident!

A *solid solution* may also undergo a transition of the first or second order. On an isobaric diagram (temperature–composition) or on an isothermal diagram (pressure–composition), the boundary between the single-phase domains α and β is formed of two lines for a first-order transition: the two single-phase domains are separated by a two-phase domain ($\alpha + \beta$). The two domains are separated by only one line when the transition is of the second order.

Order of position—See *Long-range order, Short-range order*.

Order of reaction—When the *reaction rate* of a chemical reaction:

$$aA + bB \rightarrow \text{Products}$$

may be experimentally expressed by:

$$v = d\xi/dt = k[A]^{\alpha}[B]^{\beta},$$

α (or β) represents the order of the reaction with respect to A (or B). The sum $(\alpha + \beta)$ represents the overall order of the reaction.

Oregonator—The name given to a model of the *Belousov–Zhabotinskii reaction* developed by Field in 1974 and so called in honour of his university. This model, which is straightforward because it introduces only three intermediate species, nevertheless reproduces the main observations:

$$A + Y \rightleftharpoons X \qquad [1]$$

$$X + Y \rightleftharpoons P \qquad [2]$$

$$B + X \rightleftharpoons 2X + Z \qquad [3]$$

$$2X \rightleftharpoons Q \qquad [4]$$

$$Z \rightleftharpoons fY \qquad [5]$$

A and B are the reactants, P and Q the products, and X, Y, and Z the intermediate compounds. Step [3] provides autocatalysis necessary to trigger oscillations. The most direct combination of these steps which does not introduce variations in the content of intermediate products is:

$$f[1] + f[2] + 2[3] + [4] + 2[5], \quad \text{whence globally:} \quad fA + 2B \rightleftharpoons fP + Q$$

Osmo—From the Greek $\omega\sigma\mu\sigma\varsigma$, pushing, gives *osmosis*; it must not be confused with the Greek $\sigma\sigma\mu\eta$, smell, which gives osmium.

Osmosis—Spontaneous displacement of water or any other *solvent* through a semipermeable membrane, under the influence of a *chemical potential* difference. If a membrane impermeable to the *solutes* but permeable to the solvent separates a pure solvent α from a solution β, the chemical potential of the solvent 1 in $\alpha(\mu_{1,\alpha} = \mu_1^{\circ})$ is higher than its chemical potential in β (that is, $\mu_{1,\beta} = \mu_1^{\circ} + RT \ln a_1$). The solvent will move from α to β until equality of the chemical potentials is realized. The process may be stopped by applying above β a pressure, called *osmotic pressure*, such that $\mu_{1,\beta}$ is increased and brought to the value of $\mu_{1,\alpha}$. If the pressure applied above β is higher than the osmotic pressure, the direction of transfer is reversed. This process, called 'reverse osmosis', is used in competition with distillation to desalinate sea water.

Osmotic pressure—If two solutions α and β in the same *solvent* are separated by a membrane permeable only to the molecules of solvent, the *chemical potential* of the solvent will differ on both sides of the membrane and it will migrate spontaneously in the direction of decreasing chemical potential, for instance from α to β if $\mu_{1,\alpha} > \mu_{1,\beta}$. The osmotic pressure is the pressure which must be applied above β to raise $\mu_{1,\beta}$ to the same level as $\mu_{1,\alpha}$.

Initially: $\mu_{1,\alpha} = \mu_1^\circ + RT \ln a_{1,\alpha}$ and $\mu_{1,\beta} = \mu_1^\circ + RT \ln a_{1,\beta}$. By taking for the solvent 1 the same *standard state* whatever the medium, $\mu_{1,\alpha} > \mu_{1,\beta}$ yields $a_{1,\alpha} > a_{1,\beta}$. If both solutions are initially subjected to the same atmospheric pressure P°, the osmotic pressure Π will be given by:

$$\mu_{1,\alpha} = \mu_{1,\beta} + \int_{P^\circ}^{P^\circ + \Pi} v_{1,\beta}\, dP.$$

Hence:

$$RT \ln \frac{a_{1,\alpha}}{a_{1,\beta}} = \int_{P^\circ}^{P^\circ + \Pi} v_{1,\beta}\, dP.$$

Generally, the solution α is the pure solvent: $a_{1,\alpha} = 1$, so the expressions above can be simplified by dropping the index β. Furthermore, if the solution β is dilute enough, $v_{1,\beta}$, the *partial molar volume* of the solvent 1 in the solution β may be assimilated to v_1°, the molar volume of the pure solvent, and it is possible to apply *Raoult's law*: $a_{1,\beta} = a_1 = x_1 = 1 - \sum_{i \neq 1} x_i$.

Finally, if the solvent 1 may be considered incompressible, the osmotic pressure is given by:

$$RT \ln \frac{1}{1 - \sum_{i \neq 1} x_i} = v_1^\circ \Pi.$$

$\sum_{i \neq 1} x_i$ represents the sum of all mole fractions of the species 2, 3, ... dissolved in the solution β. If, moreover, $\sum_{i \neq 1} x_i \ll 1$, the above relation simplifies and becomes:

$$RT \sum_{i \neq 1} x_i = v_1^\circ \Pi.$$

By letting $\sum_{i \neq 1} x_i = (\sum_{i \neq 1} n_i)/(\sum_i n_i)$, it follows that:

$$\left(\sum_{i \neq 1} n_i \right) RT = \left(\sum_i n_i \right) v_1^\circ \Pi,$$

a relation which can again be written: $\Pi V = n RT$.

Π is the osmotic pressure; n represents the amount of solute (do not forget the possible dissociation of electrolytes, by counting for instance two moles of solute for each mole of dissolved NaCl) and V the volume of the solution. This result is very elegant, but the countless simplifying assumptions made in obtaining it must not be lost sight of!

Otto cycle—Or 'Beau de Rochas cycle'. Nikolaus August Otto, German engineer (1832–1891), built in 1876 the first spark-ignition internal combustion engine using the four-stroke cycle patented by Alphonse Beau de Rochas in 1862. The cycle, formed of two *isochoric* steps and two *adiabatic* steps, is covered by two motor revolutions.

AB: introduction of the air–fuel mixture at low pressure
BC: adiabatic compression of the mixture
CD: combustion of the fuel
DE: adiabatic expansion of the combustion products
EBA: evacuation of burnt gases

Designating by $Q_1(>0)$ the heat arising from the combustion and by $Q_2(<0)$ the heat given to the surroundings by the exhaust gases, and assuming ideal gases with constant heat capacity, the thermal efficiency of the cycle is found as follows:

$$\eta = \frac{Q_1 + Q_2}{Q_1} = 1 - \frac{mc_V(T_E - T_B)}{mc_V(T_D - T_C)} = 1 - \frac{T_B[(T_E/T_B) - 1]}{T_C[(T_D/T_C) - 1]}.$$

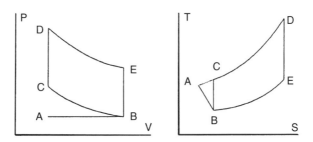

Otto cycle: P–V (left) and T–S (right) diagrams.

Using also the equations of the *isentropes* (with $\gamma = c_P/c_V$):

$$\frac{T_B}{T_C} = \left(\frac{V_C}{V_B}\right)^{\gamma-1} = \left(\frac{V_D}{V_E}\right)^{\gamma-1} = \frac{T_E}{T_D},$$

we obtain $T_E/T_B=T_D/T_C$. The expression of η is thus reduced to:

$$\eta=1-\frac{T_B}{T_C}=1-\left(\frac{V_C}{V_B}\right)^{\gamma-1}.$$

The efficiency of the Otto cycle increases with the 'compression ratio' V_B/V_C. In practice there is an upper limit because of the so-called knock phenomenon, due to the pre-ignition of the fuel by the high temperature of compression. The maximum compression ratio that can be used, fixed by the need to avoid detonation, varies between 8 and 10, which gives a theoretical efficiency of about 60%.

Overvoltage—Consider a galvanic cell in which the reaction: $Red_1+Ox_2\rightarrow Ox_1+Red_2$ proceeds. The electromotive force of the cell $\Delta\mathscr{E}=\mathscr{E}_2-\mathscr{E}_1$ is calculated from the *Nernst relation*. If a generator characterized by a voltage $\Delta\mathscr{E}' > \Delta\mathscr{E}$ is placed in opposition, the cell works as an electrolytic cell. With a reversible cell, the overall reaction observed in the electrolytic cell is the reverse of that observed in the galvanic cell. The overvoltage is the name given to the difference $\Delta\mathscr{E}'-\Delta\mathscr{E}$ which it is necessary to impose in order to observe effectively the reverse reaction.

The overvoltage is often negligible, but it can be of the order of 0.5 V when a gas is liberated at the electrode. For instance, if a cell works by the reaction $2H_2+O_2\rightarrow 2H_2O$, $\Delta\mathscr{E}=1.23$ V, whereas the electromotive force necessary to electrolyse water must be higher than $\Delta\mathscr{E}'\approx 2.2$ V.

P

Parachor—Introduced by Snuyden in 1924, the parachor is given by:

$$\mathscr{P}=M\Gamma^{0.25}/(\rho_1-\rho_v).$$

M: molar mass; Γ: surface tension at the liquid–vapour interface; ρ_1 and ρ_v: volumetric masses of liquid and vapour. The parachor is a function of temperature, is directly related to the structure of the molecule, and may be calculated by adding increments relating to atoms and bonds.

Paradigm—From the Greek παραδειγμα, example. A term borrowed from grammarians to denote an ensemble, generally reduced, of knowledge or results on which the development of a school of thought rests. *Kuhn* insists on the importance of this concept in the evolution of ideas.

Since the eighteenth century, the world has lived under the influence of the Newtonian mechanics paradigm. Newton developed mathematical tools for describing and mastering motion, which has led to a general faith in the arrival of a better world through increased knowledge and progress in techniques.

During the twentieth century the *entropy* paradigm emerged, destroying the idea of science and technology as creative of a better-ordered world. The *second law*, expressed in a language within the reach of everyone, postulates that matter and energy evolve in only one direction: from available to unavailable, from useful to useless, from ordered to disordered. Entropy is the quantity which measures, for any given system, the degree of this evolution.

Paradox—From the Greek παρα, against, and δοξα, opinion. A proposition going against accepted ideas, or whose logical consequences generate a contradiction. Thermodynamics, a science claiming to be rigorous, raises many paradoxes which tend to fault the *second law*. Most of them are false paradoxes, reducing this law to the statement 'Entropy may only increase', forgetting that it is valid only for an isolated system.

Hence all considerations about the appearance and development of complex structures in the universe or on the earth belong to this kind of fallacious paradox. The formation of stars or galaxies from a homogeneous cloud, the spread of life on the earth, would seem to express the spontaneous formation of 'order' contrary to the second law. It is easy to show that the 'order' created locally is inevitably accompanied by more considerable 'disorder' created in another part of the universe.

The formation of ordered and dissipative structures from disorder is compatible with the second law, but the reasons for their spontaneous appearance are not yet known and are the object of active research in the field of thermodynamics of irreversible processes.

Much ink has been spilt over certain paradoxes that raise more serious objections and deserve more careful investigation. Among the best known may be mentioned the *Gibbs paradox*, *Maxwell's demon*, *Poincaré's recurrence* and *speed inversion*.

Partial molar quantity—Consider a system defined by its temperature, its pressure, and $n_1, n_2, \ldots, n_k$ numbers of moles of $A_1, A_2, \ldots, A_k$. Let $Y(T, P, n_1, n_2, \ldots, n_k)$ be an integral quantity characterizing the system. At constant temperature and pressure:

$$dY = \left(\frac{\partial Y}{\partial n_1}\right)_{T,P,n_{j \neq 1}} dn_1 + \cdots + \left(\frac{\partial Y}{\partial n_k}\right)_{T,P,n_{j \neq k}} dn_k.$$

The partial molar quantity is, by definition:

$$y_i = \left(\frac{\partial Y}{\partial n_i}\right)_{T,P,n_{j \neq i}}.$$

Remark: It is important to notice the differentiation conditions; the pressure and the temperature, kept constant, are intensive quantities. It is certainly possible, using an equation of state, to express $Y - Y(T, V, n_1, \ldots n_i, \ldots)$, then to calculate $Y_i = (\partial Y/\partial n_i)_{T,V,n_{j \neq i}}$, but the new quantity Y_i so defined is no longer a partial molar quantity. In particular, it does not satisfy the *Euler identity*.

CALCULATION OF A PARTIAL MOLAR QUANTITY: If the integral quantity is given by $Y(T, P, n_1, n_2, \ldots, n_k)$, the use of the relation defining y_i raises no difficulties. On the other hand, if the integral quantity is given in its molar form: $Y_m(T, P, x_1, x_2, \ldots, x_k)$, it is necessary to be very careful when differentiating. Indeed, a partial derivative such as $(\partial Y_m/\partial x_i)_{T,P,x_{j \neq i}}$ has no meaning physically! The constraint $\sum x_i = 1$ forbids taking a derivative with respect to x_i with $x_{j \neq i}$ kept constant! Mathematicians can be convinced of this by replacing, in $Y_m(T, P, x_1, \ldots, x_i, \ldots)$, any x_i by $(1 - \sum_{j \neq i} x_j)$, to obtain various expressions of the same function Y_m, each giving a different expression of the partial derivatives $(\partial Y_m/\partial x_i)_{T,P,x_{j \neq i}}$.

To calculate y_i, it is necessary to let $x_1 = 1 - \sum_{j-2}^{n} x_j$, then to express Y in the form $Y(T, P, n, x_2, x_3, \ldots)$ with $Y = nY_m$:

$$y_i = \left(\frac{\partial Y}{\partial n_i}\right)_{n_{j \neq i}} = \left(\frac{\partial n}{\partial n_i}\right)_{n_{j \neq i}} Y_m + n\left(\frac{\partial Y_m}{\partial n_i}\right)_{n_{j \neq i}}$$

$$y_i = Y_m + n\left(\frac{\partial Y_m}{\partial n}\right)_{x_j}\left(\frac{\partial n}{\partial n_i}\right)_{n_k} + n\sum_{j=2}^{n}\left(\frac{\partial Y_m}{\partial x_j}\right)_{n,x_k}\left(\frac{\partial x_j}{\partial n_i}\right)_{n_k}.$$

Now:

$$\left(\frac{\partial Y_m}{\partial n}\right)_{x_j} = 0, \quad \left(\frac{\partial n}{\partial n_i}\right)_{n_{j \neq i}} = 1, \quad \left(\frac{\partial x_j}{\partial n_i}\right)_{n_k} = \frac{\delta_{ij} - x_j}{n},$$

whence:

$$y_i = Y_m + \sum_{j=2}^{n} (\delta_{ij} - x_j) \left(\frac{\partial Y_m}{\partial x_j} \right)_{x_{k \neq j}},$$

where δ_{ij} is the Kronecker delta, zero for $i \neq j$, and 1 for $i = j$.
We can verify that y_i obeys the Euler identity: $Y_m = \sum x_i y_i$.

APPLICATION TO UNARY SYSTEMS (pure constituent): This is trivial:

$$y_1 = Y_m = y_1^{\circ}(!!).$$

APPLICATION TO BINARY SYSTEMS: With two constituents ($i = 1, 2$), the integral molar quantity is given in the form $Y(T, P, x_2)$:

$$y_1 = Y_m - x_2 \left(\frac{\partial Y_m}{\partial x_2} \right)_{T,P}$$

$$y_2 = Y_m + (1 - x_2) \left(\frac{\partial Y_m}{\partial x_2} \right)_{T,P}.$$

These relations give a straightforward geometric construction to obtain y_1 and y_2 from the curve $Y_m = f(x_2)$. This curve, physically defined for $0 < x_2 < 1$, meets the axis $x_2 = 0$ at the ordinate y_1° and the axis $x_2 = 1$ at the ordinate y_2°. The tangent drawn from a point M on the curve (abscissa x_2, ordinate Y_m) has a slope equal to (dY_m/dx_2); it meets the verticals $x_2 = 0$ and $x_2 = 1$ at ordinates y_1 and y_2 respectively.

APPLICATION TO TERNARY SYSTEMS: With three constituents, the integral molar quantity is given by $Y_m(T, P, x_2, x_3)$:

$$y_1 = Y_m - x_2 \left(\frac{\partial Y_m}{\partial x_2} \right)_{T,P,x_3} - x_3 \left(\frac{\partial Y_m}{\partial x_3} \right)_{T,P,x_2}$$

$$y_2 = Y_m + (1 - x_2) \left(\frac{\partial Y_m}{\partial x_2} \right)_{T,P,x_3} - x_3 \left(\frac{\partial Y_m}{\partial x_3} \right)_{T,P,x_2}$$

$$y_3 = Y_m - x_2 \left(\frac{\partial Y_m}{\partial x_2} \right)_{T,P,x_3} + (1 - x_3) \left(\frac{\partial Y_m}{\partial x_3} \right)_{T,P,x_2}.$$

Geometric interpretation of these expressions is the same as for binary systems: to each point of the *Gibbs triangle* characterizing a mixture (x_2, x_3) corresponds a value of $Y_m = f(x_2, x_3)$ and a point M in the space (x_2, x_3, Y_m). The locus of point M forms a surface meeting verticals from the vertices $A_1(x_1 = 1)$, $A_2(x_2 = 1)$, and $A_3(x_3 = 1)$ of the triangle at points of altitude y_1°, y_2°, and y_3°. The tangent plane drawn

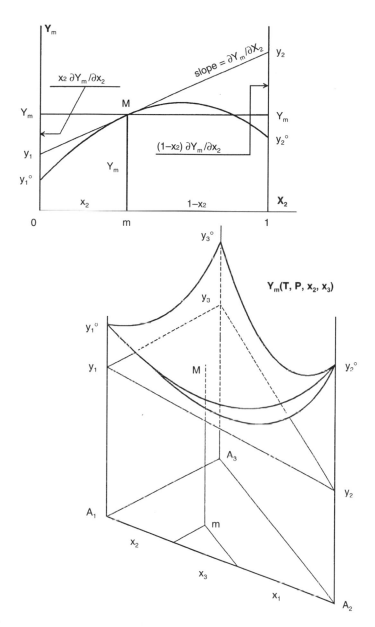

Partial molar quantity: *Geometrical construction giving, for a binary mixture (above), or a ternary mixture (below), partial molar quantities, y_i from integral molar quantity Y_m.*

P

from a point M on the surface meets these three verticals at points of altitude y_1, y_2, and y_3.

Partial molar quantity of dissolution—The name given to a *partial molar quantity* of mixing for a solute A_i. In practice, since a partial molar quantity of dissolution depends not only on the temperature and pressure but also (and mainly) on the composition, it must be measured in such conditions that the addition of a solute A_i only very slightly modifies the overall composition of the mixture. Let y_i and y_i° be respectively the values of Y for one mole of A_i in and out of the mixture; $\Delta y_i = y_i - y_i^\circ$ represents the partial molar quantity of dissolution for the constituent A_i. The quantity of mixing $\delta Y = Y_{final} - Y_{initial}$ related to the transformation:

$$\langle \text{mixture} \rangle + \delta n_i \langle A_i \rangle \rightarrow \langle \text{mixture} \rangle$$

is given by:

$$\delta Y = \left[Y + \left(\frac{\partial Y}{\partial n_i} \right)_{T,P,n_{j \neq i}} \delta n_i \right] - [Y + y_i^\circ \delta n_i].$$

Now, by definition: $y_i = (\partial Y / \partial n_i)_{T,P,n_{j \neq i}}$, whence: $\delta Y = (y_i - y_i^\circ) \delta n_i = \Delta y_i \, \delta n_i$.

$\Delta y_i = (y_i - y_i^\circ)$, the partial molar quantity of dissolution of A_i, is related to $\Delta_{mix} Y$, the integral quantity of mixing, by:

- the definition relation: $\Delta y_i = (\partial \Delta_{mix} Y / \partial n_i)_{T,P,n_{j \neq i}}$;
- the *Euler identity*: $\Delta_{mix} Y = \sum n_i \Delta y_i$;
- the *Gibbs–Duhem equation*: $\sum n_i \, \mathrm{d}(\Delta y_i) = 0$ (T and P constant).

Partial molar volume—Like every *partial molar quantity*, the partial molar volume of a constituent i in a mixture is defined by:

$$v_i = (\partial V / \partial n_i)_{T,P,n_{j \neq i}}.$$

Partial molar volumes v_i and integral volume V are linked by:

- the *Euler identiy*: $V = \sum n_i v_i$;
- the *Gibbs–Duhem equation*: $\sum n_i \, \mathrm{d}v_i = 0$ (T and P constant).

Remark: The partial molar volume represents physically not the volume occupied by one mole of i in a large quantity of mixture but, more accurately, the volume change of a large quantity of mixture after addition of one mole of i. The distinction is not a purely semantic one. Indeed, if V is by nature a positive quantity, there is nothing to prevent

v_i from being a negative one. For instance, the partial molar volume of $MgSO_4$ in water at infinite dilution is negative, which means that an addition of $MgSO_4$ to pure water leads to a decrease in the overall volume of the solution: small doubly charged Mg^{2+} ions are strongly solvated to give $[Mg(H_2O)_6]^{2+}$. The volume of the two complexes $[Mg(H_2O)_6]^{2+}$ and $[SO_4, H_2O]^{2-}$ in a dilute solution is less than the volume of seven molecules of water in the pure solvent.

Partial pressure—The partial pressure of a gas i in a gaseous mixture is the pressure measured on pure gas i occupying the whole volume available to the mixture at the temperature of the mixture. The *membrane* partial pressure of i in the mixture is the pressure measured on pure gas i in equilibrium with the mixture through a semipermeable boundary which can be crossed only by i.

Partial pressure and membrane partial pressure are equal in a mixture of ideal gases or in a mixture of real gases obeying Dalton's law of partial pressures. This law (1805), stating that the total pressure of a gaseous mixture is the sum of the partial pressures, is seldom satisfied by real gases.

The partial pressure of a gas i in a mixture is a function only of T, V, and n_i, and is independent of the composition of the mixture. In contrast, the membrane partial pressure is a function of the overall composition of the mixture. It gives access to the *fugacity* of i in the mixture, provided that an equation of state be known for pure i.

Partial volume—The partial volume of i in a gaseous mixture is the volume occupied by pure gaseous i under the total pressure P of the mixture. The partial volume of one mole of i in a mixture must not be confused with the *partial molar volume* of i in the mixture. The partial volume is known from an equation of state for pure i, and depends only on the properties of i, whereas the partial molar volume of i in a mixture, a function of the composition of the mixture, is defined by:

$v_i = (\partial V / \partial n_i)_{T, P, n_{j \neq i}}.$

The membrane partial volume of i is the volume of pure i in equilibrium with the gaseous mixture through a semipermeable boundary which can be crossed only by i. Molar partial volume and membrane molar partial volume are equal in a mixture of ideal gases or in a mixture of real gases obeying Amagat's law of partial volumes. This law (1893), stating that the volume of a gaseous mixture is the sum of the partial volumes, is generally not satisfied by a mixture of real gases.

P

Partition function—This function Z (from the German: *Zustand-summe*), introduced in distribution laws (see *Microcanonical ensemble*), is defined for a particle by the expression:

$$Z = \sum g_i \exp(-\varepsilon_i/kT),$$

where g_i is *the degeneracy* of the energy level ε_i.

For an ensemble of N particles: $Z_N = Z^N/N!$

The energy levels generally taken into account are translational, rotational, vibrational, electronic, and nuclear levels. Assuming the independence of different levels, the energy of the particle is given by:

$$\varepsilon = \varepsilon_{tr} + \varepsilon_{rot} + \varepsilon_{vib} + \varepsilon_{el} + \varepsilon_{nu},$$

whence:

$$Z = Z_{tr} Z_{rot} Z_{vib} Z_{el} Z_{nu}.$$

The calculation of Z thus requires evaluation of each contribution.

TRANSLATIONAL PARTITION FUNCTION: The translational energy of a particle whose mass is m and projections of the vector velocity are v_x, v_y, v_z, is:

$$\varepsilon(v_x, v_y, v_z) = \frac{1}{2} m(v_x^2 + v_y^2 + v_z^2).$$

The translational partition function is:

$$Z_{tr} = \sum_{v_x} \sum_{v_y} \sum_{v_z} g(v_x, v_y, v_z) \exp\left[-\frac{m}{2kT} (v_x^2 + v_y^2 + v_z^2) \right]$$

$$g(v_x, v_y, v_z) = V\left(\frac{m}{h}\right)^3 dv_x \, dv_y \, dv_z.$$

The above relation, deduced from Heisenberg's uncertainty principle, gives the number of states whose velocity lies between (v_x and $v_x + dv_x$), (v_y and $v_y + dv_y$), (v_z and $v_z + dv_z$). Since energy levels are very close together, it is possible to replace sums by integrals taken over the whole domain of velocities:

$$Z_{tr} = V\left(\frac{m}{h}\right)^3 \iiint \exp\left[-\frac{m}{2kT} (v_x^2 + v_y^2 + v_z^2) \right] dv_x \, dv_y \, dv_z$$

$$Z_{tr} = V\left(\frac{m}{h}\right)^3 \left\{ \int_{-\infty}^{+\infty} \exp\left[-\frac{m}{2kT} v_x^2 \right] dv_x \right\}^3 = V\left(\frac{2\pi mkT}{h^2}\right)^{3/2}.$$

The translational partition function depends on the volume available to the particle, whereas other partition functions depend only on

its properties. Sometimes an internal partition function is defined, as opposed to an 'external' one representing the states of translation:

$$Z_{int} = Z_{rot} Z_{vib} Z_{el} Z_{nu}.$$

ROTATIONAL PARTITION FUNCTION: The rotational energy of a molecule whose moment of inertia is I is given by:

$$\varepsilon_{rot} = J(J+1)h^2/8\pi^2 I = J(J+1)k\,\theta_{rot}.$$

$\theta_{rot} = h^2/8\pi^2 kI$: characteristic temperature of rotation.
J: quantum number of rotation ($g_{rot} = 2J+1$).

Characteristic temperatures of rotation vary from 85 K for hydrogen to a fraction of a kelvin for heavier molecules, whose moment of inertia is higher:

$$Z_{rot} = \sum_{j=0}^{\infty} (2J+1)\exp\left[-\frac{J(J+1)\theta_{rot}}{T}\right].$$

At low temperatures, θ_{rot}/T is high and each term of the sum must be calculated separately; at high temperatures, the sum may be replaced by an integral:

$$Z_{rot} = \int_0^{\infty} (2J+1)\exp\left[-\frac{J(J+1)\theta_{rot}}{T}\right]dJ.$$

$$Z_{rot} = \int_0^{\infty} \exp\left[-\frac{J(J+1)\theta_{rot}}{T}\right]d[J(J+1)]$$

$$Z_{rot} = \frac{T}{\theta_{rot}} = \frac{8\pi IkT}{h^2}.$$

P

When J takes only even or odd values (homonuclear molecules), the symmetry of the molecule divides the partition function by 2:

$$Z_{rot} = T/2\,\theta_{rot}.$$

For a polyatomic molecule, three main moments of inertia and thus three characteristic temperatures of rotation are defined. If $T \gg \theta_{rot}$:

$$Z_{rot} = \frac{\sqrt{\pi}}{\sigma h^3}(I_A I_B I_C)^{1/2}(8\pi^2 kT)^{3/2} = \frac{\sqrt{\pi}}{\sigma}\left(\frac{T^3}{\theta_A \theta_B \theta_C}\right)^{1/2},$$

with $\theta_A = h^2/8\pi^2 kI_A, \ldots$; σ is the number of symmetries of the molecule, that is, 2 for water (one binary axis), 3 for ammonia (one ternary axis), 12 for methane (four ternary axes) or benzene (six binary axes).

If the molecule has a sphere of inertia: $I_A = I_B = I_C = I$:

$$Z_{rot} = \frac{\sqrt{\pi}}{\sigma h^3} (8\pi^2 IkT)^{3/2} = \frac{\sqrt{\pi}}{\sigma} \left(\frac{T}{\theta}\right)^{3/2}.$$

VIBRATIONAL PARTITION FUNCTION: The vibrational energy is given by:

$$\varepsilon_{vib} = (v + \tfrac{1}{2})hv,$$

where v is the quantum number of vibration.

The vibration levels are not degenerate ($g_{vib} = 1$).

$$Z_{vib} = \sum_v \exp\left(-v\frac{hv}{kT}\right) = \sum_v \exp\left(-v\frac{\theta_{vib}}{T}\right).$$

To omit, as we have done, the term $hv/2$ amounts to translating the origin of energies without changing the relative population of energy levels.

Characteristic temperatures of vibration ($\theta_{vib} = hv/k$) are higher than those of rotation (for instance: 6130 K for H_2).

Z_{vib} is easy to obtain without any simplifying assumptions:

$$Z_{vib} = 1 + e^{-\theta/T} + e^{-2\theta/T} + \cdots = \frac{1}{1 - e^{-\theta/T}}.$$

There are as many vibrational partition functions as vibrational modes (theoretically: $3n - 6$ modes for a molecule with n atoms):

$$Z_{vib} = \prod_k \left[1 - \exp\left(-\frac{\theta_k}{T}\right)\right]^{-1}.$$

ELECTRONIC PARTITION FUNCTION: Electronic energy is given by $\varepsilon_{el} = hv$, where v is the frequency of the transition between the level considered and the fundamental level. The degeneracy of each level is expressed by $g_{el} = 2j + 1$, where j is an internal quantum number.

$$Z_{el} = \sum_i (2j_i + 1)\exp\left(-\frac{hv_i}{kT}\right).$$

With $\theta_{el} = hv/k$, the electronic characteristic temperature:

$$Z_{el} = g_0 + g_1 \exp(-\theta_1/T) + g_2 \exp(-\theta_2/T) + \cdots$$

Electronic characteristic temperatures are fairly often high. They may be neglected as soon as $\theta_{el} > 5T$.

NUCLEAR PARTITION FUNCTION: Transitions within nuclei are manifested by emission or absorption of γ-rays, corresponding to very high characteristic temperatures. Therefore they do not intervene in ordinary chemical reactions.

The nuclear partition function is in practice limited to its first term: $Z_{nu} = g_o = 2I_o + 1$. I_o is the quantum number of nuclear spin. Except in special cases (ortho–para hydrogen transition), it is therefore possible to omit Z_{nu} in the expression of a partition function.

The calculation of a partition function, often tedious, is not a pleasure trip! It is performed because it gives immediate access to all the thermodynamic functions:

INTERNAL ENERGY: Given by: $U = \sum n_i \varepsilon_i$. The *Boltzmann, Bose–Einstein* and *Fermi–Dirac distribution* laws give a general expression for n_i:

$$U - \frac{N}{Z} \sum g_i \varepsilon_i \exp(-\varepsilon_i / kT)$$

$$\left(\frac{\partial Z}{\partial T}\right)_V - \frac{1}{kT^2} \sum g_i \varepsilon_i \exp\left(\frac{\varepsilon_i}{kT}\right) = \frac{1}{kT^2} \frac{Z}{N} U$$

$$U = NkT^2 \left(\frac{\partial \ln Z}{\partial T}\right)_V.$$

Applied to monatomic ideal gases, particles without interaction, for which only translational energy must be taken into account, we find:

$$U = \frac{3}{2} NkT.$$

ENTROPY: In *Maxwell–Boltzmann statistics* for indistinguishable particles:

$$S = k \ln \Omega = k \ln \prod (g_i^N / n_i!).$$

With the *Stirling formula*: $\ln N! \approx N \ln N - N$:

$$S = k \sum N_i [1 + \ln(g_i / N_i)].$$

The equilibrium distribution corresponding to the maximum of $S(dS = 0)$ is the Boltzmann distribution:

$$N_i = \frac{N}{Z} g_i \exp(-\varepsilon_i / kT),$$

P

whence g_i/N_i is obtained, then inserted into the expression for S. A short calculation without any difficulties yields:

$$S = k \ln(Z_N) + U/T.$$

$Z_N = Z^N/N!$ is the partition function of N independent particles. S may be expressed as a function of Z, the partition function for one particle:

$$S = Nk\left[T\left(\frac{\partial \ln Z}{\partial T}\right)_V + \ln\left(\frac{Z}{N}\right) + 1\right].$$

By expanding the expression for Z, we get the *Sackur–Tetrode formula*. In Maxwell–Boltzmann statistics for distinguishable particles, the result would be slightly different:

$$S = k \ln \Omega = k \ln\left[N! \prod (g_i^{N_i}/N_i!)\right],$$

whence, after an analogous calculation:

$$S = Nk\left[T\left(\frac{\partial \ln Z}{\partial T}\right)_V + \ln Z\right].$$

The above expression does not assign extensive properties to the entropy, because the mixing of two identical systems makes an entropy of mixing appear:

$$S(2N) - 2S(N) = 2Nk \ln 2.$$

On the other hand, the expression of the entropy for an assembly of indistinguishable particles gives:

$$S(2N) - 2S(N) = 0.$$

When performing this last verification, it must be kept in mind that Z is proportional to the volume: on doubling the number of particles, the volume available to the system is also doubled.

HELMHOLTZ ENERGY: In the above expression of S:

$$S = k \ln(Z_N) + U/T,$$

the *Helmholtz energy* $F = U - TS$ appears. The expansion of $N!$ with the Stirling formula gives at once:

$$F = -kT \ln(Z^N/N!) = -NkT[1 + \ln(Z/N)]$$

$$F = -NkT\left(\ln\frac{Z}{N} + 1\right).$$

Since the expression of S used comes from Maxwell–Boltzmann statistics for indistinguishable particles, it follows that the derived expression of F applies to an assembly of indistinguishable particles. With distinguishable particles, the same calculation would give:

$$F = -NkT \ln Z.$$

GIBBS ENERGY: Let us express dF for a pure substance:

$$dF = -S\,dT - P\,dV + \mu\,dn.$$

μ: *chemical potential* or Gibbs energy for one mole: $G = n\mu$.
n: number of moles, not to be confused with N, number of particles:

$$dn/n = dN/N$$

$$\mu = \left(\frac{\partial F}{\partial n}\right)_{T,V} = \frac{N}{n}\left(\frac{\partial F}{\partial N}\right)_{T,V}.$$

With $F = -NkT[\ln(Z/N) + 1]$, we obtain:

$$G - n\mu = N\left(\frac{\partial F}{\partial N}\right)_{T,V} = -NkT\left(\ln\frac{Z}{N} + 1\right) - N^2 kT\left[\frac{N}{Z}\left(\frac{-Z}{N^2}\right)\right]$$

$$G = -NkT \ln\frac{Z}{N}.$$

The above expression of G is valid for an assembly of N indistinguishable particles. For an assembly of N distinguishable particles:

$$G = -NkT(\ln Z - 1).$$

PRESSURE: It is possible to obtain two expressions:

• From $G = F + PV$:

$$PV = G - F = -NkT \ln(Z/N) + NkT[\ln(Z/N) + 1] = NkT$$

$$P = NkT/V.$$

• From $dF = -S\,dT - P\,dV + \mu\,dn \Rightarrow P = -(\partial F/\partial V)_{T,n}$.
With $F = -NkT[\ln(Z/N) + 1]$, we obtain:

$$P = +NkT\left(\frac{\partial \ln Z}{\partial V}\right)_{T}.$$

Since the two expressions of P are identical:

$$\frac{1}{V} = \left(\frac{\partial \ln Z}{\partial V}\right)_T .$$

Z is therefore bound to be a function of the form: $Z = V \times \Phi(T)$, which has been verified above:

$$Z = Z_{tr} Z_{int} = V \left(\frac{2\pi mkT}{h^2}\right)^{3/2} \cdot Z_{int},$$

the internal partition function being independent of the volume. We verify on the other hand that the expression of P does not depend on the nature, distinguishable or indistinguishable, of the particles.

ENTHALPY: By the definition of the *enthalpy* function:

$$H = U + PV = NkT^2 \left(\frac{\partial \ln Z}{\partial T}\right)_V + NkT,$$

which may be expressed more straightforwardly as a function of $(\partial \ln Z / \partial T)_P$:

$$\ln Z = \ln V + \ln \Phi(T) = \ln (NkT/P) + \ln \Phi(T)$$

$$\left(\frac{\partial \ln Z}{\partial T}\right)_P = \frac{1}{T} + \frac{d \ln \Phi(T)}{dT} = \frac{1}{T} + \left(\frac{\partial \ln Z}{\partial T}\right)_V,$$

whence:

$$H = + NkT^2 \left(\frac{\partial \ln Z}{\partial T}\right)_P,$$

an expression independent of the nature, distinguishable or not, of the particles.

Pascal, Blaise (1623–1662)—French mathematician, physicist and philosopher, founder of the theory of probability, who showed that fluids transmit pressures (solids transmit forces) and demonstrated the variation of atmospheric pressure with altitude. The pascal (symbol Pa) is the name given to the unit of pressure in the SI system:

$$1 \, Pa = 1 \, N \, m^{-2} = 1 \, kg \, m^{-1} s^{-2}.$$

Path—A succession of *states* encountered by a *system* during its evolution.

Peltier effect—The *thermoelectric effect* discovered in 1834, characterized by exchange of heat with the surroundings observed when a current flows across the junction of two unlike conductors. The Peltier coefficient is defined by $\pi_{12}=dQ/dI$. The contribution of each conductor may be individualized: $\pi_{12}=\pi_2-\pi_1$. The Peltier effect is distinguishable from the Joule effect because the sign of dI changes with that of dQ.

Peritectic—From the Greek περι, around, and τηκτος, fusible. The name given to the liquid in equilibrium with a defined compound at its *incongruent* melting temperature. A peritectic transformation is characterized, on heating, by:

$$\langle\text{Solid }\beta\rangle \rightarrow \langle\text{Solid }\alpha\rangle+(\text{Liquid }L).$$

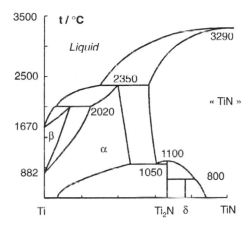

Peritectic: On this example of the titanium–nitrogen diagram, solid solutions α and β present a peritectic at 2350 and 2020°C. The compound δ dismutates with a peritectoid at 800°C Ti₂N undergoes a congruent transition at 1100°C and TiN a congruent melting at 3290°C.

The peritectic is defined by the intersection of the liquidus curves originating from α and β. On a binary temperature–composition diagram, the peritectic is a point characterized by its temperature: the incongruent melting point of the phase β, and by the compositions of the phases L, α and β. On a ternary temperature–composition diagram, the liquidus surfaces meet at a curve, the locus of peritectic points.

By analogy, an incongruent transformation in which the liquid phase is replaced by a solid solution is called peritectoid. On heating, a peritectoid transformation is characterized by:

$$\langle \text{Solid } \beta \rangle \rightarrow \langle \text{Solid } \alpha \rangle + \langle \text{Solid } \gamma \rangle.$$

Permeability—The name given to the ratio of *magnetic field* to *magnetic excitation*: $\mu \equiv \mathcal{B}/\mathcal{H}$.

In an isotropic medium, μ is a scalar; otherwise, μ is a tensor of rank 2. In the SI system, the magnetic permeability is expressed in henries per metre:

$$1\,\text{H}\,\text{m}^{-1} = 1\,\text{m}\,\text{kg}\,\text{s}^{-2}\text{A}^{-2}.$$

The permeability of vacuum is exactly: $\mu_\text{o} = 4\pi \times 10^{-7}\,\text{H}\,\text{m}^{-1}$; the ratio $\mu_\text{r} = \mu/\mu_\text{o}$ is the relative permeability of a material.

Permeable—From the Latin *permeare*, to go through. A boundary is said to be permeable when it may be crossed by matter; if it may be crossed by heat, it is better to speak of a *diathermanous* boundary.

Permittivity—The name given to the ratio of *electric excitation* to *electric field*: $\varepsilon \equiv D/E$.

ε is not the analogue of *permeability* μ, but of its inverse, μ^{-1}. In an *isotropic* medium, ε is a scalar; otherwise, ε is a tensor of rank 2. In the SI system, the permittivity is expressed in farads per metre:

$$1\,\text{F}\,\text{m}^{-1} = 1\,\text{A}^2\,\text{s}^4\,\text{kg}^{-1}\,\text{m}^{-3}.$$

In vacuum, permittivity and permeability are linked by: $\varepsilon_\text{o}\mu_\text{o} = c^{-2}$, where c is the velocity of light in vacuum:

$$\varepsilon_\text{o} = 1/\mu_\text{o}c^2 \approx 1/(36\pi \times 10^9) \approx 8.854 \times 10^{-12}\,\text{F}\,\text{m}^{-1}.$$

In a material medium, the ratio $\varepsilon_\text{r} = \varepsilon/\varepsilon_\text{o}$ is the relative permittivity, or 'dielectric constant', of the material.

Perpetual motion—The finest dream of mankind since the loss of the Garden of Eden is to have energy available without paying a price. Perpetual motion engines aim at succeeding in this. Engines said to be of the first kind work by violating the first law of thermodynamics and claim they create energy *ex nihilo*. Such an engine is nicely illustrated by a well-known lithograph of the Dutch engraver M.C. Escher (1961). Engines said to be of the second kind work by violating the second law. An ever-popular example, found in an old physics book of my grandfather, is that of a boat which takes its energy out of the sea and makes headway by dropping blocks of ice! Scientists did not wait until 1850, the date of the statement of the first two laws, to be convinced of the uselessness of such efforts. As early as 1775, the Academy of Sciences at Paris took the decision to not examine inventions concerning perpetual motion engines.

pH—Introduced in 1909 by the Danish chemist Peter Sörensen (1868–1939), the pH is defined by:

$$pH = colog_{10} a(H_3O^+) = -\log_{10} a(H_3O^+),$$

the *reference state* being the ion H_3O^+, in an infinitely dilute solution in water, with a reference concentration $c^U = 1 \, mol \, L^{-1}$. The pH, characterizing the acidity of a solution, decreases when the medium becomes more acid. In practice, in an aqueous medium, $0 < pH < 14$. An ionic concentration of H_3O^+ higher than $1 \, mol \, L^{-1}$ is not forbidden, but, taking into account the activity coefficient, the *activity* of H_3O^+ ions hardly exceeds unity.

Remark: the 'p' of pH is taken as meaning 'potential'. This choice, accepted through use, is nevertheless unfortunate because the *chemical potential* of H_3O^+ ions is low when the pH is high. This is surely why the 'p' is a small letter!

P

Phase—A phase is a *homogeneous* part of a mixture. Homogeneity is usually judged under the microscope. It is clear to the naked eye that a water–ice mixture is two-phase. The same may be true for the three phases (quartz, mica and feldspath) which make up granite, but it is generally necessary to use more advanced tools (electron microscopy, X-ray diffraction, etc.) to distinguish the different phases of a mixture. Homogeneous differs from *uniform*, because homogeneity does not preclude gradients of intensive quantities.

Phase rule—This very popular rule, stated by Gibbs, gives the *variance* of a system, defined by:

$$v = \sum (parameters) - \sum (relations).$$

The parameters describing the system are intensive parameters, that is, temperature, pressure and composition:

$$\sum(\text{parameters}) = 2 + \varphi(\varsigma - 1).$$

We must in fact know $(\varsigma - 1)$ concentrations in each of the φ phases to reconstitute the system. The ς constituents must be 'independent'. Otherwise we must subtract from c, the number of constituents, the number of chemical equilibria and the constraints imposed on the system:

$$\varsigma = c - \sum(\text{reactions}) - \sum(\text{constraints}).$$

Thermodynamics imposes at equilibrium the equality of *chemical potentials* of each constituent in all phases, that is:

$$\sum(\text{relations}) = \varsigma(\varphi - 1),$$

because each constituent needs $(\varphi - 1)$ relationships to express the equality of its chemical potential in each of the φ phases. The variance of the system is thus given by:

$$v = \varsigma + 2 - \varphi.$$

For instance, a ternary system $(\varsigma = 3)$ contains at most five phases:

• If $\varphi = 5$, $v = 0 \Rightarrow 12$ parameters $(T, P$ and 5×2 compositions) fixed
• If $\varphi = 4$, $v = 1 \Rightarrow 9$ relationships between the 10 parameters
• If $\varphi = 3$, $v = 2 \Rightarrow 6$ relationships between the 8 parameters
• If $\varphi = 2$, $v = 3 \Rightarrow 3$ relationships between the 6 parameters
• If $\varphi = 1$, $v = 4 \Rightarrow 4$ parameters $(T, P$ and 2 compositions) independent.

Phase space—When the state of a system is determined by n coordinates of position and n coordinates of velocity, it may be represented by a point in a space called 'phase space'. From this concept, introduced by Gibbs in 1901, it is possible to define two kinds of phase space:

The μ (molecular) space has a number of dimensions sufficient to determine the configuration of one molecule by one point in the space. The state of an assembly of N molecules will be given by N points.

The γ (gaseous) space has a number of dimensions such that one point in the space gives the state of N particles. With one constraint, the point travels into a $(6N - 1)$ dimensional space, called a 'surface' of the γ space of $6N$ dimensions.

Phase transition—A general expression to denote the transformation of a macroscopic ensemble of thermodynamic objects, or of an ensemble of shapes, into an another distinct ensemble. Such a definition includes allotropic transitions and change from a homogeneous structure to a heterogeneous one (precipitation, *unmixing*, etc.). The analogy with first- or second-order transitions may be found in the kinetic description of the *phase space*.

Phenomenological coefficient—During an irreversible process, the *entropy production* is expressed by a sum of products of *generalized forces* (or potential gradients, or generalized affinities) X_i by the corresponding *fluxes* (or *generalized currents*) J_i:

$$\dot{\sigma} = \frac{d\sigma}{dt} = \sum_i J_i X_i > 0.$$

At equilibrium, each flux becomes zero together with the forces which produce it. Generally, relationships between fluxes and forces are complex. However, it is possible to use a limited expansion of fluxes as a function of forces and to imagine, close to equilibrium, that the expansion may be limited to the first term. Relations between fluxes and forces are then linear:

$$J_i = \sum_k L_{ik} X_k.$$

The coefficients L_{ik} are called phenomenological coefficients. Coefficients L_{ii} are the direct, or proper, coefficients. They express the direct relationship between conjugated fluxes and forces; for instance, a temperature gradient causes a heat flux. Coefficients $L_{ik}(i \neq k)$ are cross, or mutual, or coupling coefficients; they express coupling of two irreversible processes i and k; for instance, the formation of a concentration gradient may be caused by a temperature gradient.

The evolution criterion $\sum J_i X_i > 0$ applied to only one process yields $L_{ii} > 0$. Proper coefficients are positive. The same criterion applied to two processes i and k yields:

$$L_{ii} X_i^2 + (L_{ik} + I_{ki}) X_i X_k + L_{kk} X_k^2 > 0.$$

Coupling coefficients may be positive or negative, subject to the above condition, which implies: $(L_{ik} + L_{ki})^2 < 4 L_{ii} L_{kk}$.

If a magnetic process is not involved, the cross coefficients L_{ik} and L_{ki} obey the *Onsager reciprocity* relations: $L_{ik} = L_{ki}$.

P

The linear thermodynamics of irreversible processes, in which relations between fluxes and forces are linear by means of phenomenological coefficients, applies close to equilibrium. The range of validity may be large (Fourier's law of thermal conductivity, Fick's laws of diffusion), but it is often very narrow (chemical reactions for instance).

Phenomenon—From the Greek φαινομενον, that which is shown, φαινειν, to show. An observable, experimental fact or event.

Phlogiston—From the Greek φλογιστος, flammable. The theory of phlogiston was developed in 1697 by the German chemist Georg Ernst Stahl (1660–1734) in order to explain combustion. Phlogiston is a massless substance present in every combustible material, the substances richest in phlogiston burning best. According to Stahl, metals are formed of calx and phlogiston, combustion being a loss of phlogiston. In 1756, Lomonosov rejected the theory with the argument that metals increase in weight during combustion. In 1783, Lavoisier, by showing that combustion always involved a reaction with oxygen, overthrew the theory definitively.

pK—By analogy with the definition of pH, pK is given by:

$$pK = \operatorname{colog}_{10} K = -\log_{10} K.$$

The concept of pK may be applied to acids and bases:

$$AH + H_2O \rightarrow A^- + H_3O^+ \Rightarrow K_a = [a(H_3O^+)a(A^-)]/[a(AH)a(H_2O)]$$

$$A^- + H_2O \rightarrow AH + OH^- \Rightarrow K_b = [a(AH)a(OH^-)]/[a(A^-)a(H_2O)].$$

pK_a and pK_b are linked by:

$$pK_a + pK_b = pK_e = 14 (\text{at } 24\,^\circ C).$$

Owing to the above relation, the two concepts of pK_a and pK_b are redundant. Acids are thus classified according to their pK_a. Bases are classified according to the pK_a of their conjugated acid. In an aqueous medium:

- p$K_a < 0$: strong acids and negligible bases
- $3 < pK_a < 11$: weak acids and weak bases
- p$K_a > 14$: negligible acids and strong bases.

Planck function—The function $Y = -G/T$. Introduced in 1869, together with the *Massieu function* ($J = -F/T$), and popularized by Planck, it is no longer used today.

Planck, Max (1858–1947)—German physicist, who in 1900 introduced the constant that now bears his name to account for the thermal radiation emitted by a black body. By showing that light absorption or emission necessitates the implementation of discrete quantities of energy, he gave birth to the quantum theory and earned the Nobel prize for physics in 1918.

Plasma—A plasma is an electrically conducting medium, generally made up of electrons, cations, and neutral particles. Its properties depend on the collective behaviour of charged particles. Each type of particle may be characterized by its kinetic temperature and by its kinetic pressure:

$$u = 3kT/2, \qquad P = NkT/V.$$

At thermodynamic equilibrium, all the temperatures are the same. The most important parameter in a plasma is the *Debye length l*, the distance beyond which a positive ion no longer has any influence on an electron:

$$l = \left(\frac{\varepsilon_0 kT_e}{n_e q_e^2}\right)^{1/2} \approx 69 \sqrt{\frac{T_e}{n_e}},$$

where ε_0, k, T_e, n_e and q_e represent respectively the permittivity of vacuum, the Boltzmann constant, the electron temperature, the electron density (number of electrons per m^3) and the elementary charge. The Debye length allows a more accurate definition of a plasma: 'A conducting medium is called a plasma when its dimensions are larger than the Debye length'. Plasmas are also characterized by the angular frequency of plasma oscillations:

$$\omega/\text{rad s}^{-1} = \left(\frac{n_e q_e^2}{m_e \varepsilon_0}\right)^{1/2} \approx 56 \sqrt{n_e}.$$

The plasma state, sometimes described as the fourth state of matter, is by far the commonest in the Universe. It is encountered:

- at the centre of stars ($T_e \approx 2 \times 10^7$ K, $n_e \approx 10^{31}$ m^{-3})
- on the surface of stars ($T_e \approx 10^4$ K, $n_e \approx 10^{20}$ m^{-3})
- in interstellar clouds ($T_e \approx 10^2$ to 10^4 K, $n_e \approx 10^2$ to 10^6 m^{-3})
- in intergalactic space ($T_e \approx 3$ K, $n_e \approx 1$ m^{-3})
- inside metals ($T_e \approx 10^4$ K, $n_e \approx 10^{28}$ m^{-3})
- in plasma torches, lightning strokes ($T_e \approx 2 \times 10^4$ K, $n_e \approx 10^{22}$ m^{-3}).

Poincaré's recurrence—From Latin *recurrens*, running back. Poincaré's theorem [Henri Poincaré, French mathematician, (1854–1912)] states that if a mechanical system passes through a succession of states between two times t_1 and t_2, it will repass after a finite time through a succession of states arbitrarily close to the succession observed between the times t_1 and t_2.

Let us imagine a box with two compartments, and a gas confined in one of them. At time $t = 0$, communication is established and the gas expands into the whole box. Poincaré's theorem states that after a finite time, all the molecules will be found in the same compartment, which seems to contradict the *second law*, because it would imply a spontaneous decrease of the *entropy* of the system.

It must be kept in mind that the second law has a probabilistic character. It does not preclude an isolated system, characterized by its entropy S_1, from spontaneously achieving a state whose entropy is $S_2 < S_1$, but gives the probability P of such an event:

$$P = \exp[(S_2 - S_1)/k],$$

k being the *Boltzmann constant*. When an isolated system contains a large number of particles, the probability of a significant decrease in its entropy becomes very low. If the above system holds 10^{24} particles, the time needed for all the particles to return spontaneously to the same compartment is of the order of $10^{10^{24}}$ seconds, a time which must be compared with the 5×10^{17} seconds representing the widely accepted age of the Universe!

Poisson distribution—The name given to the *binomial distribution* in the limiting case where, the number of trials N tending towards infinity, the probability p of an event A tends towards zero, but the product $N \times p$ tends towards a finite value $\lambda \ll N$. The most probable value of n_A is then:

$$\langle n_A \rangle = Np = \lambda.$$

The probability of an event A in the binomial distribution is given by:

$$P_N(n_A) = \frac{N!}{n_A! n_B!} p^{n_A} q^{n_B}.$$

The application of the *Stirling formula* to $N!$ and $n_B!$, because n_A is not a large number: $N! \approx \sqrt{2\pi N}\,(N/e)^N$, yields:

$$P_N(n_A) = \lambda^{n_A} \frac{e^{-\lambda}}{n_A!}.$$

We can verify the normalization condition:

$$\sum_{n_A = 0}^{\infty} \lambda^{n_A} \frac{e^{-\lambda}}{n_A!} = e^{\lambda} e^{-\lambda} = 1.$$

The Poisson distribution is determined by a knowledge of only the first *moment*: $\langle n_A \rangle = \lambda$.

Polarization—The presence of a dielectric material in an *electric field* has the effect of decreasing the field. The electric polarization is defined as the *electric moment* per unit volume induced by the electric field in the material. Using the example of a plane capacitor with a charge $\mathcal{Q}$, volume V, and distance between electrodes d, we have, in vacuum: $\varepsilon_o E = \mathcal{Q}d/V$, the electric moment per unit volume of the charged capacitor. In the presence of a dielectric material, $\varepsilon_o E$ has two contributions, that of the charge between the electrodes and that of the polarization P of the material:

$$\varepsilon_o E = \frac{\mathcal{Q}d}{V} - P.$$

Polytropic—A transformation is called polytropic when it proceeds at constant *heat capacity*: $c = dQ/dT = \text{const.}$ *Isothermal* $(c = \infty)$ and *adiabatic* $(c = 0)$ transformations represent two particular cases of polytropic transformations. *Isochoric* or *isobaric* transformations are polytropic only with the approximation c_V or c_P constant. If we take one of the expressions of dQ from *calorimetric coefficients*:

$$dQ = c_V \, dT + (c_P - c_V) \left(\frac{\partial T}{\partial V} \right)_P dV,$$

we obtain, with $dQ/dT = c$ for a polytropic transformation:

$$\frac{c - c_V}{c_P - c_V} = \left(\frac{\partial T}{\partial V} \right)_P \left(\frac{\partial V}{\partial T} \right)_c.$$

This relationship, integrated for an ideal gas with c_P and c_V constant, yields the equation of polytropic curves:

$$PV^{\gamma'} = \text{const.} \quad \text{with:} \quad \gamma' = \frac{c_P - c}{c_V - c}.$$

We naturally obtain the equation for adiabatic curves with $c = 0$ and the equation for isothermal curves with $c = \infty$.

Pomeranchuk effect—The name given to the decrease in temperature observed below 0.35 K during the solidification of liquid helium-3 by compression. Such paradoxical behaviour is explained by the shape of the liquid–solid equilibrium curve for helium-3, which shows a minimum at $T = 0.35$ K and $P = 29.5$ bar. Recall the *Clapeyron equation*:

$$\Delta_{fus} H = T(v_l - v_s) \left(\frac{\partial P}{\partial T} \right)_{eq}.$$

As $(v_l - v_s) > 0$ whatever the temperature, the enthalpy of melting of helium-3 becomes zero at 0.35 K and negative below this temperature. Heating the liquid below 0.35 K causes it to solidify! When the solidification is induced by compression, the temperature decreases.

Popper, Karl Raimund (1902–1994)—Austrian philosopher, who introduced the concept of falsifiability (*Logik der Forschung*, 1935). According to Karl Popper, a hypothesis is truly scientific only if it is *falsifiable*.

Postulate—From the Latin *postulatum*, request. An unproved statement taken as the premise of an argument. A postulate is not as evident as an *axiom* and may be more easily questioned. The postulates of quantum mechanics underlie any presentation of statistical thermodynamics. Also, the thermodynamics of irreversible process rests on the postulate of *local equilibrium*.

Potential—From the Latin *potentialis*, existing in possibility. A *field* of vectors F derives from a potential V if it is irrotational (curl $F = 0$), otherwise, if it is possible to write $F = \text{grad} V$:

$$F = \left(\frac{\partial V}{\partial x} \right)_{y,z} i + \left(\frac{\partial V}{\partial y} \right)_{x,z} j + \left(\frac{\partial V}{\partial z} \right)_{x,y} k.$$

V is a scalar potential. If a field of vectors B may be described by a rotation: $B = \text{curl} A$, A is a vector potential.

Potential energy—To raise a mass m by a height dz in a gravitational field g, one must provide it with *work*: $dW = mg\, dz$. If g is constant, the *energy* of a system transferred from a height $z = 0$ to a height z changes by mgz. With respect to the level $z = 0$, the system acquires a potential energy:

$$E_p = mgz.$$

The potential energy of a system placed in an external field contributes to its *external energy* provided that the field has no effect on its microscopic properties, which is the case with a gravitational field. With a *magnetic field* inducing a *magnetization* or an *electric field*, inducing *polarization* of the material, a fraction of the potential energy contributes to the *internal energy* of the system.

Pourbaix diagram—Or 'potential–pH diagram', in which *electrode potential*, given by the *Nernst relation*, is expressed as a function of pH at a constant temperature, generally 25 °C. Such a diagram allows the stability domains of various species in solution to be visualized. Consider, for instance:

$$Ox + zH_3O^+ + ne^- \rightarrow Red.$$

The electrode potential at 25°C is given by:

$$E = E^\circ + \frac{0.059}{n} \log_{10} \frac{[Ox]}{[Red]} - 0.059 \frac{z}{n} pH.$$

If the Ox and Red species are dissolved, it is natural to consider that the boundary separating the two stability domains is defined by the condition [Ox] = [Red], and thus by the straight line of the equation:

$$E = E^\circ - 0.059 \cdot (z/n) \cdot pH.$$

On a Pourbaix diagram, two important straight lines are those delimiting the stability domain of water. Water may indeed be oxidized at high potential or reduced at low potential:

- the line of oxidation of water: $6H_2O \rightarrow O_2 + 4H_3O^+ + 4e^-$ has the equation: $E = 1.23 - 0.059$ pH (equilibrium H_2O/O_2 under $P_{O_2} = 1$ bar);
- the line of reduction of water: $2H_3O^+ + 2e^- \rightarrow H_2 + 2H_2O$ has the equation: $E = -0.059$ pH (equilibrium H_2O/H_2 under $P_{H_2} = 1$ bar).

Between these two lines, species in solution may neither oxidize nor reduce water; the *overvoltage* phenomenon, important for gases, may slow down the reaction of water when the solution potential is less than 0.5 V from the above lines. It is thus possible to retain oxidizers such as MnO_4^- or reducing agents such as Ni or Fe, temporarily in water.

The position of a boundary may be somewhat arbitrary. Consider, for instance, the equilibrium Cu/Cu^{2+} ($E^\circ = +0.34$ V). It is possible to say that metallic copper is stable when it is in equilibrium with Cu^{2+}

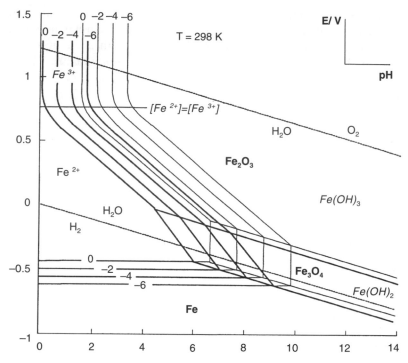

Pourbaix diagram: *Visualization of the stability domain of water, between the two parallel dashed lines, of the metastable diagram of iron (thin lines), and of the stable diagram of iron (bold lines). Hydroxides $Fe(OH)_2$ and $Fe(OH)_3$ are metastable phases, whereas oxides Fe_2O_3 and Fe_3O_4 are stable phases. Numbers (0, −2, −4, −6) represent $log_{10}([Fe^{2+}]+[Fe^{3+}])$. The horizontal $E = +0.77V$ separates the domains of predominance of Fe^{2+} and Fe^{3+} ions.*

ions whose *activity* is less than 10^{-2}, 10^{-4}, or 10^{-6}. Depending on the choice, the boundary separating the stability domain of Cu from that of Cu^{2+} ions will be the horizontal $E = +0.31$, $E = +0.28$, or $E = +0.25\,V$.

Power—Energy exchanged during unit time. In the SI system, power is expressed in watts:

$$1\,W = 1\,J\,s^{-1} = 1\,\Omega\,A^2 = 1\,kg\,m^2\,s^{-3}.$$

Poynting effect—The name given to the increase in saturated vapour pressure with the total pressure imposed. This effect was demonstrated by John Henry Poynting (1852–1914), English physicist, author of works on the flow of electric energy, radiation pressure, the density of the earth, and the gravitational constant. Let $p°$ be the saturated vapour pressure of a substance subjected to this pressure alone:

$$\mu_l° = \mu_v° + RT \ln p°.$$

Now, if the liquid is subjected to the pressure P of an inert gas, its *chemical potential* increases. That of the vapour in equilibrium must also increase in order to maintain the equilibrium with the liquid. Let p be the new vapour pressure. The equality of chemical potentials for the new equilibrium yields:

$$\mu_l° + \int_{p°}^{P} v_l\,dP = \mu_v° + RT \ln p.$$

By subtracting the two relations, we get: $\int_{p°}^{P} v_l\,dP = RT \ln p/p°$, a relation which may be expressed in differential form: $v_l\,dP = v_v\,dp$, v_l and v_v being respectively the molar volumes of the liquid and of the vapour.

ppb—Parts per 10^9 (billion in the USA and now generally elsewhere in the English-speaking world; milliard or thousand million otherwise); always represents *mass fractions*.

ppm—Parts per million, always a *mass fraction*:

$$1000\,\text{ppm} - 1\% = 0.1\,\text{mass}\%.$$

Pressure—Pressure, a quantity of *tension* whose conjugated *extensity* is volume, is force per unit surface. In the SI system, pressure is expressed in pascals:

$$1\,\text{Pa} = 1\,\text{N m}^{-2} = 1\,\text{kg m}^{-1}\text{s}^{-2}.$$

Pressure–temperature diagram—This diagram generally represents the stability domains of various polymorphic phases for a pure substance. The *bivariant* domains are separated by *monovariant* equilibrium curves whose equation is given by integration of the *Clapeyron equation*:

$$(\partial P/\partial T)_{eq} = \Delta_{tr} H/T\Delta_{tr}V.$$

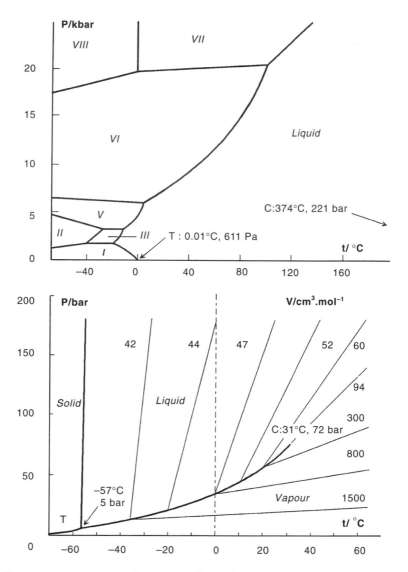

Pressure–temperature diagram: *Above, diagram of water showing seven stable varieties of ice. Ice IV is metastable. At the scale of the diagram, the vapour domain is confused with the abscissa. Below, diagram of CO_2 showing a few isochoric curves. The pressure of the triple point (5 bars) is exceptionally high. The saturating vapour pressure of dry ice (solid CO_2) is 1 bar at $-78°C$.*

A liquid–vapour equilibrium curve stops at a critical point, for which $\Delta_{vap}V = 0$ and $\Delta_{vap}H = 0$. Theoretically, there are no critical points for transitions between condensed phases, because they generally do not present the same elements of symmetry.

Three monovariant equilibrium curves meet at a point of *invariant* equilibrium, or triple point. The tangents to any two curves drawn from a triple point meet at an angle less than 180°. To be convinced of this, it is sufficient to extend any of the three curves into its metastable domain, beyond the triple point. The extended α–β equilibrium curve can be neither in the α domain nor in the β domain, because that would mean that α and β could share not only a common boundary but also a common stability domain, which contradicts the *phase rule*. The extended α–β equilibrium curve must therefore penetrate into the γ domain.

Remark 1: Whatever the nature of the diagram, the junction at a triple point of two curves at an angle greater than 180° leads to a particularly ugly drawing. Now what is ugly is nearly always wrong. Naturally, the aesthetic criterion, which is entirely subjective, is not a scientific one, but it helps in asking pertinent questions.

Remark 2: On a pressure–temperature diagram, we can verify the following statements, which are easy to prove directly:

- the saturated vapour pressure of a *metastable* phase is always higher than that of a stable phase;
- the melting temperature of a metastable phase is always lower than that of a stable phase.

Pressure–volume diagram—See *Clapeyron diagram* (P vs V) and *Amagat diagram* (PV vs P).

Prigogine, Ilya—Belgian thermodynamicist, born in 1917, who has made an important contribution to the non-linear thermodynamics of irreversible processes and developed mathematical models of what he calls *dissipative structures*. Nobel prize for chemistry in 1977.

Principle—A principle is a general statement, unproved, but justified *a posteriori* by verification of its consequences. There is a difference in nature between *axiom, postulate*, and principle. A principle is more general than a postulate and its refutation would lead to the collapse of a whole edifice. Thermodynamics rests on four principles, stated for the first time by Clausius (first and second law, 1850), by Maxwell (zeroth law, 1872), and by Nernst (third law, 1918):

• The *zeroth law* asserts that it is possible to measure a temperature.
• The *first law* formulates energy conservation.
• The *second law* provides the key to the evolution of a system.
• The *third law* states the inaccessibility of absolute zero.

A structure, satisfying though it may appear, can always be improved. Indeed, some epistemologists think that 'with four principles, there are three too many!'; several attempts have been made to reduce the basis of thermodynamics to only one principle. The most convincing result has been obtained by Hatsopoulos and Keenan, who in 1972 stated the *law of stable equilibrium*. Sometimes one comes across, as a *fourth law*, statements which are often only conjectures.

Probability—In mathematics, the probability of an event E is the limit towards which tends the ratio of the number Ω of cases in which E is observed to the number Σ of experiments realized when $\Sigma \rightarrow \infty$. To evaluate a probability without doing any experiment, it is possible to calculate the ratio of the number Ω of favourable cases of the occurrence of E to the number Σ of possible cases. Such a calculation implies naturally that every possible event be equally probable. Probabilities are doubtless the most attractive and the easiest branch of mathematics, as shown by the passion of our society for horse races and games of chance!

The mathematical probability $P = \Omega/\Sigma$ of an event E is a number between 0 (impossible event) and 1 (certain event). In thermodynamics, the probability of a macroscopic state E equals the number Ω of *complexions* or *microscopic states* allowing its realization. A thermodynamic probability is thus a number greater than 1. The elimination of the denominator Σ is justified because, for large values of Σ (in thermodynamics, Σ is easily of the order of $10^{10^{24}}$), near equilibrium Ω is very close to Σ.

It is easy to be convinced of this by the simple example of a box divided into two identical compartments, each holding 10^{24} molecules of the same gas. If communication is established between the two compartments, the most probable distribution will be 10^{24} molecules in each one, with a *standard deviation* of 7×10^{11}. Now the probability that an actual distribution differs from the most probable one by more than 100 times the standard deviation, that is, that the pressures in the two compartments differ by more than $7 \times 10^{13}/2 \times 10^{24} \approx 3.5 \times 10^{-11}$ in relative value, is of the order of 10^{-2173}!

As a consequence, equipartition of the molecules (that is, $10^{24} \pm 7 \times 10^{13}$ molecules per compartment) corresponds to near-certainty. Any other distribution has such a low probability that it may be considered as unlikely, even if in rigorous terms its probability is not exactly zero.

Let us calculate now the probability $P = \Omega/\Sigma$ of the above equilibrium state ($N = 10^{24}$ molecules in each compartment):

$$P = \frac{\Omega}{\Sigma} = \frac{C_N^{2N}}{2^{2N}} = \frac{1}{2^{2N}} \cdot \frac{(2N)!}{N! \, N!}.$$

The use of the *Stirling formula*: $\ln (N!) \approx N \ln N - N$, yields:

$$\ln \Omega \approx \ln \Sigma = 2N \ln 2, \text{ or } P \approx 1.$$

This result justifies neglecting Σ in defining a thermodynamic probability. The equilibrium state, given by seeking the maximum of Ω, corresponds to a state for which $\Omega \approx \Sigma$.

Remark: If we had continued the Stirling expansion further, the result of the calculation would have been $P \approx 0$. This result is logical because the probability of strict equipartition of the molecules ($10^{24} \pm 1$ molecules in each compartment) is effectively vanishingly small. However, knowledge of the exact value of such a probability is devoid of significance. If our perception of the outer world allows us to consider that the equality $N_A = N_B$ is satisfied when $N_A \approx N_B \approx 10^{24}$ within about $\pm 10^{18}$, and if calculation shows us that the state $N_A \approx N_B \approx 10^{24}$ to $\pm 10^{14}$ represents a near-certainty, logic dictates that the approximate result $P \approx 1$ has a greater significance than the more rigorous result $P \approx 0$!

Probability density—When a variable X is continuous, it is not possible to define a *probability* for X to take exactly the value x, because in the strictest sense this probability equals zero. It is nevertheless possible to define a probability density function $P(x)$. In such conditions, $P(x) \, dx$ represents the probability for the variable X to take a value included between x and $x + dx$ with $\int P(x) \, dx = 1$.

Process—A process (from the Latin *procedere*, to proceed) is defined by an initial state, a final state, and the path followed. A process must not be mistaken for a *transformation*, which is defined by only its initial and final states.

Product—A general name to designate a defined compound, used instead of 'chemical product'. In a *chemical reaction* $\sum v_i M_i = 0$, the 'initial products', or 'reactants', refer to the constituents M_i on the left-hand side ($v_i < 0$) whereas the 'final products', or simply 'products', refer to the constituents M_i on the right-hand side ($v_i > 0$).

Proper—In thermodynamics, this adjective has no universally accepted meaning. Some authors, with Guggenheim, use it with the meaning of 'divided by amount of substance'. Others, with Prigogine, apply it to a pure substance and mark it with an asterisk, for instance:

$$\mu^*(T, P) = \mu^\circ(T) + \int_{P^\circ}^{P} V \, dP.$$

Psychrometry—From the Greek $\psi v \chi \rho o \varsigma$, cold. The name given to the study of air–water vapour mixtures in conditions close to those prevailing over the earth's surface. The measure of degree of humidity by psychrometry is associated with the decrease in temperature caused by water evaporation. When the ambient atmosphere is dry, evaporation is rapid, so the decrease in temperature is large. In a medium saturated with humidity, water does not evaporate, so the temperature does not decrease.

Pure substance—A material formed from only one constituent, as opposed to a *mixture*. A physical quantity related to a pure substance is symbolized by an asterisk, for instance:

$$\mu^*(T, P) = \mu^\circ(T) + \int_{P^\circ}^{P} V \, dP.$$

From a strict thermodynamic point of view, a pure substance does not exist! The *Gibbs energy* of the transformation:

$$\langle A \rangle + \delta n \langle B \rangle \rightarrow \langle \text{Mixture} \rangle$$

is given by:

$$\delta G = \delta n \cdot \Delta \mu_B = \delta n \cdot RT \ln a_B.$$

δn being fixed, $\delta G \rightarrow -\infty$ when $a_B \rightarrow 0$, that is, when B becomes infinitely dilute. It is thus impossible to imagine a physical process separating a mixture into its strictly pure constituents.

Q

Quantity of mixing—This expression may designate an *extensive* quantity Y, an *intensive* one y_i, characterizing a mixture, or their change $\Delta_{\text{mix}}Y$ or Δy_i during a mixing process.

Some extensities Y, such as volume or entropy, may be experimentally determined, which gives access to $\Delta_{\text{mix}}Y$ by straightforward subtraction. When Y is defined in relative value, which is the case for most of thermodynamic functions, experiment gives only $\Delta_{\text{mix}}Y$. Thermodynamicists, even beginners, must not feel annoyed by this distinction, because they may often arbitrarily put $Y° = 0$!

On the other hand, to mix water and alcohol with the aim of preparing a product with attractive organoleptic properties and to put some bleach into a swimming pool to improve its bacteriological quality represent two processes of a very different nature.

In the first case, the properties of the product (brandy, perhaps) will differ from those of its constituents: the quantities of mixing of interest will be integral quantities of mixing, symbolized by $\Delta_{\text{mix}}Y$. In the second case, the state of the swimming pool after bleaching will differ little from its initial state and the integral quantities of mixing will be of little consequence. On the other hand the bleach, concentrated before mixing and dilute after bleaching, will have seen its state modified. The quantities of interest will be *partial molar quantities* of mixing related to bleach i, defined by:

$$\Delta y_i = (\partial \Delta_{\text{mix}}Y/\partial n_i)_{T,P,n_{j \neq i}}.$$

There is a close relationship between integral quantities and partial molar quantities. It is the *Euler identity*:

$$\Delta_{\text{mix}}Y = \sum n_i \Delta y_i$$

Classical relations between thermodynamic functions apply naturally to the integral quantities as well as to the partial molar quantities. For instance, from $\Delta_{\text{mix}}G$ or $\Delta\mu_i$, we derive:

$$\Delta_{\text{mix}}S = -\left(\frac{\partial \Delta_{\text{mix}}G}{\partial T}\right)_P \qquad \Delta s_i = -\left(\frac{\partial \Delta\mu_i}{\partial T}\right)_P$$

$$\Delta_{\text{mix}}V = +\left(\frac{\partial \Delta_{\text{mix}}G}{\partial P}\right)_T \qquad \Delta v_i = +\left(\frac{\partial \Delta\mu_i}{\partial P}\right)_T$$

$$\Delta_{mix}H = -T^2\left[\frac{\partial}{\partial T}\left(\frac{\Delta_{mix}G}{T}\right)\right]_P \qquad \Delta h_i = -T^2\left[\frac{\partial}{\partial T}\left(\frac{\Delta\mu_i}{T}\right)\right]_P.$$

We verify: $\Delta_{mix}G = \Delta_{mix}H - T\Delta_{mix}S$ and $\Delta\mu_i = \Delta h_i - T\Delta s_i$.

It is quite easy to obtain experimentally a curve of $\Delta_{mix}G$ (or $\Delta\mu_i$) as a function of composition. However, the above functions, although rigorous, are inaccurate for providing access to volumes, enthalpies, or entropies of mixing. It is better to use them to verify the coherence of experimental measurements of volumes or enthalpies of mixing. Entropies of mixing, hard to measure directly, are obtained from: $\Delta_{mix}S = (\Delta_{mix}H - \Delta_{mix}G)/T$.

Quantum state—Schrödinger's equation, which defines the wave function associated with a particle, has real solutions only for discrete values of physical quantities characterizing the particle, such as *energy*, *angular momentum*, or *magnetic moment*. With each of these values is associated a quantum number. The quantum state of a particle is defined by the ensemble of its quantum numbers.

In thermodynamics, the quantity of interest is energy. The quantum state of an assembly of particles is defined when the distribution of particles among the various energy levels is known.

Quantum statistics—A name given to *Fermi–Dirac statistics*, which govern the behaviour of *fermions*, and to *Bose–Einstein statistics*, which govern that of *bosons*. Expressions of entropy established for both statistics can be represented by a single form:

$$S = k\ln\Omega = k\sum_i\left[N_i\ln\left(\frac{g_i}{N_i} - \delta\right) - \delta g_i\ln\left(1 - \delta\frac{N_i}{g_i}\right)\right].$$

$\delta = +1$: Fermi–Dirac statistics; $\delta = -1$: Bose–Einstein statistics.

When $g_i \gg N_i$, it is possible to put $\delta = 0$, and we obtain the expression of entropy for the *Maxwell–Boltzmann statistics* of indistinguishable particles. Such an approximation is generally satisfactory. However, some applications (*radiation*, gases at low temperatures, electrons in metals, etc.) need the use of quantum statistics.

Quantum theory—The first statement was given by Max Planck in 1900: 'Electromagnetic radiations are emitted or absorbed in quanta of energy':

$$W = h\nu.$$

v: frequency of the radiation (s^{-1})
h: Planck constant, $6.626\,075\,5 \times 10^{-34}\,\text{J s}$.

Quasi-static—A *process* is called quasi-static when it follows a succession of equilibrium states; the *surroundings* may be irreversibly altered during the process so that after a return path, the universe ends up in a final state which differs from its initial state.

Quasi-static is thus not synonymous with *reversible*. A quasi-static process obeys the first two criterions of reversibility, but does not obey the third one, because over a cycle, $\oint dW \neq 0$. The progressive magnetization of a ferromagnetic bar followed by its demagnetization is a quasi-static process, owing to the hysteresis phenomenon.

R

Radiation—Radiation may be considered as a collection of photons, each being characterized by its energy $h\nu$ and its *momentum hv/c* (v: frequency of the radiation; h: Planck constant; c: velocity of light in vacuum). Radiation in equilibrium with its surroundings is in thermal equilibrium. Its temperature is then, by definition, that of the matter which surrounds it.

Photons being *bosons*, thermodynamic properties of radiation follow from the application of *Bose–Einstein statistics*:

$$S = k \ln \Omega.$$

With bosons:

$$\Omega = \prod_i \frac{(N_i + g_i - 1)!}{N_i!(g_i - 1)}.$$

By applying the *Stirling formula*: $(\ln N! \approx N \ln N - N)$:

$$S = k \sum \left[N_i \ln\left(\frac{g_i}{N_i} + 1\right) + g_i \ln\left(1 + \frac{N_i}{g_i}\right) \right],$$

where g_i is the *degeneracy* of the energy level ε_i. More precisely, $g_i \, dv_i$ represents the number of photons in a volume V, whose frequency is included between v_i and $(v_i + dv_i)$ or whose energy $(\varepsilon = hv_i)$ is included between ε_i and $(\varepsilon_i + d\varepsilon_i)$:

$$g_i = 8\pi V c^{-3} v_i^2.$$

N_i is the number of photons with energy ε_i in a volume V. The total energy of the system is then given by $E = \sum N_i \varepsilon_i$. The volume V and the energy E being given, the equilibrium condition is obtained by seeking the maximum of S. The differentiation of S with the constraints E and V constant, i.e. g_i constant, yields:

$$dS = k \sum \left[\ln\left(\frac{g_i}{N_i} + 1\right) dN_i \right] = 0$$

$$dE = \sum \varepsilon_i dN_i = 0,$$

which implies, after introducing a *Lagrange multiplier* λ:

$$\frac{dS}{k} - \lambda \, dE = \sum \left[\ln\left(\frac{g_i}{N_i} + 1\right) - \lambda \varepsilon_i \right] dN_i = 0.$$

Now:

$$\left(\frac{\partial S}{\partial E}\right)_V = k\lambda = \frac{1}{T}.$$

The Lagrange multiplier λ is thus identified with $1/kT$ and the equilibrium condition $dS = 0$ implies:

$$\ln\left(\frac{g_i}{N_i} + 1\right) = \frac{\varepsilon_i}{kT}$$

$$N_i = g_i \left[\exp\left(\frac{\varepsilon_i}{kT} - 1\right) \right]^{-1},$$

whence the energy of the radiation:

$$E = \sum N_i \varepsilon_i = \sum g_i \varepsilon_i \left[\exp\left(\frac{\varepsilon_i}{kT} - 1\right) \right]^{-1}.$$

The energy of the radiation in a volume V is obtained by using the expression for g_i, then integrating over all values of v_i:

$$g_i \, dv_i = 8\pi V c^{-3} v_i^2 \, dv_i$$

$$E = 8\pi V c^{-3} \int_0^\infty h v^3 \left[\exp\left(\frac{\varepsilon_i}{kT} - 1\right) \right]^{-1} dv.$$

The integral is calculated by using a series expansion of the bracket, which yields the relation:

$$E = aT^4V,$$

where a is a universal constant defined by:

$$a = 8\pi^5 k^4 / 15 c^3 h^3 = 7.5646 \times 10^{-16} \, \text{J} \, \text{m}^{-3} \, \text{K}^{-4}.$$

By introducing into the expression of the radiation entropy:

$$S = k \sum \left[N_i \ln\left(\frac{g_i}{N_i} + 1\right) + g_i \ln\left(1 + \frac{N_i}{g_i}\right) \right],$$

the equilibrium condition: $\ln[(g_i/N_i) + 1] = (\varepsilon_i / kT)$, we get:

$$S = \sum N_i \frac{\varepsilon_i}{kT} - \sum g_i \ln\left[1 - \exp\left(-\frac{\varepsilon_i}{kT}\right)\right],$$

whence:

$$F = E - TS = kT \sum g_i \ln\left[1 - \exp\left(-\frac{\varepsilon_i}{kT}\right)\right].$$

F is calculated by the same method as E:

$$F = 8\pi V c^{-3} kT \int_0^\infty v^2 \ln\left[1 - \exp\left(-\frac{hv}{kT}\right)\right] dv$$

$$F = -\frac{8\pi^5 k^4}{45 c^3 h^3} T^4 V = -\frac{1}{3} aT^4 V.$$

From the differential: $dF = -S \, dT - P \, dV$, there follows immediately:

- the radiation entropy: $S = 4aT^3V/3$
- the radiation pressure: $P = aT^4/3$
- the radiation energy: $E = F + TS = aT^4V$
- the radiation enthalpy: $H = E + PV = 4aT^4V/3$
- the radiation Gibbs energy: $G = H - TS = 0$.

Thermodynamic functions are simpler for radiation than for a gas, even an ideal one. Indeed, for a gas there is no relation between pressure and temperature, the variables T, P and V being related only by an equation of state $f(T, P, V) = 0$, whereas for radiation there is a direct relationship between temperature and pressure.

If radiation is confined in an enclosure with perfectly white walls (that is, which reflect 100% of the incident radiation), a volume change due to the displacement of a piston will cause a temperature change.

If the transformation is isentropic (that is, adiabatic and reversible): $S =$ const. From the preceding relationships, it follows that:

$$VT^3 = \text{const.} \quad PT^{-4} = \text{const.} \quad PV^{4/3} = \text{const.}$$

The analogy between the last expression and the equation of an adiabat for an ideal gas with $\gamma = c_P/c_V = 4/3$ is merely fortuitous, because for radiation: $c_V = (\partial U/\partial T)_V = 4aT^3V$, but $c_P = (\partial H/\partial T)_P = \infty$.

Radiation pressure—*Radiation* is a collection of photons, each being characterized by its energy $h\nu$ and its *momentum* $h\nu/c$. Photons exert on the unit surface of a wall a force which is called radiation pressure. Its existence may be proved by a thought experiment.

Let H and C be two black bodies (absorbing all incident radiation) at constant temperatures T_H and T_C ($T_H > T_C$) enclosed by a white boundary (reflecting all incident radiation). In the enclosure, always kept under vacuum, it is possible to introduce, move and withdraw a thermally insulating mobile piston without destroying the vacuum. If the piston is near H, the enclosure is in thermal equilibrium with C at the temperature T_C. Let us withdraw the piston to introduce it near the body C, then shift it inside the enclosure to place it in contact with the body H. Through this operation, the whole energy emitted by the body C will be absorbed by the body H. Since the *second law* forbids the transfer without energy compensation from a cold body C to a hot body H, the shift from C to H of the piston inside the enclosure therefore requires the supply of work from the surroundings. This work, done against the radiation pressure, increases with $\Delta T = (T_H - T_C)$.

Radiation pressure was forecast by Maxwell in 1874. However, owing to its very low value, it was not demonstrated experimentally until 1901, by Lebedev. It is possible to infer the radiation pressure from Stefan's law giving the *Helmholtz energy* of radiation:

$$F = -\frac{1}{3} aT^4 V,$$

where a is a universal constant (see *Radiation*):

$$a = 8\pi^5 k^4/15c^3h^3 = 7.5646 \times 10^{-16} \, \text{J m}^{-3} \, \text{K}^{-4},$$

whence:

$$P = -\left(\frac{\partial F}{\partial V}\right)_T = \frac{1}{3} aT^4.$$

Radiation pressure cannot be neglected when $T > 10^4 \, \text{K}$.

Rankine cycle—Proposed in 1859 by William John Macquorn Rankine, Scottish physicist and engineer (1820–1872), this reversible cycle, formed of two *adiabatic* and two *isobaric* steps, is the most straightforward of the cycles used in a steam engine. As opposed to the *Brayton cycle*, it involves a condensable gas.

Without superheating of the vapour, the system follows, on a *temperature–entropy diagram*, the cycle ABCDJA:

AB: adiabatic compression of the liquid in a pump
BCD: heating then vaporization of the liquid at constant pressure
DJ: adiabatic expansion with partial condensation of the vapour
JA: exothermic condensation of the vapour.

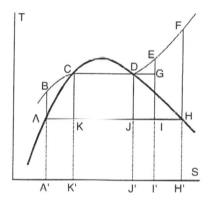

Rankine cycle: *Representation of the cycle ABCDJA on a temperature–entropy diagram. The liquid–vapour equilibrium curve is in bold. The line BCDEF is an isobaric curve.*

For a motor, the efficiency of the Rankine cycle is given by:

$$\eta = -W/Q_2 - (Q_1 + Q_2)/Q_2.$$

$Q_1(<0)$ and $Q_2(>0)$ represent the heat exchanged by the fluid with the surroundings, in the condenser and in the boiler respectively. As the transformations are reversible, $Q = \int T \, dS$. The efficiency is thus given, on a T–S diagram, by $\eta = \text{area}(ABCDJ)/\text{area}(A'BCDJ')$.

The efficiency of the Rankine cycle is lower than that of the *Carnot cycle*, for instance the cycle KCDJ:

$$\frac{\text{area (ABCDJ)}}{\text{area (A'BCDJ')}} < \frac{\text{area (KCDJ)}}{\text{area (K'CDJ')}}.$$

R

A Carnot cycle would need the use of a pump able to compress a liquid–vapour mixture to follow the path KC, which raises technical problems. The Rankine cycle is thus preferred, in spite of its lower efficiency. It may however be improved by superheating the vapour, always keeping the pressure constant. The fluid then follows the path ABCDEI, whence the efficiency η = area (ABCDEI)/ area (A′BCDEI′).

In a Carnot cycle, the vapour may be superheated following the isotherm DG, which implies a drop in pressure. The Rankine cycle JDEI here has an efficiency higher than that of the Carnot cycle JDGI, because the temperature of point E is higher than that of point G.

If the Rankine cycle is used in a refrigerator or a heat pump, the most efficient path must then be HFDCK:

HF: adiabatic compression
FDC: condensation with heat supplied to the hot source
CK: expansion through a valve or a pressure drop
KH: evaporation by means of heat supplied by the cold source.

There is a striking difference between the operation of a refrigerator and that of a steam engine. In a refrigerator, the expansion CK by pressure drop is irreversible, whereas in a steam engine, the path KC performed with a pump is reversible.

Raoultian activity—The *activity* of a constituent i is called Raoultian when the *standard state* selected is the constituent i pure in the same *aggregation state* as the solution. Compositions are then expressed in *mole fractions*. An activity expressed by the relation $a_i = \gamma_i x_i$ is Raoultian if $\gamma_i \to 1$ when $x_i \to 1$.

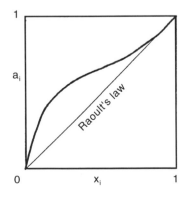

Raoultian activity: Equal to the mole fraction near pure constituent i. Its value for a given solution is independent of the unit selected to express the composition.

Remark: As concentrated solutions obey *Raoult's law*, it follows that, if $x_i \to 1$, not only Raoultian activity $a_i \to 1$, and *activity coefficient* $\gamma_i \to 1$, but also $da_i/dx_i \to 1$ and $d\gamma_i/dx_i \to 1$.

Note: Raoultian activity must not be confused with Raoult's law. Raoultian activity implies a well-specified choice of a standard state, whereas Raoult's law characterizes the behaviour of solutions. Such behaviour is common to all concentrated solutions and hence cannot be dependent on an arbitrary choice by the experimenter.

Raoult's law—François-Marie Raoult (1830–1901), French physicist, in 1882 invented *tonometry*, *cryometry*, and *ebulliometry*. A *solvent* obeys Raoult's law when its *activity* is proportional to its *mole fraction*: $a_1 = kx_1$. When selecting as *standard state* the pure solvent, Raoult's law is expressed by $a_1 = x_1$. If Raoult's law is satisfied over the whole composition domain $(0 < x_1 < 1)$, the solution is called *ideal*. The validity of *Henry's law* for a solute implies that of Raoult for the solvent. Indeed, it is shown (see *Ideal solution*) that, if the ideality condition is satisfied for one constituent, it is also satisfied for all constituents of the mixture.

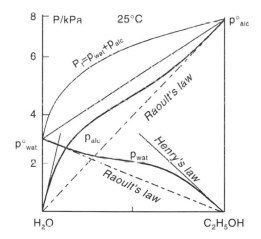

Raoult's law: *Application to the water–alcohol mixture at 25°C. If p_i is the saturated vapour pressure of i pure, and x_i its mole fraction in the solution, Raoult's law allows the vapour pressure of i to be expressed in the form: $p_i = p_i^\circ x_i$; it is satisfied all the more accurately as x_i approaches 1.*

Ratio—The quotient of two physical quantities with the same dimensions. A ratio, also called factor or index, is thus a dimensionless number. In contrast to a *fraction*, lower than unity, a ratio may be higher or lower than 1. The main ratios encountered are the mass ratio, the quotient of the mass of constituent i over the mass of solvent, and the mole ratio, the quotient of amount (number of moles) of constituent i over amount of solvent.

Reactants—The name given to the constituents on the left-hand side in a *chemical reaction*, as opposed to 'products' on the right-hand side.

Reaction rate—The *extent* ξ of a *chemical reaction* $\sum v_i M_i = 0$ is defined by: $d\xi = dn_i/v_i$, dn_i being the change in the amount (number of moles) of M_i in the time interval dt. The rate v of the reaction is defined by: $dv = d\xi/dt$, independent of the constituent M_i selected. The *entropy production* is related to the rate v and the *affinity* $\mathscr{A}$ of the reaction by:

$$\dot\sigma \equiv \frac{d\sigma}{dt} = \frac{1}{T} \mathscr{A} v > 0.$$

Real gas—For the behaviour of real gases, see: *Acentric factor, Amagat diagram, Clapeyron diagram, Corresponding states, Enthalpy–entropy diagram, Enthalpy–pressure diagram, Fugacity, Gas mixture, Partial pressure, Partial volume, Pressure–temperature diagram, Temperature–entropy diagram, Virial, Volume–temperature diagram.*
The main equations of state encountered for a real gas are:

- the *virial* expansion:

$$P = \frac{nRT}{V}\left[1 + \frac{n}{V}B(T) + \left(\frac{n}{V}\right)^2 C(T) + \cdots\right],$$

- the *van der Waals* equation:

$$\left[P + a\left(\frac{n}{V}\right)^2\right](V - nb) = nRT,$$

- the Beattie–Bridgman equation. This well known equation, proposed in 1928, contains five parameters, tabulated for many substances:

$$P = \frac{RT(1-\varepsilon)}{V^2}(V + B) - \frac{A}{V^2}$$

$$A = A_0(1 - a/V)$$

$$B = B_0(1 - b/V)$$

$$\varepsilon = C/VT^3.$$

The Beattie–Bridgman equation is quite accurate as long as the fluid density does not exceed 0.8 times the critical density. For higher densities, it is possible to use an equation with eight parameters proposed in 1940 by Benedict, Webb, and Rubin, or better, an equation with 16 parameters given in 1962 by Stobridge.

Reduced extensity—*Extensity* divided by amount of substance. Sometimes it may be convenient to reduce an extensity to unit mass or to unit volume. It must be kept in mind that language may be misleading: a reduced extensity has nothing of an extensity, but everything of an *intensity*! So a volumetric mass is not a mass, a molar enthalpy is not an enthalpy, just as power is not work!

Reference—This word, from the Latin *referre*, to carry back, is often defined by its context, as an object for comparison. In practice a *reference state* must not be confused with a *standard state*.

In a solvent–solute mixture, it is possible to describe the solvent by reference to the pure constituent and the solute by reference to the constituent in an infinitely dilute solution. To specify the nature of the reference, or object of comparison, it is the custom to define a standard state for the solvent, a state which generally identifies with the reference state, and a reference state for the solute, a state which never identifies with the standard state.

A constituent i is in a standard state when its *activity* a_i equals unity; it is in a reference state when its *activity coefficient* equals unity. The activity coefficient being defined by: $\gamma_i = a_i c_i^{\ominus}/c_i$, the definition of a reference state necessitates that of a reference composition $c_i^{\ominus}$.

Reference state—The state of a constituent i, arbitrarily selected, for which its *activity coefficient* is made equal to unity. To select a reference state, it is necessary to specify:

- the pressure
- the *aggregation state*
- the nature (i pure or dissolved in a *solvent*)
- the solvent (if i is a *solute*)
- the concentration or the titre
- the reference *composition*: $c_i^{\ominus}$.

Remark 1: This last point is necessary. Indeed, the *activity*, defined by $a_i = f_i/f_i^{\circ}$, does not depend on the unit selected to define composition;

on the other hand, the activity coefficient, defined by $\gamma_i = a_i c_i^{\ominus}/c_i$, depends on the choice of $c_i^{\ominus}$.

Remark 2: The reference state generally differs from the *standard state*. It is clear that, if $a_i = 1$ also implies $\gamma_i = 1$, both states are identical. This is the situation when the reference state selected is the pure condensed constituent, with titres expressed in mole fractions, which amounts to letting $x_i^{\ominus} = 1$.

It is often awkward to select such a reference state when i is a solute. It is then more convenient to select the solute in infinitely dilute solution, the concentrations being expressed (for instance) in moles per litre, which is equivalent to letting $c_i^{\ominus} = 1 \, \text{mol L}^{-1}$. The standard state implied by such a reference state is generally hard to realize with accuracy, and may sometimes correspond to a state without any physical meaning.

Refrigeration—Transfer of *heat* from a cold reservoir (source) to a hot reservoir (sink) with the aim of heating the sink (heat pump) or cooling the source (refrigerator). According to the *second law*, such an operation is not cost-free but requires the supply of *work*. The most efficient cycle for the job is the *Carnot cycle*, reversible working between two isothermal and two adiabatic curves. However, the easiest to use is the *Brayton cycle*, followed in the reverse direction, naturally.

Regular solution—A binary solution is called regular when its *excess Gibbs energy of mixing* is given by:

$$\Delta_{\text{mix}} G^{\text{xs}} = \alpha \, x_1 x_2.$$

The concept of regular solution, coined by Hildebrand, may be generalized to solutions with n components:

$$\Delta_{\text{mix}} G^{\text{xs}} = \sum_{i=1}^{n-1} \sum_{j=i+1}^{n} \alpha_{ij} x_i x_j,$$

which yields, for a ternary solution:

$$\Delta_{\text{mix}} G^{\text{xs}} = \alpha_{12} x_1 x_2 + \alpha_{13} x_1 x_3 + \alpha_{23} x_2 x_3.$$

Energy parameters α are generally a linear function of temperature: $\alpha = \eta - \sigma T$, η being the enthalpic term and σ the entropic one. *Activity coefficients* γ_i are easily obtained from:

$$\Delta_{\mathrm{mix}} G^{\mathrm{xs}} = \sum x_i \Delta \mu_i^{\mathrm{xs}} = RT \sum x_i \ln \gamma_i.$$

For a binary system $(i = 1, 2)$, we get (see *Partial molar quantity*):

$$\gamma_i = \exp\left[\frac{\alpha}{RT}(1 - x_i)^2\right].$$

Construction of the curve $\Delta_{\mathrm{mix}} G = f(x_2) = RT \sum x_i \ln a_i$ $(i = 1, 2)$ is unproblematical. It is easy to verify that $\Delta_{\mathrm{mix}} G \to 0$ with a vertical tangent when $x_2 \to 0$ or 1. The curve presents only one minimum for $x_2 = 0.5$ when $\alpha/RT < 2$; when $\alpha/RT > 2$, it presents three extrema: a maximum for $x_2 = 0.5$ and two minima symmetrical with respect to 0.5. The solution is then unstable and shows *unmixing* between the two minima.

Remark: Some authors define a regular solution as a solution for which the excess entropy is zero, with, as consequences: $\Delta_{\mathrm{mix}} G^{\mathrm{xs}} = \Delta_{\mathrm{mix}} H$ and $\Delta_{\mathrm{mix}} S = \Delta_{\mathrm{mix}} S^{\mathrm{id}} = -R \sum x_i \ln x_i$. Taking into account the relation existing between entropy and Gibbs energy: $S = -(\partial G/\partial T)_P$, the coherence of the definition implies $\Delta_{\mathrm{mix}} S^{\mathrm{xs}} = (\Delta_{\mathrm{mix}} G^{\mathrm{xs}}/\partial T)_P = 0$, thus $\Delta_{\mathrm{mix}} G^{\mathrm{xs}}$ is independent of temperature. To identify excess Gibbs energy with enthalpy is not mentally satisfying. The existence of an enthalpy of mixing $(\Delta_{\mathrm{mix}} H \neq 0)$ implies interactions, attractive or repulsive, between the constituents, interactions incompatible with an ideal entropy of mixing.

Relativity—The theory of relativity, elaborated by Einstein in 1905, is based on Lorentz transformations. These transformations give the measurement of space (x, y, z) and time (t) made by an observer moving with a velocity v in the direction of the x-axis as a function of the measurements (x_0, y_0, z_0, t_0) made by an observer at rest.
Let $\beta = \sqrt{1 - v^2/c^2}$ (c: velocity of light in vacuum):

$$x = \frac{x_0 + vt_0}{\beta}, \quad y = y_0, \quad z = z_0, \quad t = \frac{1}{\beta}\left(t_0 - \frac{vx}{c^2}\right).$$

The most important results of the theory relate to the energy of a mass m moving with a velocity v:

$$E = \frac{mc^2}{\beta} = mc^2 + \frac{1}{2}mv^2 + \frac{3}{8}m\frac{v^4}{c^2} + \cdots$$

R

Length, volume, mass, and energy are transformed according to:

$$l = l_o \beta$$
$$V = V_o \beta$$
$$m = m_o / \beta$$
$$E = E_o / \beta.$$

How are thermodynamic quantities transformed in a system moving with a velocity v? Planck, as early as 1907, provided an answer to this legitimate question by proposing the following transformations:

$$n = n_o \qquad P = P_o \qquad S = S_o$$
$$T = T_o \beta \qquad Q = Q_o \beta.$$

The invariance of entropy arises directly from its definition in statistical thermodynamics: $S = k \ln \Omega$. The number of *microscopic states* Ω which realizes a given *macroscopic state*, as well as the amount of substance, does not depend on the velocity of the system. In relativistic thermodynamics, the function of interest is the *enthalpy*, because it is expressed as a function of three invariants: amount of substance, pressure, and entropy:

$$dH = T\, dS + V\, dP + \mu\, dn.$$

Some months later, Einstein confirmed these results by different reasoning. These relationships raised no objection for half a century, until another school appeared and proposed:

$$T = T_o / \beta \qquad Q = Q_o / \beta.$$

Both points of view retain entropy invariance. Since no scientific truth is decided by a majority of votes, for the moment let us wisely refrain from taking any position.

Renversable—A term coined by Duhem in 1893 to describe a process which satisfies the first two criteria of *reversibility* but not the third. Renversable is not synonymous with quasi-static, because a renversable process is not performed as a succession of equilibrium states.

Reservoir—A *system* that is able to exchange with another an unlimited quantity of heat, work, or matter without changes in the intensive quantities (pressure, temperature, chemical potential, etc.) which characterize it. A reservoir is of interest for the thermodynamicist only if its capacity is infinite; it may behave as a *source* or as a *sink*, depending on the direction of exchange.

Resistivity—The name given to the reciprocal of *conductivity*:

$$\text{Resistivity} = \frac{1}{\text{Conductivity}} = \frac{\text{Gradient of intensity}}{\text{Flux of extensity}}.$$

In the SI system, electrical resistivity is expressed in $\Omega\,\text{m}$, or $\text{m}^3\,\text{kg}\,\text{s}^{-3}\,\text{A}^{-2}$.

Retrograde—From the Latin *retro*, backward, and *gradus*, step. In common parlance, the adjective 'retrograde' is pejorative: retrograde behaviour represents deterioration. A scientist perceives in such behaviour, contrary to generally accepted ideas, a fine subject for reflection and a salutary reminder to beware obviousness!

A retrograde solubility is a solubility varying in a direction opposite to the expected one: the solubility of a liquid in a solid is called retrograde if it increases with temperature; that of a solid in a liquid is called retrograde if it decreases when the temperature increases.

Vapour→liquid condensation is called retrograde if it occurs as a result of a decrease in pressure or an increase in temperature. Conversely, vaporization is called retrograde if it occurs when the temperature falls or the pressure rises.

Reversibility—A *process* is called reversible if it takes place as a succession of equilibrium states, the reverse process leaving the *surroundings* unmodified. A reversible *transformation* satisfies three criteria:

1. the transformation must follow a perfectly defined path, which is the definition of a process;
2. the reverse transformation, following the same path, must be possible at any moment;
3. if the system is made to undergo the forward process and then the reverse process following the same path, the work exchanged with the surroundings must be zero: $\oint dW = 0$.

The first two criteria, not very constraining, characterize a *renversable* transformation. The third one means that the transformation must occur without any friction; its presence, more or less unavoidable, prevents actual transformations from being reversible.

Root mean square deviation—See *Standard deviation*.

S

Sackur–Tetrode equation—Proposed in 1912, it allows one to evaluate, in *Maxwell–Boltzmann statistics*, the entropy of an ideal gas containing N indistinguishable particles with a mass m, occupying a volume V at temperature T. Using the expression of S calculated from the *partition function* Z:

$$S = Nk \left[T \left(\frac{\partial \ln Z}{\partial T} \right)_V + \ln \left(\frac{Z}{N} \right) + 1 \right],$$

and taking into account only the contribution of translation:

$$Z_{\mathrm{tr}} = V \left(\frac{2\pi m k T}{h^2} \right)^{3/2},$$

we get the Sackur–Tetrode formula giving the entropy of a monatomic ideal gas made up of indistinguishable particles without interaction:

$$S = Nk \left[\ln \left(\frac{V}{N} \right) + \frac{3}{2} \ln \left(\frac{2\pi m k T}{h^2} \right) + \frac{5}{2} \right],$$

k being the Boltzmann constant and h the Planck constant. With a polyatomic ideal gas, it is possible to obtain a more accurate expression of S by introducing other contributions:

$$Z = Z_{\mathrm{tr}} Z_{\mathrm{int}}.$$

The Sackur–Tetrode formula gives satisfactory results at high temperatures, but it cannot be applied at low temperatures, owing to the term $\ln T$, which makes $S \to -\infty$ when $T \to 0$.

The underlying reason for this behaviour comes from the fact that at low temperatures, the particles, which can be either *fermions* or *bosons*, do not follow the Maxwell–Boltzmann statistics. Quantum effects arise when the density of particles n becomes higher than a quantum density n_Q:

$$n = \frac{N}{V} > n_Q = \left(\frac{2\pi m k T}{h^2} \right)^{3/2}.$$

There is thus a limiting temperature $T_\circ = (h^2 / 2\pi m k)(N/V)^{2/3}$, below which the gas is called degenerate. When $T \to 0$, the particles, bosons or fermions, occupy a fundamental and single state, the consequence of which is to make the entropy of the gas tend towards zero.

Saturated vapour pressure—Or 'saturated vapour tension', the pressure of the vapour above a constituent i in equilibrium with the condensed phase. The vapour pressure of a pure substance depends on the temperature according to the *Clapeyron equation*; it also increases with the total pressure: the *Poynting effect*. More generally, the equilibrium condition of a constituent i between a condensed phase and its vapour is obtained from the equality of *chemical potentials*: $\mu_{ic} = \mu_{iv}$, whence:

$$\mu_{ic}^{\circ} + RT \ln a_{ic} = \mu_{iv}^{\circ} + RT \ln(f_i/f_i^{\circ})$$

$$(f_i/f_1^{\circ}) = a_{ic} \exp(-\Delta_{vap} G^{\circ} / RT).$$

If the compound i is pure ($a_{ic} = 1$), we obtain the Clapeyron equation.

Second—One of the seven base units of the SI system, the second (symbol s) was defined in 1968 by the 13th Conférence Générale des Poids et Mesures as the duration of 9 192 631 770 periods of the radiation corresponding to the transition between the two hyperfine levels of the ground state of the cesium-133 atom.

Second law—This fundamental law of thermodynamics is a principle of evolution; there exist many statements of it, the best known being those of *Clausius*, *Planck*, and *Caratheodory*. Each amounts to affirming the existence of *stable equilibrium states*:

Among all states available to an isolated system, there is one state, and only one, that is a stable state. This state may be reached from any state available to the system, taking into account the constraints imposed.

S

Corollary 1: An *isolated* system being characterized by its *energy,* its *composition*, and the imposed physical or experimental constraints, it follows that any property of a system in stable equilibrium may be expressed as a function of its energy, its composition, and the constraints imposed. This corollary, also known by the name of *state principle,* may be presented as a statement of the second law.

Corollary 2: In an isolated system, a *transformation* from any given state to an equilibrium state can only be *irreversible*. Indeed, if it were *reversible*, it would be possible, contrary to the second law, to make a system evolve from a stable equilibrium state to any state without modification of the *surroundings*, because the third criterion of reversibility implies $\oint dW = 0$.

Corollary 3: A system in a stable equilibrium state may receive, but cannot produce, *work*; otherwise it would be possible to use the work produced to bring the system into a state out of equilibrium, and this, without influence on the surroundings.

Corollary 4: If two systems A and B are in *mutually stable equilibrium*, they are, when taken separately, in a stable equilibrium state. Otherwise it would be possible by separating the two systems to obtain work, which contradicts corollary 3.

Corollary 5: When a system exchanges work with a *reservoir*, evolving from a well-defined state i towards a well-defined state f, work exchanged during a reversible transformation is less than work exchanged during an irreversible transformation:

$$\int_i^f (dW)_{rev} < \int_i^f (dW)_{irr}.$$

Indeed, the existence of a reversible *path* i → f supposes that states i and f are equilibrium states according to corollary 2.

On the other hand, if the transformation i → f is effected by an irreversible path, followed by the reverse transformation f → i by a reversible path, the difference:

$$\int_i^f (dW)_{irr} - \int_i^f (dW)_{rev} = \oint (dW)_{irr}$$

can only be positive according to corollary 3.

Corollary 6: The work exchanged between a system and a reservoir is the same for every reversible transformation making the system evolve from a well-defined state i towards a well-defined state f. The differential $(dW)_{rev}$ is thus *exact*; the function W exists and is symbolized by F. It is a state function called *Helmholtz energy*, or available work:

$$\int_i^f (dW)_{rev} = F_f - F_i = \Delta F.$$

Corollary 7: The Helmholtz energy F and the *energy E* are *extensive* state functions; their difference is thus an extensive state function.

This corollary allows a new function of state S to be defined, called *entropy*:

$$dS = \beta(dE - dF) = (dE - dF)/T.$$

The entropy, defined by its differential, is thus known, apart from an additive constant. The coefficient β, or its reciprocal $T = 1/\beta$, which is called the *thermodynamic temperature* of the reservoir, is defined except for a multiplicative constant. It is possible to fix this constant so that the thermodynamic temperature identifies with the temperature defined by an *ideal gas* thermometer.

Corollary 8: For an *adiabatic* reversible *process*, $dS = 0$. Indeed, dS has been defined according to corollary 7 by $dS = \beta(dE - dF)$. Now, for an adiabatic process, $dE = dW$, whereas for a reversible process, $dF = dW$ according to corollary 6. Hence $dS = \beta(dE - dF) = 0$.

Corollary 9: For an adiabatic irreversible process, $dS > 0$. In fact, for an adiabatic process, $dE = dW$; moreover, according to corollary 5, $(dW)_{irr} > (dW)_{rev}$, whence:

$$dS = \beta(dE - dF) = \beta[(dW)_{irr} - (dW)_{rev}] > 0.$$

Corollary 10: For every transformation in an isolated system, $dS > 0$. In fact, the transformation is adiabatic because the system is isolated; it is also irreversible according to corollary 2, whence $dS > 0$ owing to corollary 9. As a consequence, the state of equilibrium for an isolated system is obtained by seeking the state for which S reaches its maximum.

Corollary 11: During a *diabatic* reversible process, $dS = dQ/T$. In fact, $dS = \beta(dE - dF)$ with $dE = dQ + dW$. If the transformation is reversible, $dW = dF$ owing to corollary 6, whence:

$$dS = \beta \, dQ_{rev} = dQ_{rev}/T.$$

Corollary 12: During a diabatic irreversible process, $dS > dQ/T$. In fact, $dS = \beta(dE - dF)$ with $dE = dQ_{irr} + dW_{irr}$ and $dF = dW_{rev}$, whence $dS = \beta(dQ_{irr} + dW_{irr} - dW_{rev})$.
Since $dW_{irr} > dW_{rev}$ according to corollary 5, it follows that:

$$dS > \beta \, dQ_{irr} = dQ_{irr}/T.$$

Remark 1: From the statement of the second law, we infer the existence of two state functions, namely, Helmholtz energy (corollary 6) and entropy (corollary 7). Corollary 10, expressing that the entropy of an isolated system may only increase ($\Delta S \geqslant 0$), is often presented as a statement of the second law. Such a presentation is quite legitimate; however, so as not to 'go round in circles', it is necessary to define entropy without any reference to the second law and to show beforehand that entropy so defined is really a state function.

Remark 2: Corollaries 11 and 12 may be expressed as:

$$\mathrm{d}S = \frac{\mathrm{d}Q}{T} + \mathrm{d}\sigma.$$

$\mathrm{d}\sigma \geqslant 0$ represents the *entropy production* due to the irreversibility of the transformation or (for those who like fancy words) the entropy change of the universe: $\mathrm{d}\sigma > 0$ for an irreversible transformation and $\mathrm{d}\sigma \to 0$ when the transformation tends towards reversibility. This relation is often used to calculate an entropy change. However, it is more easily used with $\mathrm{d}\sigma = 0$; hence the need to find a reversible transformation which leads the system from the same initial state towards the same final state, a search which is sometimes impossible.

Remark 3: The above expression may also be presented in an equivalent form:

$$\mathrm{d}S = \mathrm{d}_\mathrm{e}S + \mathrm{d}_\mathrm{i}S.$$

The entropy change is then separated into two contributions:

- $\mathrm{d}_\mathrm{e}S = \mathrm{d}Q/T$ is the external contribution to the entropy change, due to the heat quantity $\mathrm{d}Q$ exchanged by the system with its surroundings. $\mathrm{d}_\mathrm{e}S$ represents a flux term.
- $\mathrm{d}_\mathrm{i}S = \mathrm{d}\sigma > 0$ is the internal contribution to the entropy change, entropy produced inside the system by the irreversibility of the transformation. $\mathrm{d}_\mathrm{i}S$ represents a source term.

The source term $\mathrm{d}_\mathrm{i}S$ corresponds to entropy creation in the universe, as opposed to the flux term $\mathrm{d}_\mathrm{e}S = \mathrm{d}Q/T$ because the entropy of the surroundings changes during the transformation by $(-\mathrm{d}_\mathrm{e}S) = -\mathrm{d}Q/T$.

Seebeck effect—*The thermoelectric effect* discovered in 1822 by the Estonian physicist Thomas Johann Seebeck (1770–1831), which is manifested by the generation of an electromotive force in a loop consisting of two dissimilar conductors 1 and 2 when the two junctions are maintained at different temperatures. The Seebeck coefficient, or 'relative thermoelectric power', is defined by $\alpha_{12} = \mathrm{d}\mathscr{E}/\mathrm{d}T$. It is possible to characterize the contributions of each conductor: $\alpha_{12} = \alpha_2 - \alpha_1$. Coefficients α_1 and α_2 are the Seebeck coefficients, or 'thermoelectric powers' of the conductors 1 and 2. The Seebeck coefficient α is related to the *Peltier* coefficient π by the Kelvin relationship: $\pi = \alpha T$.

Semipermeable—A semipermeable boundary, also called a *membrane*, is a wall which allows the passage of certain chemical species, molecules or ions, under the influence of a *chemical potential* difference. The membrane is thus permeable to these species and impermeable to the others. A colloidal precipitate of copper ferrocyanide constitutes a membrane for an osmometer; water can pass through the membrane, but not ions or dissolved species. The membrane separating the two compartments of a *galvanic cell* is an ionic conductor, through which only certain ions may flow.

Short-range order—In a solid solution A–B, short-range order is established when the probability of finding constituent B as nearest neighbour to constituent A is higher than the statistical probability. Short-range order is the manifestation of attractive interactions between species A and B:

$$\varepsilon_{AB} < (\varepsilon_{AA} + \varepsilon_{BB})/2.$$

The enthalpies of formation of the bonds: ε_{AA}, ε_{BB} and ε_{AB} are, by nature, negative. The solid solution is characterized by:

N_A and N_B: number of atoms, or atom groups A and B.

z: coordination number, or number of nearest neighbours of a given site.

P_{AA}, P_{BB}, and P_{AB}: numbers of bonds AA, BB, and AB.

These values are not independent but linked by two relationships. With the simplifying assumption that z is the same for all sites:

$$N_A = (P_{AB} + 2P_{AA})/z \qquad N_B = (P_{AB} + 2P_{BB})/z.$$

The only independent parameter is thus P_{AB}, but it is possible to introduce an 'order parameter' $\mathscr{P}$, whose physical meaning is more direct:

$$\mathscr{P} = \frac{(P_{AB})_{obs} - (P_{AB})_{sta}}{(P_{AB})_{max} - (P_{AB})_{sta}}.$$

$\mathscr{P} = 1$ for maximal order and $\mathscr{P} = 0$ for complete disorder. A negative order parameter would mean that the bonds A–B are not attractive but repulsive. Then there would no longer be order, but formation of *clusters*. With attractive interactions, the order parameter takes an equilibrium value which is obtained by seeking the minimum of the *Gibbs energy G*.

There is necessarily a minimum, because an increase in order decreases both the enthalpic (the number of A–B bonds increases) and entropic terms. In the expression $G = H - TS$, terms H and S work in opposite directions. It is necessary beforehand to calculate separately the enthalpy and the entropy of the solution. To obtain enthalpy presents no great difficulty:

$$H = P_{AA}\varepsilon_{AA} + P_{BB}\varepsilon_{BB} + P_{AB}\varepsilon_{AB}$$

$$H = \frac{1}{2}zN_A\varepsilon_{AA} + \frac{1}{2}zN_B\varepsilon_{BB} + P_{AB}\left[\varepsilon_{AB} - \frac{1}{2}(\varepsilon_{AA} + \varepsilon_{BB})\right].$$

By selecting as *standard state* for A and B the pure constituents A and B in the same *aggregation state* as the solid solution, the integral enthalpy of mixing is given by:

$$\Delta_{mix}H = P_{AB}\left[\varepsilon_{AB} - \frac{1}{2}(\varepsilon_{AA} + \varepsilon_{BB})\right] = P_{AB}\,\omega.$$

The entropy of mixing, whose accurate calculation is trickier, contains two contributions. The first, called non-configurational (it is generally an entropy of vibration), is hard to evaluate. Like the enthalpy of mixing, the non-configurational entropy is proportional to the number of A–B bonds:

$$\Delta_{mix}S^{nc} = P_{AB}\eta.$$

The *configurational entropy* is calculated by evaluating the multiplicity of a solution containing P_{AA}, P_{BB}, and P_{AB} pairs AA, BB, and AB, Ising's problem, which nobody knows how to solve in three-dimensional space! Faced with such a situation, it is necessary to make reasonable assumptions. Using the law of least intellectual effort, it is possible to put:

$$\Delta_{mix}S^{conf} = \Delta_{mix}S^{id} = -k(N_A\ln x_A + N_B\ln x_B).$$

The above expression is unsatisfactory, because it is hard to imagine that attractive interactions between A and B can lead to a perfectly disordered solution with a statistical distribution of bonds! A more realistic hypothesis consists in supposing the $Nz/2$ pairs of the crystal distributed at random over $Nz/2$ positions, which amounts to writing:

$$\Delta_{mix} S^{conf} = \frac{1}{2} kz(P_{AA} \ln x_{AA} + P_{BB} \ln x_{BB} + P_{AB} \ln x_{AB}),$$

where x_{AA}, x_{BB}, x_{AB} are the mole fractions of the pairs AA, BB, AB.

This more satisfactory expression nevertheless yields too great a number of configurations. Guggenheim suggested a correction by pointing out that, for a statistical distribution:

$$x_{AA} = x_A^2 \quad x_{BB} = x_B^2 \quad x_{AB} = 2x_A x_B,$$

which led him to propose:

$$\Delta_{mix} S^{conf} = -k(N_A \ln x_A + N_B \ln x_B)$$

$$-\frac{kz}{2}\left(P_{AA} \ln \frac{x_{AA}}{x_A^2} + P_{BB} \ln \frac{x_{BB}}{x_B^2} + P_{AB} \ln \frac{x_{AB}}{2x_A x_B}\right).$$

The equilibrium concentration is obtained by solving:

$$\partial(\Delta_{mix} H - T\Lambda_{mix} S)/\partial x_{AB} = 0.$$

Hence:
$$\frac{x_{AB}^2}{x_{AA} x_{BB}} = 4 \exp\left(-4 \frac{\omega - \eta T}{zRT}\right).$$

The above expression looks like an equilibrium constant for the reaction $AA + BB \rightarrow 2AB$ between bonds instead of compounds, whence the name 'quasi-chemical model' generally given to this kind of treatment. By applying this relation to an equimolecular solution, the 'mass action constant' is easily expressed as a function of the order parameter:

$$(P_{AB})_{sta} = zN/4, \quad (P_{AB})_{max} = zN/2,$$

whence: $P_{AB} = zN(\mathscr{P}+1)/4, \quad P_{AA} = P_{BB} = zN(1-\mathscr{P})/8.$

Substitution gives: $\dfrac{1+\mathscr{P}}{1-\mathscr{P}} = \exp\left(\dfrac{4\eta}{zR}\right) \exp\left(-\dfrac{4\omega}{zRT}\right).$

If $(\omega - \eta T) < 0, \Rightarrow \mathscr{P} > 0$: appearance of order.
If $(\omega - \eta T) > 0, \Rightarrow \mathscr{P} < 0$: formation of clusters.
$\mathscr{P} \rightarrow 0$ when $T \rightarrow \infty$: the order persists at high temperatures.

Siemens—The name given to the unit of electrical conductance in the SI system (symbol S), from Ernst Werner von Siemens (1816–1892), German electrical engineer. $1S = 1\Omega^{-1} = 1A\,V^{-1} = 1m^{-2}kg^{-1}s^3 A^2$.

Sievert's law—A diatomic gas dissolved in a liquid obeys Sievert's law when its solubility is proportional to the square root of its pressure:

$$(X_2)_{gas} \rightarrow 2[X]_{dissolved} \qquad K = (a_X)^2/f_{X_2}.$$

Sievert's law (1909) assumes that three conditions are fulfilled:

- the diatomic gas is wholly dissociated in solution
- the gas is considered as ideal: $f_{X_2} = P_{X_2}$
- the solution obeys *Henry's law*: $a_X = \gamma_X^\infty c_X$.

To a first approximation, Sievert's law is satisfied well by diatomic gases dissolved in liquid metals under pressures lower than 1 bar.

Sign convention—The thermodynamicist always looks from the point of view of the *system*: what the system receives is counted positive; what the system loses is counted negative. When a *process* leads a system from an initial state i to a final state f, the change in a *state function Y* is:

$$\Delta Y = \int_i^f dY = Y_f - Y_i,$$

according to the sign convention. On the other hand, if dY is not an *exact differential*, the function Y does not exist. However, it is always possible to calculate $Y = \int_i^f dY$. The result, a function of the path followed, is positive if the system receives the quantity Y, or negative if the system provides the quantity Y to the surroundings.

Simple substance—A material made up of identical atoms, as opposed to a *compound*. When the simple substance is in a molecular form, the molecules may comprise one atom (rare gases, most metallic vapours), two atoms (H_2, O_2, N_2, Cl_2, etc.), three atoms (O_3), or more (As_4, S_8, C_{60}, etc.).

Sink—A *reservoir* which receives heat, work or matter from a *system* without alteration in its characteristic intensive properties.

Smectic—From the Greek $\sigma\mu\varepsilon\kappa\tau\iota\kappa\sigma\varsigma$ ($\sigma\mu\varepsilon\gamma\mu\alpha$, soap). A mesomorphic state characterized by the arrangement of the centres of gravity of the molecules in parallel planes.

Solid—A state of matter in which atoms occupy a fixed position in space: the *kinetic energy* of the atoms is very low compared with their potential energy of interaction. Solids in their stable state appear in crystalline form.

Solid solution—A single-phase solid mixture. In order that two constituents give a solid solution in all proportions, they must have the same crystal structure and the sizes of ions or atoms capable of occupying the same crystal sites must be similar. In practice, the difference between their ionic radii must not exceed 15%. Thus the silicates Fe_2SiO_4 and Mg_2SiO_4 give a solid solution in all proportions because both conditions are fulfilled; KCl and NaCl at low temperatures have a limited mutual solubility, but above 600°C they give a solid solution in all proportions.

These two conditions are not always sufficient to give a solid solution. For instance, FeO and NiO have the same structure, NaCl type, and satisfy both criteria, but give only a limited solid solution, because they have no common stability domain on an *Ellingham diagram*. FeO in presence of NiO dismutates with the formation of an iron–nickel alloy and a spinel solid solution. Spinels Fe_3O_4 and $NiFe_2O_4$, on the other hand, give a solid solution in all proportions, because their stability domains possess a common zone.

Two compounds with different structure may exhibit only a limited solid solubility. In order that B, with structure β, be dissolved in A, with structure α, it is necessary that B undergoes the transition $\beta \rightarrow \alpha$ ($\Delta_{tr}G > 0$), then dissolves following B(α, pure) $\rightarrow$ B(α, dissolved) ($\Delta_{mix}G < 0$). The solubility of B in A will be very low if the transition $\beta \rightarrow \alpha$ is difficult ($\Delta_{tr}G \gg 0$). So, face-centred cubic Al dissolves up to 60 at% of hexagonal Zn, whereas the reverse solubility, that of Al in Zn, does not exceed 2 at%.

Solid solutions in which a site may be occupied by different atoms are substitutional solid solutions. There are also solid solutions of insertion, in which atoms, generally of small size, such as boron, carbon, or nitrogen, occupy a fraction of the interstitial sites of a lattice. Austenite, for instance, is a solid solution in which the carbon atoms occupy interstitial sites of γ-iron, face-centred cubic.

Solidus—On a composition–temperature binary diagram, the solidus is a boundary line separating two domains: on the low-temperature side, no liquid phase can be stable; on the high-temperature side, a liquid phase may be stable, alone or in equilibrium with a solid.

On a ternary diagram, where the composition is defined by two parameters (x_1, x_2, with $x_3 = 1 - x_1 - x_2$), it is possible to use the third dimension for the temperature. The solidus then appears as a surface, or an assembly of surfaces, below which no stable liquid phase may exist.

Solubility product—The name given to the *equilibrium constant* for the dissolution reaction of an insoluble or slightly soluble compound:

$$\langle A_m B_p \rangle_{sol} \rightarrow m(A^{p-})_{dis} + p(B^{m+})_{dis}.$$

By selecting as *standard state* for $A_m B_p$ the pure solid, and as *reference state* for ionic species their infinitely dilute solution in the solvent, the reference concentration being $c_i^{\ominus} = 1\,\text{mol}\,L^{-1}$, we obtain:

$$K(T) = [a(A^{p-})]^m \cdot [a(B^{m+})]^p \quad \text{with} \quad a_i = \gamma_i c_i / c_i^{\ominus}.$$

If $c_i \rightarrow 0$, $\gamma_i \rightarrow \gamma_i^{\infty} = \text{const.}$ The constant equals unity when the *interaction parameters* equal zero. The *activity* of ionic species may then be replaced by the composition, expressed in $\text{mol}\,L^{-1}$.

The concept of solubility product may be applied to various fields, for instance to the solubility of an oxide in a liquid metal:

$$\langle Al_2 O_3 \rangle_{sol} \rightarrow 2[Al]_{dis} + 3[O]_{dis}.$$

By selecting as standard state for $Al_2 O_3$ the pure solid oxide, and as reference state for dissolved species their infinitely dilute solution in the metal, the reference concentration being $c_i^{\ominus} = 1\,\text{mass}\%$, we get:

$$K(T) = (a_{Al})^2 \cdot (a_O)^3 \quad \text{with} \quad a_i = \gamma_i [\%i]/[\%i]^{\ominus}.$$

If $[\%i] \rightarrow 0$, $\gamma_i \rightarrow \gamma_i^{\infty} = \text{const.}$ The constant equals unity when the interaction parameters equal zero. The activity of dissolved species may then be replaced by the composition, expressed in mass per cent.

Solute—The minority species in a solution, as opposed to the *solvent*. More often than not, a pure solute does not appear in the same state as the solvent (for instance: sugar in coffee, salt in water, gas in a metal), which poses serious problems in the selection of a *standard state*. The choice of a standard concentration c° raises difficulties, because c° is generally too high for the solute to reasonably obey *Henry's law*. It is then more convenient to select a *reference state*, generally the solute in an infinitely dilute solution in the solvent, the reference concentration being $c^{\ominus} = 1\,\text{mol}\,L^{-1}$. If the solute deviates little from Henry's law, $c^{\circ} \approx c^{\ominus} = 1\,\text{mol}\,L^{-1}$, but such a situation is seldom encountered. Often the standard concentration corresponds to a solution that is metastable because it is supersaturated, or is devoid of physical meaning if there is no concentration corresponding to unit activity.

Solution—A mixture consisting of only one condensed phase. A solution may be solid or liquid, but its thermodynamic treatment remains the same. *Solid solutions*, of great practical importance, form in restrictive conditions.

Liquid solutions are more common, the constraint of structure having disappeared. Nevertheless, two liquids giving a solution must not be too different chemically. Water and alcohol mix very well ; water and phenol show unmixing below 67°C ; water, oil, and mercury do not mix significantly at all.

Gas mixtures, at low pressures, are single-phase. Above the critical point, there is only one fluid state, within which it is no longer possible to draw a boundary between liquid and vapour states. In such conditions, nothing precludes several gaseous phases from coexisting at equilibrium.

Solvent—The majority or potentially majority species in a solution, as opposed to *solute*. The expression 'potentially majority' takes into account a mixture of solvents. For instance, both constituents of an alcohol–water mixture may be considered as solvents because both may become majority, even if they cannot both be so together! In a solution composed of one solvent and several solutes, the solvent is traditionally numbered 1 and the solutes 2, 3, etc.

The need to distinguish between solvent and solute arises from the fact that, when pure, solutes do not appear in the same state as in the solution, which is of consequence for the selection of *standard states*. For the solvent, the commonest standard state is the pure solvent. This choice, together with the fact that a solvent in a dilute solution obeys *Raoult's law*, allows its *activity* to be equal to its *mole fraction*.

Solvus—On a temperature–composition binary diagram, the solvus is the boundary line separating a single-phase domain $\langle \alpha \rangle$ from a two-phase domain $\langle \alpha + \beta \rangle$. The equilibrium of a constituent, for instance B, between the phases α and β is obtained by equating the *chemical potentials*:

$$\mu_{B,\beta}^{\circ} + RT \ln a_{B,\beta} = \mu_{B,\alpha}^{\circ} + RT \ln a_{B,\alpha}.$$

This relationship is very general, but to be useful (to seek the equation of the solvus is legitimate curiosity), it is necessary to select a *standard state* for constituent B. The experimenter is free to choose what he pleases, the best choice depending on the nature of the problem to be solved. In this particular case, to choose only one standard state for B would not be convenient, because this would amount to letting:

$$\mu_{B,\beta}^{\circ} = \mu_{B,\alpha}^{\circ}, \quad \text{whence:} \quad a_{B,\beta} = a_{B,\alpha} \text{ (deadlock!)}.$$

In the particular case where the solution β is rich in B and the solution α rich in A, it is of interest to select as a standard state βB (that is, pure B with the structure β) when B is in solution in β, and αB (that is, pure B with the structure α) when B is in solution in α. It follows that:

$$a_{B,\beta} = \gamma_{B,\beta} x_{B,\beta} \quad \text{with} \quad \gamma_{B,\beta} \to 1 \quad \text{when} \quad x_{B,\beta} \to 1$$

$$a_{B,\alpha} = \gamma_{B,\alpha} x_{B,\alpha} \quad \text{with} \quad \gamma_{B,\alpha} \to \gamma_{B,\alpha}^{\infty} \quad \text{when} \quad x_{B,\alpha} \to 0.$$

$\mu_{B,\alpha}^{\circ} - \mu_{B,\beta}^{\circ} = \Delta_{\beta B}^{\alpha B} \mu^{\circ} = \Delta_{tr}\mu_B^{\circ}$, the *standard Gibbs energy* for the transition $\beta B \to \alpha B$. The equilibrium of B between the phases α and β is then expressed by:

$$\frac{a_{B,\alpha}}{a_{B,\beta}} = \exp\left(-\frac{\Delta_{tr}\mu_B^{\circ}}{RT}\right),$$

a relationship which may be simplified with the approximations:

$a_{B,\beta} = x_{B,\beta}$ (*Raoult's law* applied to the solution β rich in B)

$a_{B,\alpha} = \gamma_{B,\alpha}^{\infty} x_{B,\alpha}$ (*Henry's law* applied to the solution α rich in A).

$$\Delta_{tr}\mu_B^{\circ} = \Delta_{tr}H_B^{\circ}\left(1 - \frac{T}{T_{tr}}\right), \quad \text{neglecting } \Delta c_P = c_{P,\alpha B}^{\circ} - c_{P,\beta B}^{\circ}.$$

T_{tr}: equilibrium transition temperature: $\beta B \to \alpha B$. Hence:

$$\frac{x_{B,\alpha}}{x_{B,\beta}} = \frac{1}{\gamma_{B,\alpha}^{\infty}} \exp\left[\frac{\Delta_{tr}H_B^{\circ}}{R}\left(\frac{1}{T_{tr}} - \frac{1}{T}\right)\right].$$

A symmetrical expression may be obtained by inverting A and B on the one hand, α and β on the other hand. Moreover, it is possible to introduce $\Delta h_{\beta B,\alpha}^{\infty}$, the dissolution enthalpy of pure βB in pure αA. Then:

$$RT \ln \gamma_{B,\alpha}^{\infty} = \Delta\mu_{\alpha B,\alpha}^{xs} = \Delta h_{\alpha B,\alpha}^{\infty} - T\Delta s_{\alpha B,\alpha}^{xs}.$$

$\Delta h_{\alpha B,\alpha}^{\infty} = \Delta h_{\alpha B,\alpha}^{xs}$ and $\Delta s_{\alpha B,\alpha}^{xs}$ represent respectively the enthalpy and the excess entropy of dissolution of pure αB in pure αA. On substitution:

$$RT \ln \frac{x_{B,\alpha}}{x_{B,\beta}} = -\Delta h_{\alpha B,\alpha}^{\infty} + T\Delta s_{\alpha B,\alpha}^{xs} + \Delta_{tr} H_B^{\circ} \frac{T}{T_{tr}} - \Delta_{tr} H_B^{\circ}.$$

Now $\Delta h_{\beta B,\alpha}^{\infty} = \Delta_{tr} H_B^{\circ} + \Delta h_{\alpha B,\alpha}^{\infty}$, whence:

$$RT \ln \frac{x_{B,\alpha}}{x_{B,\beta}} = -\Delta h_{\beta B,\alpha}^{\infty} + T(\Delta s_{\alpha B,\alpha}^{xs} + \Delta_{tr} S_B^{\circ}).$$

Soret effect—A *thermodiffusion effect* demonstrated in 1879 and characterized by the appearance of a concentration gradient under the influence of a temperature gradient.

Source—A reservoir able to supply *heat*, *work*, or matter to a *system* without modifying the intensive parameters (temperature, pressure, *chemical potentials*, etc.) which characterize it.

Species—A widely used term denoting particles, atoms, defined compounds, ions, or molecules present in a medium.

Specific—An extensive quantity is called specific when it is divided by another one. (*Remark*: the ratio of two extensities is no longer an *extensity*, but an *intensity*.) This adjective 'specific' is quite imprecise; it is better to speak of volumetric mass or mass volume, rather than specific mass or specific volume. The *specific heat*, a concept introduced by Lavoisier, now denotes the *heat capacity* per unit mass, but it is often wrongly used with the meaning of molar heat capacity.

Specific gravity—A dimensionless number, the ratio of the mass of a substance to the mass of the same volume of another substance taken as reference. When the substance is gaseous, it is common to select air as the reference, whereas for condensed matter, water is generally selected. It is necessary to specify the temperature and pressure used to calculate the ratio. Specific gravity is a number easy to visualize; nevertheless, wisdom dictates that its use be avoided in calculations and that density (mass per unit volume) or specific volume (volume per unit mass) be preferred.

Specific heat—This expression, encountered in old and respectable books, now represents the *heat capacity* reduced to unit mass. Its use

S

must be proscribed, as well as that of adjective '*specific*', which is quite imprecise.

Speed inversion—As a consequence of the *second law*, an isolated system evolves spontaneously in the direction of an increase in its *entropy*. Let us imagine that, after a time t, all the velocities of the particles constituting the system are reversed. Owing to the reversibility of the laws of mechanics, the system will evolve spontaneously in the reverse direction and after a time $2t$, will again be in its initial state, whence a decrease in its entropy between the times t and $2t$, which contradicts the second law.

This paradox, of the same nature as *Poincaré's recurrence*, finds a solution in the fact that the second law has a statistical character and is justified only by the large number of particles. It may be violated, but the probability is minute for a large system, as is the probability of spontaneous inversion of each velocity.

Spinodal—In a binary solution, spinodal points are points characterized by a *stability* equal to zero:

$$St = (\partial^2 G / \partial x^2)_{T,P} = 0.$$

It amounts to the same thing to define spinodal points as points for which the second derivative of the integral *Gibbs energy of mixing* becomes zero:

$$(\partial^2 \Delta_{mix} G / \partial x^2)_{T,P} = 0.$$

When they exist, there are generally two spinodal points. Between spinodal points, $(\partial^2 \Delta_{mix} G / \partial x^2)_{T,P} < 0$: the solution is unstable with respect to a small composition fluctuation. *Unmixing* proceeds following a mechanism called 'spinodal decomposition'. Outside the spinodal points, $(\partial^2 \Delta_m G / \partial x^2)_{T,P} > 0$: the solution is stable with respect to a small composition fluctuation. Nevertheless it may be metastable, but unmixing, in order to develop, needs seeds. The decomposition mechanism proceeds by 'germination and growth'.

Stability—A system is in a state of *stable* (or *metastable*) equilibrium when a small fluctuation of a parameter makes the system react in the direction which brings it back to its initial state. Mathematically, this is expressed by writing that the second derivative of the *thermodynamic potential* with respect to the variable that fluctuates must be positive.

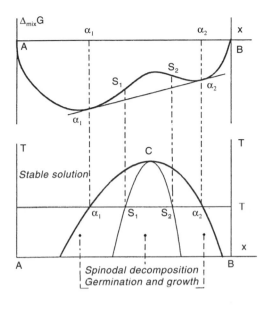

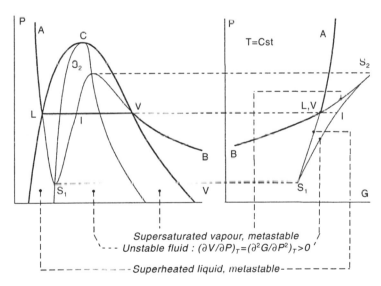

S

Spinodal, Stability: *Above, the spinode, locus of points S_1 and S_2, lies inside the binode, locus of points α_1 and α_2. The two curves meet at a critical point C. The same phenomenon is found again below, during the separation of a fluid phase, below the critical point, into a liquid and a vapour.*

282 Stability

Let G be the thermodynamic potential (if the constraints are P and T constant, G is then the *Gibbs energy*) and x be a variable likely to fluctuate. By expanding in Taylor series the function G around x:

$$G(x+\varepsilon) = G(x) + \varepsilon\frac{\mathrm{d}G}{\mathrm{d}x} + \frac{\varepsilon^2}{2}\frac{\mathrm{d}^2G}{\mathrm{d}x^2} + \cdots$$

At equilibrium: $\mathrm{d}G/\mathrm{d}x = 0$. If $\mathrm{d}^2G/\mathrm{d}x^2 > 0$, a fluctuation ε of the variable x will have the effect of increasing G and the system, which will react so that G tends towards a minimum, will regain its initial equilibrium position: the equilibrium is then stable (or metastable). On the other hand, if $\mathrm{d}^2G/\mathrm{d}x^2 < 0$, a fluctuation ε of the variable x will tend to decrease G and the system will depart from its initial equilibrium position to evolve towards a new equilibrium.

When x represents a composition parameter, it is clear that a spontaneous fluctuation of x in an elementary volume will be accompanied by a fluctuation of opposite sign in an adjacent elementary volume. Consider two neighbouring volumes δV, whose composition fluctuates around a mean value x, say $(x-\varepsilon)$ in one volume δV and $(x+\varepsilon)$ in the other volume δV. δG, the change in the thermodynamic potential, will be given by:

$$\delta G = G(x+\varepsilon) + G(x-\varepsilon) - 2G(x) = \varepsilon^2(\mathrm{d}^2G/\mathrm{d}x^2).$$

If $\mathrm{d}^2G/\mathrm{d}x^2 > 0$, the composition fluctuation will be damped and the system will recover its initial state; if $\mathrm{d}^2G/\mathrm{d}x^2 < 0$, the fluctuation will increase and the system will evolve towards a new equilibrium: there will be *unmixing*. Darken in 1953 proposed to define the stability function of a solution by:

$$St = (\partial^2G/\partial x^2)_{T,P}.$$

Generally, the stability is not unique and it is necessary to specify the direction of differentiation. On the other hand, a binary solution has only one composition parameter. Integration of the *Gibbs–Duhem equation* yields, for a binary (see *Partial molar quantity*):

$$\mu_i = G + (1-x_i)\frac{\mathrm{d}G}{\mathrm{d}x_i} \quad (i=1,2)$$

$$St = \frac{\mathrm{d}^2G}{\mathrm{d}x_i^2} = \frac{1}{1-x_i}\cdot\frac{\mathrm{d}\mu_i}{\mathrm{d}x_i} = -2\frac{\mathrm{d}\mu_i}{\mathrm{d}[(1-x_i)^2]}.$$

Stability may also be expressed as a function of the *activities*:

$$\mu_i = \mu_i^\circ + RT \ln a_i$$

$$St = \frac{d^2 G}{d x_i^2} = -2RT \frac{d \ln a_i}{d[(1-x_i)^2]}.$$

For an *ideal solution*: $St^{id} = RT/x_1 x_2$, whence the excess stability:

$$St^{xs} = St - St^{id} = -2 \frac{d\mu_i^{xs}}{d[(1-x_i)^2]} = -2RT \frac{d \ln \gamma_i}{d[(1-x_i)^2]}.$$

Remark: The stability of a binary solution does not depend on the constituent i selected. Indeed, it is easy to verify, using the Gibbs–Duhem equation, the equality:

$$\frac{d\mu_1}{d[(1-x_1)^2]} = \frac{d\mu_2}{d[(1-x_2)^2]},$$

a relationship also satisfied when replacing μ_i by $\ln a_i$ or by $\ln \gamma_i$.

PHYSICAL MEANING: A negative stability implies an unstable solution leading to unmixing; it would seem logical to imagine that high stability expresses a particular behaviour of the solution. Indeed, in drawing for a solution the curves of $\ln a_i$ vs $(1-x_i)^2$, one or more inflection points may be observed, inflections corresponding to a maximum of the stability function.

Experimentally, these maxima of the stability function, when they exist, correspond to particular compositions of the solution. For instance, with liquid slags ($SiO_2 + CaO$), the stability maxima correspond to the compositions $CaSiO_3$ and Ca_2SiO_4, for which very stable definite compounds exist in the solid state. Although these mixed oxides undergo decomposition on melting, their 'memory' remains in the liquid. The same remark may be made of the compositions Ni_3Ti and $NiTi$ of the titanium–nickel liquid alloy, as well as the composition Ni_3Fe of the iron–nickel solid solution, at a temperature higher than that of disappearance of order around this composition.

Stable equilibrium—A *system* is said to be in a state of stable equilibrium when it is impossible to modify its state permanently without simultaneously also modifying its *surroundings* permanently. The *second law* states that a *constrained* system evolves spontaneously towards a state of stable equilibrium which is unique. In order to find it, we have to look for the minimum of a *thermodynamic potential* function, whose nature depends on the constraints imposed on the system.

Standard—From the Old French *estendard*, a flag to mark a rallying place (*Chanson de Roland*, twelfth century), a standard is an element of comparison.

In thermodynamics, the introduction of standard quantities, or the definition of a *standard state*, arises from the fact that most thermodynamic functions are defined by a differential whose integration introduces a constant. The standard state of a constituent *i* is a state, arbitrarily defined, for which its *activity* is put equal to unity.

Standard affinity—The *affinity* of a *chemical reaction* is called 'standard' when each constituent of the reaction is taken in its *standard state*. At constant temperature and pressure:

$$\mathscr{A}^\circ = -\Delta_r G^\circ = -\sum v_i \mu_i^\circ .$$

The standard affinity $\mathscr{A}^\circ$ is a quantity characterizing the reaction, independent of its *extent*, but a function of its stoichiometric numbers and, of course, a function of the standard state selected for each constituent. It is related to the *affinity* of the reaction by:

$$\mathscr{A} = \mathscr{A}^\circ - RT \ln \prod_i (a_i)^{v_i} .$$

Standard deviation—Still called 'root mean square deviation', the standard deviation σ_x of a stochastic variable X equals the square root of the variance:

$$\sigma_x = \sqrt{\langle x^2 \rangle - \langle x \rangle^2} .$$

$\langle x \rangle$: *moment* of order 1, or mean of the values taken by X.

$\langle x^2 \rangle$: moment of order 2, or mean of the squares.

The standard deviation characterizes the width of the *distribution* for the variable X. A low standard deviation gives a sharp peak around the mean value $\langle x \rangle$. In the case of a *normal distribution*, where the mean value of X is also the most probable value, the standard deviation represents the difference between the most probable value of X and the value of X corresponding to the inflection point on the Bernoulli bell curve. The probabilities of observing a difference between the result of an experiment and the expected mean value greater than

1, 2, 3, 4, 5, and 100 times the standard deviation are respectively 32%, 5%, 0.25%, $7 \times 10^{-3}\%$, $5 \times 10^{-5}\%$, and $3 \times 10^{-2171}\%$!

Standard element reference—This notion is applied only to *elements*. When tables of data give the *enthalpy* or the *Gibbs energy* of a component, the values cannot be *absolute*, but only relative, given with respect to a reference, for which $H=0$ or $G=0$ has been arbitrarily set. The standard element reference, SER for short, has been introduced with the aim to define an enthalpy or a Gibbs energy of formation without ambiguity.

By convention, the standard element reference is the most stable *aggregation state* of an element at a given temperature under *standard pressure*. As a consequence, the standard element reference is independent of pressure (because the pressure is standard) but changes for each allotropic transition, or for each change of state of an element.

Standard enthalpy of reaction—The *enthalpy* of a *chemical reaction* is said to be standard when each constituent is in its *standard state*. The standard enthalpy $\Delta_r H^0$ of a reaction $\sum v_i \langle M_i \rangle = 0$ may be calculated from the standard enthalpies of formation, $\Delta_f H^\circ(M_i)$, generally tabulated for a temperature $T_0 = 298.15\,\text{K}$:

$$\Delta_r H^\circ(T_0) - \sum v_i \Delta_f H^\circ(M_i, T_0).$$

The standard enthalpy of the reaction for a temperature T is then obtained from the *heat capacities* $c_P(M_i)$:

$$\Delta_r H^\circ(T) = \Delta_r H^\circ(T_0) + \int_{T_0}^{T} \Delta c_P \, dT,$$

with $\Delta c_P = \sum v_i c_P(M_i)$. This relationship, due to Kirchhoff, is valid only if between T_0 and T no constituent M_i undergoes a phase transition.

If the constituents M_i undergo n phase transitions between T_0 and T, each being characterized by an enthalpy $\Delta_{tr} H^\circ$, the curve $\Delta_r H^\circ$ vs T shows n discontinuities, needing the calculation of $(n+1)$ integrals, each corresponding to a different expression of Δc_P. In fact, when a constituent M_i undergoes a transition $\alpha \to \beta$, there is no relationship between $c_P(\alpha - M_i)$ and $c_P(\beta - M_i)$:

$$\Delta_r H^\circ(T) = \Delta_r H^\circ(T_0) + \int_{T_0}^{T} \Delta c_P \, dT + \sum v_i \Delta_{tr} H^\circ(M_i).$$

S

The sum is performed only over the transitions observed between T_0 and T. For a second-*order* transition, $\Delta_{tr}H^\circ = 0$, but the curve $c_P(M_i)$ vs T shows a singularity in the shape of a λ, and the integration is more arduous.

Remark 1: It is often possible, to a first approximation, to let $\Delta c_P \approx 0$, but it is dangerous to neglect transitions, especially vaporizations. For instance, the standard enthalpies of melting (at $0°C$), vaporization (at $100°C$), and formation (at $25°C$) of the water are respectively 6, 40, and $-245\,kJ\,mol^{-1}$.

Remark 2: Is it necessary to point out that the standard enthalpies of formation of the elements taken in their standard state are zero? (!):

$$\Delta_f H^\circ(Fe) = 0, \ \Delta_f H^\circ(H_2) = 0, \ \text{but } \Delta_f H^\circ(H^\bullet) = +218\,kJ\,mol^{-1} \text{ at } 25°C.$$

Remark 3: When enthalpies of *combustion* $\Delta_c H^\circ(M_i)$ are given, which is often the case with organic compounds, the standard enthalpy of a chemical reaction is then expressed by:

$$\Delta_r H^\circ = -\sum v_i \Delta_c H^\circ(M_i).$$

One must fall into the trap at least once to get this into one's head! The minus sign arises from the fact that, in a formation reaction, the compounds synthesized lie on the right-hand side, whereas in a combustion reaction the compounds burnt lie on the left-hand side!

Standard entropy—The value taken by the *entropy* of a compound in its standard state. Standard entropies, generally given in tables at a temperature of $298.15\,K$, are calculated from the *third law*, which allows us to put $S(0\,K) = 0$ for every substance.

For a gaseous substance at $T_o = 298.15\,K$ which does not undergo any polymorphic transition in the solid state:

$$S^\circ(T_o) = \int_0^{T_f} \frac{c_{P,s}}{T}\,dT + \frac{\Delta_{fus}H}{T_f} + \int_{T_f}^{T_b} \frac{c_{P,l}}{T}\,dT + \frac{\Delta_{vap}H}{T_b} + \int_{T_b}^{T_o} \frac{c_{P,v}}{T}\,dT.$$

$\Delta_{fus}H$: enthalpy of fusion at the melting temperature T_f;

$\Delta_{vap}H$: enthalpy of vaporization at the boiling temperature T_b under the *standard pressure*, which must in principle equal 1 bar.

Remark: When the molar entropy of a gas is quoted for the old standard pressure of 1 atm, we get the molar entropy of this gas under the new standard pressure of 1 bar by adding $R \ln 1.01325$, that is,

0.1094 J. For condensed substances, the correction to be applied is about 1000 times smaller, i.e. less than experimental uncertainty.

Standard entropy of reaction—An *entropy of reaction* is said to be standard when each constituent is in its *standard state*. The standard entropy $\Delta_r S°$ of a reaction $\sum v_i M_i = 0$ may be calculated from the *standard entropies* of the constituents M_i, generally given in tables at $T_0 = 298.15\,K$:

$$\Delta_r S°(T_\circ) = \sum v_i S_i°(T_\circ).$$

The standard entropy of reaction at a temperature T is then calculated from the *heat capacities* $c_P(M_i)$:

$$\Delta_r S°(T) = \Delta_r S°(T_\circ) + \int_{T_\circ}^{T} \frac{\Delta c_P}{T}\,dT,$$

with $\Delta c_P = \sum v_i c_P(M_i)$. This relationship, similar to that which Kirchhoff established for the *standard enthalpy of reaction*, is valid only if between $T_\circ$ and T no constituent M_i undergoes a phase transition. Otherwise it is necessary to take such transitions into account and the remarks made about the standard enthalpy of reaction, always valid, lead to the relationship:

$$\Delta_r S°(T) = \Delta_r S°(T_\circ) + \int_{T_\circ}^{T} \frac{\Delta c_P}{T}\,dT + \sum v_i \frac{\Delta_{tr} H°}{T_{tr}}.$$

The sum is performed only over the transitions observed between $T_\circ$ and T. This expression represents as many branches of curve plus one as the number of transitions between T_0 and T. A first-*order* transition is characterized by a discontinuity in the curves $\Delta_r H°$ and $\Delta_r S°$ vs T, which is not the case for a second-order transition. For a second-order transition, $\Delta_{tr} H° = 0$, and the curve $c_P(M_i)$ vs T shows a singularity in the shape of a λ, which makes the integration more arduous.

S

Standard fugacity—The *fugacity* of a constituent taken in its *standard state*.

For a gas, if the selected standard fugacity is 1 bar, the corresponding standard pressure is, strictly speaking, slightly different from 1 bar, and, moreover, a function of temperature! This difficulty is usually ignored, owing to the small error introduced by assimilating pressure and fugacity in the region of 1 bar. One must nevertheless be conscious that the error may exceed 1% with organic vapours, even those of low complexity.

The standard fugacity of a condensed substance equals the vapour pressure of the compound taken in its standard state, usually the pure condensed compound. This vapour pressure may not be accessible to experiment, but there are numerous techniques allowing direct access to the *activity*: $a_i = f_i / f_i^\circ$.

Standard Gibbs energy of reaction—A *Gibbs energy of reaction* is said to be standard when each constituent is in its *standard state*.

The difference in nature between $\Delta_r G$, the Gibbs energy of a *chemical reaction* $\sum v_i M_i = 0$, and $\Delta_r G^\circ$, the standard Gibbs energy of the reaction, must be emphasized, because it is not a purely mathematical one:

$$\Delta_r G = \Delta_r G^\circ + RT \ln \prod a_i^{v_i}.$$

$\Delta_r G^\circ$ is indeed standard, because it is the value taken by $\Delta_r G$ when all the constituents M_i are in their standard state, that is, when their *activities* a_i equal unity. At fixed temperature, $\Delta_r G$ depends only on the state of the system (pressure and composition), whereas $\Delta_r G^\circ$ is a function only of the choice of standard states. $\Delta_r G^\circ$ varies with temperature, because it is convenient not to choose a standard temperature; $\Delta_r G^\circ$ does not vary with pressure, because the choice of a standard state implies that of a standard pressure.

$\Delta_r G^\circ$ and the *equilibrium constant* K are related by:

$$\Delta_r G^\circ \equiv -RT \ln K.$$

The sign of $\Delta_r G$ indicates the direction of evolution of a system; that of $\Delta_r G^\circ$ denotes only the position of K with respect to unity.

Standard state—The state of a constituent i, arbitrarily defined by the thermodynamicist, for which the *activity* is put equal to unity. To choose a standard state, it is necessary to specify:

- the pressure
- the *aggregation state*
- the nature (pure constituent or *solute* dissolved in a *solvent*)
- the solvent (if i is a solute)
- the concentration or the titre.

A standard state must not be confused with a *reference state*. Since ces of the two states are related, it may be more convenient, depending on the situation, to define either a standard state or a reference state. Generally, if a compound is considered whose titre may vary between 0 and 100% in a single phase, the best choice will be to specify a standard state, whereas in the case of a solute, it will be preferable to specify a reference state.

Standard temperature—A standard temperature is a monstrosity which does not exist in thermodynamics literature. Suppose for a moment that the scientific community, in a praiseworthy effort to leave nothing to chance, were to propose the adoption of a standard temperature $T°$. The consequences would be troublesome. Indeed, such a decision would amount to binding one hand and foot by forbidding the realization of a standard state at another temperature. This would lead logically to expressing the *activity* of a pure condensed substance as a function of temperature, except at the cost of unnatural mental gymnastics. A happy consequence would be to yield a truly constant *'equilibrium constant'*, but at what a price!

Thermodynamic tables are often given at 298.15 K. It must be clear in everyone's mind that such a temperature is not a standard temperature! In the relationship $\Delta_r G° = \Delta_r H° - T\Delta_r S°$, the change in the standard Gibbs energy $\Delta_r G°$ is explicitly a function of T. Rigorously, $\Delta_r H°$ and $\Delta_r S°$ also depend on T.

State—The state of a *system* is defined by the ensemble of physical quantities: temperature, pressure, composition, etc., which characterize the system, but neither by its *surroundings* nor by its history.

State function—A function of *state variables*. For a function Φ to be a state function, it is necessary and sufficient that the *differential* $d\Phi$ be an *exact differential*. The following statements are equivalent; if one of them is satisfied, the other three are also satisfied:

- Φ is a state function;
- $d\Phi$ is an exact differential;
- $\oint d\Phi = 0$;
- $\int_i^f d\Phi = \Phi(f) - \Phi(i)$, independent of the path followed.

State of matter—Tradition acknowledges three states of matter: *solid*, *liquid*, and *gaseous*. The condensed states (solid or liquid) contrast with the gaseous state; the fluid states (liquid or gas) contrast with the solid state. *Mesomorphous* states are intermediate between the solid and liquid states; the *plasma* state is sometimes considered as a fourth state of matter.

State principle—See *Law of stable equilibrium, corollary 4*. The state principle is sometimes presented as a statement of the *second law*.

State space—Space containing enough dimensions to allow the description by one point of the *macroscopic state* of a dynamic system. During its evolution with time, this representative point describes a trajectory in the state space. Attractors are mathematical objects described in the state space.

State variable—A physical quantity, a knowledge of which gives information about the state of a *system*. Among the variables that describe the state of a system, some are not independent. It is then possible to choose arbitrarily certain quantities as state variables and to consider others as functions of these state variables.

Stationary state—An *open* system is said to be in a stationary state when all transformations that occur, or all elementary steps of the same overall reaction, take place at the same rate, independent of time; this distinguishes the stationary state from the *steady state*.

In a stationary state, as well as in an *equilibrium state*, parameters characterizing the system do not change with time. However, there is a striking difference between the two states: in an equilibrium state, the *entropy production* is zero and the surroundings remain unmodified, whereas in a stationary state, entropy production exists, which obviously modifies the state of the surroundings. The theorem of *minimum entropy production* states that in a stationary state, the entropy production is a minimum.

Steady state—An *open* system is said to be in a steady state when all transformations that occur, or all elementary steps of the same overall reaction, take place at the same rate, dependent on time; this distinguishes the *stationary state* from the steady state. A system may tend towards equilibrium through a succession of steady states. In a steady state, as well as in a stationary state, the *entropy production* is a minimum.

Stefan–Boltzmann law—Discovered experimentally in 1879 by Josef Stefan, Austrian physicist (1835–1893), it was given a thermodynamic proof by *Boltzmann* five years later. It states that the radiation energy density of a black body, that is, an object which absorbs all radiation falling on it, is proportional to the fourth power of the temperature. The proof given here under the entry *radiation*, arising from statistical considerations, is not that of Boltzmann:

$$E/V = aT^4$$

$$a = 8\pi^5 k^4/15c^3h^3 = 7.5646 \times 10^{-16}\,\mathrm{J\,m^{-3}\,K^{-4}},$$

where aT^4 represents the radiation energy per unit volume. If we drill a hole through the wall of an enclosure which contains radiation at equilibrium at a temperature T, the energy emitted through the hole per unit time and per unit surface is also proportional to the fourth power of the temperature:

$$E/t\mathscr{S} = \sigma T^4,$$

σ being the Stefan–Boltzmann constant, defined by:

$$\sigma = ac/4 = 5.67051 \times 10^{-8}\,\mathrm{W\,m^{-2}\,K^{-4}}.$$

Stirling cycle—The first external combustion engine working by this cycle, composed of two *isochoric* and two *isothermal* curves, was demonstrated in 1818 by Robert Stirling (1790–1878), Scottish clergyman. The Stirling engine is trickier to use than the traditional internal combustion engine or than the Brayton gas turbine, but it is low-polluting, not demanding as regards the nature of the necessary heat supply, and has a better efficiency. Although obsolete, it may still have a future.

S

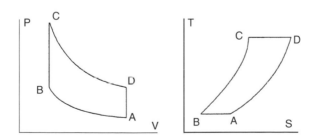

Stirling cycle: P–V (left) and T–S (right) diagrams.

Stirling formula—Incorrectly attributed to James Stirling (1692–1770), Scottish mathematician who made important advances in the theory of infinite series, it was first originated by Abraham de Moivre (1667–1754), pioneer in the theory of probability. It gives an approximation for $N!$ when N is a large number:

$$N! = N^N e^{-N} \sqrt{2\pi N}(1+\varepsilon)$$

or:

$$\ln N! = N \ln N - N + \frac{1}{2}\ln(2\pi N) + \frac{\Theta(N)}{12N}.$$

In most cases, it is possible to make do with the simplified expansion:

$$\ln N! \approx N \ln N - N.$$

Proof: Consider the integral: $\Gamma(N) = \displaystyle\int_0^\infty x^N e^{-x}\,dx$.

Integration by parts yields:

$$\Gamma(N) = N \int_0^\infty x^{N-1} e^{-x}\,dx = N\Gamma(N-1).$$

By recurrence, we show that if N is a whole number, $\Gamma(N) = N!$

We get a limited expansion of $\Gamma(N)$ with change of variables:

$$x = N + y\sqrt{N} = N(1 + y/\sqrt{N})$$

$$x^N e^{-x}\,dx = \sqrt{N}\exp(N\ln N - N)\exp[\ln(1+y\sqrt{N}) - y\sqrt{N}]\,dy.$$

By expanding: $\ln(1+y\sqrt{N}) = y\sqrt{N} - Ny^2/2 + \ldots$

Then, substituting: $N! \approx \sqrt{N} e^{-N} N^N \displaystyle\int_{-\sqrt{N}}^\infty e^{-y^2/2}\,dy.$

When $N \to \infty$, the integral tends towards: $\displaystyle\int_{-\infty}^\infty e^{-y^2/2}\,dy = \sqrt{2\pi}$,

whence:

$$N! \approx \sqrt{2\pi N}\, e^{-N} N^N.$$

Remark: We must be well aware of the approximations made when using the Stirling formula. Consider for instance the sharing of N balls between two bags, both having the same probability (0.5) of receiving a ball. The probability of finding $N/2$ balls in each bag is:

$$P = \frac{1}{2^N} \cdot \frac{N!}{(N/2)!\,(N/2)!}.$$

On using, when $N \to \infty$, the Stirling formula in its simplified form: $\ln N! \approx N \ln N - N$, we get $P = 1$, whereas on using the more complete expansion: $\ln N! \approx N \ln N - N + \ln \sqrt{2\pi N}$, we get: $P = \sqrt{2/\pi N}$, a different result, because the probability of equipartition, far from being a certainty, tends on the contrary to zero when N increases indefinitely.

Mathematically, the reason arises from the fact that the two expressions $\ln N!$ and $(N \ln N - N)$ are equivalent because their ratio tends towards 1 when N tends towards infinity, but their difference also tends towards infinity. Physically (see *Probability*), it is clear that for large values of N (in practice, $N \approx 10^{24}$), the probability of strict equipartition becomes vanishing small, whereas the probability of equipartition within ε in relative value becomes a near-certainty. In fact, the ratio between the *standard deviation* $\sigma = \sqrt{0.25N}$ and the most probable distribution ($N/2$) tends towards 0 when $N \to \infty$.

Stoichiometry—In 1794, the German chemist Jeremias Benjamin Richter (1762–1807) suggested that substances react with one another in definite proportions. To characterize this behaviour, he coined the term stoichiometry, from the Greek $\sigma\tau o\iota\chi\varepsilon\iota o\nu$, well-ordered. Although ionic (NaBr) or molecular (C_6H_6) compounds are often stoichiometric, non-stoichiometry nevertheless remains widespread. Thus, the composition of iron protoxide 'FeO', or wüstite, may vary from $Fe_{0.95}O$ to $Fe_{0.87}O$. In a *chemical reaction* $\sum v_i M_i = 0$, the v_i are the stoichiometric numbers, sometimes incorrectly called stoichiometric coefficients.

Stress—Force per unit surface developed within a material owing to the application of a force, a permanent strain, a temperature gradient, etc. Stress has the dimensions of pressure. It may be normal when applied perpendicularly to the cross-section of the material, by pulling or squeezing; a shear stress is exerted in the plane of the cross-section.

Sublimation—A direct transition from the solid state to the vapour state without intermediate melting. Sublimation is an *endothermic* phenomenon.

Surface tension—When a molecule lies in the bulk of a material, far from an interface, it is subjected to interactions from each molecule which surrounds it, over a solid angle of 4π steradians. Near an interface, it is also subjected to interactions from each molecule which surrounds it, but these molecules are no longer symmetrically distributed in the space. Since interactions between molecules are attractive, at equilibrium and without any other interaction a system will take a shape which gives it the minimum surface. In order to increase the surface of a system, it is necessary to provide work.

The surface tension Γ represents the ratio between dW, the work exchanged by the system with the surroundings, and $d\mathscr{S}$, the observed surface change. In the absence of any other interaction, the change in *internal energy* of a system whose surface varies by $d\mathscr{S}$ is given by:

$$dU = \Gamma\, d\mathscr{S}.$$

In the SI system, surface tension is expressed in $J\,m^{-2}$ or in $N\,m^{-1}$; it is often lower than $1\,J\,m^{-2}$, depending on the temperature, pressure, and nature of the phases in contact:

- liquid water–ice interface at $0°C$: $\Gamma = 22\,mN\,m^{-1}$;
- liquid water in equilibrium with its vapour at $25°C$: $\Gamma = 72\,mN\,m^{-1}$;
- grain boundary in an austenitic steel: $\Gamma \approx 850\,mN\,m^{-1}$;
- liquid copper in equilibrium with its vapour at $1100°C$: $\Gamma = 1.25\,mN\,m^{-1}$.

The surface tension at a liquid–vapour interface decreases when T increases and becomes zero at the critical temperature. The influence of surface tension forces is often negligible. It increases with the surface/volume ratio, and becomes important for emulsions or very small particles. Owing to surface tension forces, the vapour pressure of a *droplet* is a reciprocal function of its radius of curvature.

Surroundings—The surroundings, or environment, represent the part of the universe which does not belong to the *system* defined by the thermodynamicist. The surroundings are of interest to him only if they can have some influence on the evolution of the system. Unless specified to the contrary, the surroundings are considered as a *reservoir* whose capacity is infinite and whose intensive properties (temperature, pressure, chemical potentials, etc.) do not vary, whatever the exchange of matter and energy with the system.

Symmetry principle—Stated by Pierre Curie in 1894, this 'principle' states that an effect must have the same symmetries as the causes which produce it. The appellation 'principle' is a misuse, because this is actually a result deduced from tensor calculus. In thermodynamics, it finds expression in the fact that *coupling* occurs only between two processes whose tensor order is the same. For instance, no coupling can be observed between a chemical reaction and a heat transfer. Indeed, the *affinity*, the cause of the reaction, is a scalar (tensor of order zero) whereas the thermal gradient, the cause of the heat transfer, is a vector (tensor of order 1).

Syncrystallize—From the Greek $\sigma\upsilon\nu$, together. To syncrystallize is to produce a solid *solution*, either by precipitation from a liquid solution or by cooling from a liquid phase. A syncrystallization gap for a solid solution, or miscibility gap, is a zone where *unmixing* of the solution occurs.

Syntectic—From the Greek $\sigma\upsilon\nu$, together, and $\tau\varepsilon\kappa\tau\sigma\varsigma$, fusible. The name given to the transformation observed on heating:

$$\langle \text{Solid } \gamma \rangle \rightarrow (\text{Liquid L}_1) + (\text{Liquid L}_2).$$

The solid melts with formation of two immiscible liquids whose compositions differ. Alkali metals (Na, K, Rb) give easily with zinc an intermetallic compound MZn_{13} which undergoes a syntectic decomposition:

$$NaZn_{13} \xrightarrow{557^\circ C} L_1(0.8 \text{ at\% Zn}) + L_2(94 \text{ at\% Zn}).$$

System—The object of closest attention by the thermodynamicist, a system is a fraction of the *universe* which must be carefully defined. A system is separated from the remainder of the universe by a *boundary* which may be material or not, but which, by convention, delimits a finite volume. The possible exchanges of work, heat or matter between the system and the *surroundings* take place across this boundary.

T

Temperature—The concept of temperature, a *tension* associated with *entropy*, follows from the *zeroth law* of thermodynamics. In the SI system, temperature is expressed in kelvins (symbol K); a temperature difference may be expressed indifferently in kelvins or in degrees Celsius (symbol °C), with $T/K = t/°C + 273.15$.

Temperature–entropy diagram—Sometimes called the Mollier diagram together with the *enthalpy–entropy diagram*, it serves the same purposes. In the two-phase domain, the equilibrium lines have zero slope; the loci of the L and V points meet at the critical point with a horizontal tangent. On this diagram, *isenthalpic*, *isobaric*, and *isochoric* curves are drawn. From the expressions of dS obtained from *Maxwell's equations:*

$$dS = c_P \frac{dT}{T} - \left(\frac{\partial V}{\partial T}\right)_P dP; \quad dH = c_P dT + \left[V - T\left(\frac{\partial V}{\partial T}\right)_P\right]dP,$$

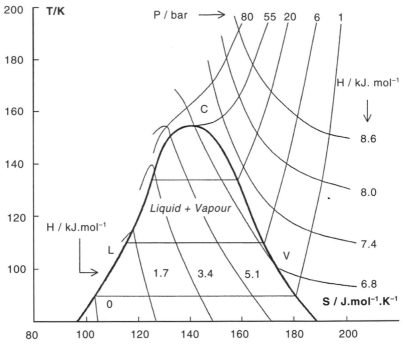

Temperature–entropy diagram: On this diagram for dioxygen, the origin of the isenthalpic curves (H=0) has been arbitrarily chosen at the point corresponding to the liquid in equilibrium with its vapour at 90 K under 1 bar.

we obtain:

- $H = \text{const.} \Rightarrow \left(\dfrac{\partial T}{\partial S}\right)_H = \dfrac{T}{c_P}\left[1 - \dfrac{T}{V}\left(\dfrac{\partial V}{\partial T}\right)_P\right] = \dfrac{T}{c_P}(1 - \alpha_P T);$

- $P = \text{const.} \Rightarrow \left(\dfrac{\partial T}{\partial S}\right)_P = \dfrac{T}{c_P};$

- $V = \text{const.} \Rightarrow \left(\dfrac{\partial T}{\partial S}\right)_V = \dfrac{T}{c_V}.$

The relative order of the slopes of the isochoric, isobaric, and isothermal curves is the same as on an enthalpy–entropy diagram:

$$\left(\frac{\partial T}{\partial S}\right)_V = \frac{T}{c_V} > \left(\frac{\partial T}{\partial S}\right)_P = \frac{T}{c_P} > \left(\frac{\partial T}{\partial S}\right)_T = 0.$$

The slope of the isenthalpic curves may be positive or negative, that is, greater or less than that of the isotherms, which is zero by construction. However, it is always less than that of the isobars. It is easily shown that the sign of the slope of the isenthalps is opposite to the sign of the Joule–Thomson coefficient.

$$\mu_J = \left(\frac{\partial T}{\partial P}\right)_H = \frac{1}{c_P}\left[T\left(\frac{\partial V}{\partial T}\right)_P - V\right].$$

By drawing on a temperature–entropy diagram the locus of the extrema of the isenthalpic curves, we get a parabolic curve which divides the plane into two domains. Inside the curve, near the critical point, the slope of the isenthalps is negative: an isenthalpic expansion induces cooling of the fluid. Outside the curve, the slope of the isenthalps is positive: an isenthalpic expansion induces heating of the fluid. For an *ideal gas*, the slopes of the isenthalpic and isothermal curves are zero, the enthalpy of an ideal gas being dependent only on the temperature; the slopes of the isobars and isochors remain respectively equal to T/c_P and T/c_V.

The equilibrium curve between two phases is formed of two branches, whose slopes can be obtained by means of the differential of dS:

$$\left(\frac{\partial S}{\partial T}\right)_x = \frac{c_P}{T} - \left(\frac{\partial V}{\partial T}\right)_P\left(\frac{\partial P}{\partial T}\right)_x = \frac{c_x}{T} \Rightarrow \left(\frac{\partial T}{\partial S}\right)_x = \frac{T}{c_x},$$

where c_x is the *heat capacity* along the saturated vapour curve.

The above expression may be applied to both liquid and vapour phases. For the liquid, $c_x > 0$; for the saturated vapour, c_x is generally negative. If they are two inversion points for which c_x becomes zero, the curve related to the saturated vapour has two vertical tangents between which $c_x > 0$, that is, $(\partial T/\partial S)_x > 0$. An *adiabatic* compression of the saturated vapour takes the system into the two-phase (liquid + vapour) domain. On the other hand, when $c_x < 0$, an adiabatic compression of the saturated vapour takes the system into the single-phase domain of the unsaturated vapour.

Tension—In thermodynamics, a tension is the partial derivative of the *internal energy* with respect to an *extensity*, other extensive quantities being kept constant. Hence with any extensity x_i it is always possible to associate a tension variable $X_i = \partial U/\partial x_i$, which is called conjugate. According to the *first law*, the change in internal energy of a *system* is given by:

$$dU = \sum X_i \, dx_i.$$

When two systems A and B interact, the assembly (A + B) being *isolated*, the energy exchanged under the influence of a difference in tension is manifested by a transfer of extensity. The greater the difference in tension between A and B, the greater is the flux of extensity crossing the boundary. The definition of a tension has two direct consequences:

- a tension is an intensive quantity which may be defined at each point of the space;
- tensions measured on both sides of the boundary separating two systems at equilibrium are equal.

Remark 1: Be careful not to confuse tension and *intensity*. All tensions are intensive quantities, because they are defined punctually as the derivative of an extensity (internal energy) with respect to another extensity, but the reverse is not true. Not all intensive quantities are necessarily equal on both sides of the boundary separating two systems at equilibrium. A concentration for instance is an intensive quantity but not a tension, as opposed to a *chemical potential*.

Remark 2: The word 'tension' is easily put to all sorts of uses; somewhat affected expressions such as 'electrical tension' and 'vapour tension' are strictly speaking not incorrect, because an electrical potential or a vapour pressure is in fact a tension, but they lack preciseness.

Tensor—Some physical quantities, called 'invariant', are not modified by a change in the axes of coordinates. Tensor calculus is the branch of mathematics dealing with the invariant description of systems. A tensor is a mathematical object which is to be regarded with respect. For the average thermodynamicist, it is enough to know that:

• a tensor of order 0 is a scalar;
• a tensor of order 1 making two points correspond is a vector;
• a tensor of order 2 making two vectors correspond is a matrix.

Tesla—The unit of *magnetic field* in the SI system (symbol T), from Nikola Tesla (1857–1943), Croatian engineer:

$$1\,T = 1\,kg\,s^{-2}A^{-1} = 1\,Wb\,m^{-2} = 1\,V\,s\,m^{-2}.$$

Thermal expansion coefficient—See *Expansivity*.

Thermochemistry—The branch of thermodynamics dealing with heat exchange accompanying transformations: mixing, phase transitions, chemical reactions, etc. The laws of thermochemistry rest on two statements:

• Lavoisier and Laplace's law (1780): 'The heat exchange accompanying a transformation is equal and opposite to the heat exchange accompanying the reverse transformation'.
• Hess's law (1840): 'The heat exchange accompanying a transformation is the same whether the process occurs in one or several steps'.

Both laws preceded the statement of the *first law* (1850), but it is easy to be convinced that they are a direct consequence of it!

Thermodiffusion—A *coupling* effect between diffusion and heat transfer. Recall the general expression for *entropy production*:

$$\dot\sigma \equiv \frac{d\sigma}{dt} = \sum_i J_i X_i.$$

If $\dot\sigma$ is the entropy production per unit volume, J_i is a *flux* and X_i a *generalized force*. Within the framework of the linear thermodynamics of irreversible processes:

$$J_i = L_{ii}X_i + L_{iQ}X_Q = L_{ii}\left(-\frac{1}{T}\frac{\partial\mu_i}{\partial x}\right) + L_{iQ}\left(\frac{\partial(1/T)}{\partial x}\right)$$

$$J_Q = L_{Qi}X_i + L_{QQ}X_Q = L_{Qi}\left(-\frac{1}{T}\frac{\partial\mu_i}{\partial x}\right) + L_{QQ}\left(\frac{\partial(1/T)}{\partial x}\right).$$

T

$J_i = L_{ii}[-(1/T)(\partial\mu_i/\partial x)]$ expresses Fick's law relating to diffusion (1856), the driving force being the *chemical potential* gradient.

$J_Q = +L_{QQ}[\partial(1/T)/\partial x]$ expresses Fourier's law relating to thermal conduction (1818), the driving force being the $(1/T)$ gradient.

In the absence of any initial temperature gradient, the presence of a chemical potential induces heat transfer, whence the build-up of a temperature gradient: the *Dufour effect*. This effect is appreciable only with gases.

In the absence of any initial chemical potential gradient, the presence of a temperature gradient induces displacement of matter, whence the build-up of a concentration gradient: the *Soret effect*.

Thermodynamic potential—The name given to a function whose minimum gives the equilibrium state of a system subjected to *constraints*. It is possible to describe a thermodynamic potential corresponding to one constraint or to many constraints. Recall for instance the expression for the *internal energy*:

$$dU = dQ + dW = T(dS - d\sigma) + dW = -T\,d\sigma + \sum X_i\,dx_i.$$

Now $d\sigma \geqslant 0$ according to the *second law*; X_i represents a quantity of *tension* and x_i a quantity of *extensity*. It is clear that if the extensive quantities are kept constant, there remains $dU = -T\,d\sigma \leqslant 0$. The function U can only decrease and tend towards a minimum which will be reached at equilibrium. The internal energy is thus the thermodynamic potential of a system subject to the constraints $x_i = \text{const.}$

From the state function U, it is possible to introduce other state functions by using Legendre transformations. For instance:

$$\Omega = U - X_j x_j, \text{ hence: } d\Omega = -T\,d\sigma - x_j\,dX_j + \sum_{1 \neq j} X_i\,dx_i.$$

The function Ω is the thermodynamic potential of a system subjected to the constraints $X_j = \text{const.}$ and $x_{i \neq j} = \text{const.}$ Among the most often encountered thermodynamic potentials may be cited *negentropy* (isolated system), *Helmholtz energy* (temperature, volume, and amount of substance constant), and *Gibbs energy* (temperature, pressure, and amount of substance constant). Of course, as well as internal energy already discussed, *enthalpy* is also the thermodynamic potential of a system in which entropy, pressure, and amount of substance are kept constant.

Thermodynamics—Thermodynamics did not arise like Aphrodite in all her glory out of the sea-foam, but blossomed gently, developing during the nineteenth century and reaching its full flowering during the first half of the twentieth century. The word itself was coined by Joule in 1858 to designate the science of relations between heat (Greek $\theta\varepsilon\rho\mu\eta$) and power (Greek $\delta\upsilon\nu\alpha\mu\iota\varsigma$). Of course, a classical phenomenon in the history of science, its subject matter quickly expanded, and thermodynamics now designates the science of all transformations of matter and energy. It is convenient to divide it into two distinct branches:

THERMODYNAMIC OF EQUILIBRIUM, whose aim is to provide an answer to a plain question: given a system in a well-defined initial state, subject to accurately specified constraints, what will be the state of the system once it has reached equilibrium? This branch developed between 1824 and 1918 with Carnot, Clapeyron, Joule, Clausius, Kelvin, Maxwell, Gibbs, Planck, Nernst, and so on. It is possible to divide it into two branches again, depending on the point of view from which the system is considered.

What is called 'classical', or better, 'macroscopic' thermodynamics considers matter as a continuous medium and does not make any assumption about its detailed structure, as opposed to 'microscopic' thermodynamics, which views matter as an assembly of particles, atoms, or molecules whose behaviour it analyses. Such thermodynamics, also called 'statistical', because it uses the theory of probability, was born with Boltzmann in 1870 and underwent considerable expansion at the beginning of the twentieth century with the birth of atomic physics and the development of quantum mechanics.

THERMODYNAMICS OF IRREVERSIBLE PROCESSES, which truly began with De Donder, then Onsager in 1931. It is concerned with the description of the state of a system out of equilibrium and its evolution with time. The study of systems close enough to equilibrium for a linear relationship to be assumed between observed fluxes and the forces that cause them gave rise to the 'linear thermodynamics of irreversible processes'. The linear hypothesis having revealed limitations, for instance in the study of chemical reactions, a 'non-linear thermodynamics of irreversible processes' was developed with Prigogine, which is still in progress.

T

Thermodynamic temperature—Since corollary 7 of the *second law* lays down that both *energy* and *Helmholtz energy* are state functions, their difference must also be a state function, which allows us to put:

$$dS = \frac{1}{T}(dE - dF),$$

introducing at the same time the *entropy* S and the thermodynamic temperature T. Corollary 11 shows that during a *reversible diabatic process* $dS = dQ_{rev}/T$. Finally, according to *Carnot's theorem*, another consequence of the second law, all reversible engines working between two temperatures T_1 and T_2 have the same efficiency:

$$|Q_1|/|Q_2| = -Q_1/Q_2 = f(T_1, T_2).$$

There are many functions $f(T_1, T_2)$ which satisfy the above relationship. As it is not useful to complicate when it is possible to simplify, it is agreed to put:

$$-Q_1/Q_2 = T_1/T_2.$$

This choice characterizes the thermodynamic scale of temperature. The application of the relationship $dS = dQ_{rev}/T$ to an engine exchanging over a reversible cycle a heat quantity Q_1 with a reservoir at temperature T_1 and Q_2 with a reservoir at temperature T_2 yields:

$$\Delta S = \oint \frac{dQ_{rev}}{T} = \frac{Q_1}{T_1} + \frac{Q_2}{T_2} = 0.$$

The above thermodynamic temperature is defined apart from constant. To fix this constant, we must select the thermodynamic temperature of a particular reservoir. In 1954, the thermodynamic temperature of the triple point of water was fixed at exactly 273.16 K. Once this temperature is defined, which is easily done, it is theoretically possible to obtain that of any other reservoir by measuring the efficiency of a reversible engine working between that temperature and 273.16 K.

The thermodynamic temperature, based on the concept of a reversible engine, is thus an absolute temperature. Unfortunately, a reversible engine belongs to 'thermotopia' and it is easier to define the thermodynamic temperature from the ideal-gas thermometer. Fortunately, by making a reversible engine work over a *Carnot cycle*, it is shown that thermodynamic temperature identifies with the absolute temperature defined from an ideal-gas thermometer.

Thermoelastic—A system is called thermoelastic when the only work exchanged with the surroundings is that of pressure forces. In such a system, the basic relationships are simplified:

$$dU = dQ + dW = T(dS - d\sigma) - P\,dV.$$

Thermoelectric effect—A thermoelectric effect results from *coupling* between thermal and electric phenomena, generally observed at the junction between two metals.

The *entropy production* is expressed by:

$$\dot{\sigma} = \sum J_i X_i = J_Q X_Q + J_e X_e,$$

where J_Q and J_e represent the *fluxes* of heat and electricity through the junction. *Generalized forces* associated with these fluxes are:

- the reciprocal temperature gradient: $X_Q = \Delta(1/T) = -\Delta T/T^2$;
- the *electrochemical potential* gradient: $X_e = -\Delta\tilde{\mu}/T$.

The phenomenological relationships are written (with $L_{Qe} = L_{eQ}$):

$$J_Q = -L_{QQ}\frac{\Delta T}{T^2} - L_{Qe}\frac{\Delta\tilde{\mu}}{T}$$

$$J_e = -L_{eQ}\frac{\Delta T}{T^2} - L_{ee}\frac{\Delta\mu}{T}.$$

From the above expressions, it is possible to deduce three effects:

- the *Peltier effect*: $(J_Q/J_e)_{\Delta T=0} = L_{Qe}/L_{ee}$;
- the *Seebeck effect*: $(\Delta\tilde{\mu}/\Delta T)_{J_e=0} = -L_{eQ}/TL_{ee}$;
- the *Thomson effect*: $(J_e/J_Q)_{\Delta\tilde{\mu}=0} = L_{eQ}/L_{QQ}$.

Taking into account the *Onsager reciprocity*: $L_{eQ} = L_{Qe}$, we obtain the Kelvin equation relating π, the Peltier coefficient, and α, the Seebeck coefficient: $\pi = \alpha T$.

Thermomechanical effect—The name given to the appearance of a heat flux under the influence of a difference in pressure between two systems maintained at constant temperature. The heat flux between the two systems A and B is obtained by measuring the heat quantity which must be supplied to enclosure A or extracted from enclosure B in order to keep the temperature constant. In the *steady state*, the two quantities must be equal in absolute value. The *entropy production* is expressed by:

$$\dot{\sigma} = \sum J_i X_i = J_Q X_Q + J_m X_m,$$

where J_Q and J_m represent the *fluxes* of heat and matter through the junction. *Generalized forces* associated with these two fluxes are:

- the reciprocal temperature gradient: $X_Q = \Delta(1/T) = -\Delta T/T^2$;
- the *chemical potential* gradient: $X_m = -\Delta\mu/T = -V\cdot\Delta P/T$.

The phenomenological relationships are written (with $L_{Qm} = L_{mQ}$):

$$J_Q = -L_{QQ}\frac{\Delta T}{T^2} - L_{Qm}\frac{V}{T}\Delta P$$

$$J_m = -L_{mQ}\frac{\Delta T}{T^2} - L_{mm}\frac{V}{T}\Delta P.$$

From the above expressions, it is possible to deduce two effects:

- the thermomechanical effect: $(J_Q/J_m)_{\Delta T = 0} = L_{Qm}/L_{mm}$;
- the *thermomolecular effect*: $(\Delta P/\Delta T)_{J_m = 0} = -L_{mQ}/L_{mm}VT$.

Thermomolecular effect—The name given to the pressure gradient which develops between two fluids separated by a capillary or a porous boundary in the absence of a flux of matter under the influence of a temperature gradient. The name thermomolecular effect is sometimes reserved for the difference in pressure developed between gaseous systems, whereas for the liquid systems the term 'thermo-osmosis' is used. From the phenomenological equations given under *thermomechanical effect*, the thermomolecular effect is characterized by:

$$(\Delta P/\Delta T)_{J_m = 0} = -L_{mQ}/L_{mm}VT.$$

Third law—The functions U, H, F, and G are defined apart from a constant, which in practice raises no problem, because thermodynamics deals only with the changes in these functions. The *entropy* function is also defined apart from a constant, but it must necessarily be fixed, otherwise the calculation of a change ΔG by integration of an expression such as $dG = -S\,dT + V\,dP$ would give a result depending on the value attributed to the constant defining S, which is unacceptable physically. The third law of thermodynamics results from that necessity.

There are several equivalent statements, proposed between 1906 and 1918 (see *Nernst theorem*). Is the third law truly a *principle*, or must it be reduced to the level of a theorem? The debate, which is still not closed, is generally ignored by textbooks.

THE FIRST STATEMENT is that of the inaccessibility of *absolute zero*:

It is impossible to reach absolute zero in a finite number of operations.

Nernst gave this statement in 1912, and in 1918 proposed a proof which may be summarized in few lines: the work W that must be supplied to a system in order to transfer a heat quantity Q from a cold source (temperature T_c) to a hot sink (temperature T_h) is:

$$W = -Q[1 - (T_h/T_c)].$$

The above relationship shows that $W \to \infty$ when $T_c \to 0$. It is also possible to say that the efficiency of a *Carnot cycle*:

$$\eta = (T_h - T_c)/T_h \to 1 \text{ when } T_f \to 0.$$

A reversible engine working by a Carnot cycle between a hot reservoir and a cold one at 0 K would take heat only from the hot source to transform it entirely to work, which contradicts the *second law*.

All these demonstrations have been criticized by Einstein, who showed that it is impossible to exchange heat at 0 K and hence to make a reversible engine work between a finite temperature T_h and a temperature $T_c = 0$. Indeed, since the isotherm $T = 0$ identifies with an adiabat (reversible isentrope), it is impossible to conceive an adiabat leaving the isotherm $T = 0$. Thus, the third law is not a consequence of the second law.

Without going into details of the academic debate surrounding this serious question, let us indicate that it is possible in some cases to prove the inaccessibility of absolute zero from the second law.

THE SECOND STATEMENT, also called the 'Nernst principle' is the following:

For any isothermal and reversible transformation of a system between two stable states, $\Delta S \to 0$ when $T \to 0$.

Nernst made the above proposition in 1906, and in 1918 showed that it is a consequence of the inaccessibility of absolute zero.

Indeed, if the entropies $S(\alpha)$ and $S(\beta)$ of the system in these two states were different at 0 K [let for instance $S(\beta, 0) < S(\alpha, 0)$], it would be possible, by starting from the system in its state β at a finite temperature T, such that $S(\beta,T) < S(\alpha,0)$, to bring the temperature down to $0\,\mathrm{K}$ in only one operation by means of an isentropic process. During this process, only a fraction of the system would undergo the transformation $\beta \to \alpha$.

Conversely, if $S(\alpha,T) \to S(\beta,T)$ when $T \to 0$, it would be impossible, whatever the initial temperature of the system in its state β, to bring it down to $0\,\mathrm{K}$ by means of an isentropic process in which the system would undergo the transformation $\beta \to \alpha$.

Thus the entropy of a system tends, when $T \to 0$, towards the same value, whatever its initial state, provided that this state is a stable one. In 1911, Planck proposed, for every system in a stable equilibrium state, to put $S = 0$ for $T = 0$. This convention allows an absolute value to be assigned to the entropy function.

Corollary 1: When $T \to 0$, $c_V \to 0$ and $c_P \to 0$. Indeed:

$$\text{at } V = \text{const.:} \quad \Delta S = \int_{T_\circ}^{T} \frac{c_V}{T}\,\mathrm{d}T; \quad \text{at } P = \text{const.:} \quad \Delta S = \int_{T_\circ}^{T} \frac{c_P}{T}\,\mathrm{d}T.$$

In order for the integrals to tend towards a finite value when $T \to 0$, it is necessary that c_V and c_P tend towards 0 and be proportional to $T^n (n \geqslant 1)$. Generally, experiment gives $n = 3$, in good agreement with the Debye model; for metals, $n = 1$, because to the general term due to the lattice, one must add a linear term due to the heat capacity of the conduction electrons.

Corollary 2: When $T \to 0$, the *expansivity* α_P and the *bulk expansion coefficient* β_V tend towards 0. Indeed, from *Maxwell's equations*:

$$\alpha_P = \frac{1}{V}\left(\frac{\partial V}{\partial T}\right)_P = -\frac{1}{V}\left(\frac{\partial S}{\partial P}\right)_T; \quad \beta_V = \frac{1}{P}\left(\frac{\partial P}{\partial T}\right)_V = \frac{1}{P}\left(\frac{\partial S}{\partial V}\right)_T.$$

As the entropy of every substance tends towards the same limit when $T \to 0$, this limit depends neither on the volume nor on the pressure: $(\partial S/\partial P)_T$ and $(\partial S/\partial V)_T$ thus tend towards zero when $T \to 0$; the same is therefore true for α_P and β_V. We must beware the trap of excessive generalization: the *compressibility*, expressed as a function of the two preceding quantities by $\chi_T = \alpha_P/P\beta_V$, tends towards a finite, non-zero value when $T \to 0$.

Corollary 3: An enthalpy of transition tends towards zero when $T \to 0$. Indeed, at equilibrium: $\Delta_{\mathrm{tr}}S = \Delta_{\mathrm{tr}}H/T_{\mathrm{tr}}$. When $T_{\mathrm{tr}} \to 0$, $\Delta_{\mathrm{tr}}S \to 0$ because the entropy of a substance at $0\,\mathrm{K}$ is independent of its crystal structure. The enthalpy of transition thus tends towards zero and is proportional to T^n (with $n > 1$) when $T \rightarrow 0$. Experiment shows generally that $n = 2$.

Furthermore, the third law has several limitations:

Limitation 1: The third law cannot be applied to metastable states (glasses, etc.) or to transformations out of equilibrium. In fact, some solids may show a residual entropy at 0 K because they retain some disorder owing to the low rate of diffusion processes at low temperatures. The best-known example is that of molecular crystals such as CO, NO, N_2O, etc. The fact that the relative positions of the C and O atoms in a molecule of CO may be easily reversed, an inversion retained at low temperatures, explains the presence of a residual entropy at 0 K. The entropy calculated by spectrometry is indeed slightly higher than that measured by calorimetry on the basis of the third law.

Limitation 2: The third law cannot be applied to gases, real or ideal. The entropy change of an ideal gas whose pressure varies from P_1 to P_2 is given by $\Delta S = -R \ln(P_2/P_1) \neq 0$, which contradicts the third law. This difficulty may be overcome in two ways: the first consists in emphasizing that at 0 K, since all substances are condensed with a vapour pressure equal to zero, there are no longer any gases, ideal or not. This conjuring trick maybe allows the problem to be eliminated elegantly, but surely not solved.

A more satisfying analysis consists in noting that the behaviour of an ideal gas is explained by a model of particles without interactions. This model, realistic at low pressures and above 0 K, does not take into account the quantum effects that are important when approaching absolute zero (see for instance *Sackur–Tetrode formula*). The application of quantum statistics to the ideal gas allows it to come back to order again, rescuing the third law (phew!).

Limitation 3: The third law seems not to apply to the solid–liquid transition of helium-3. Indeed, the equilibrium curve $P = f(T)$ goes through a minimum at $T = 0.35$ K and $P = 29.5$ bar. Below 0.35 K, the equilibrium pressure increases and tends towards 35 bar when $T \rightarrow 0$ with a seemingly vertical tangent. Now the *Clapeyron equation*: $(dP/dT)_{eq} = \Delta_{tr}S/\Delta_{tr}V$ would imply $\Delta_{tr}S \rightarrow -\infty$ when $T \rightarrow 0$, because $\Delta_{tr}V$ remains finite, which contradicts the third law. Here the problem is a real one, because quantum mechanics does not, as in the above case, come to play *deus ex machina* and save the situation. It is always possible to hope that $(dP/dT)_{eq}$ condescends at very low temperatures, still not available experimentally, to tend towards zero, as expected from the theory!

T

Thomson, Sir William, Lord Kelvin of Largs (1824–1907)— British engineer, mathematician and physicist who in 1852 discovered the cooling of gases by expansion and contributed to the establishment of an absolute scale of temperature. His considerable work, more than 600 papers, also relates to electromagnetism, hydrodynamics, underwater transmission, and determination of the age of the earth.

Thomson effect—A *thermoelectric effect* discovered in 1857 and related to the *Seebeck* and *Peltier effects*. However, it is harder to demonstrate. It is characterized by an exchange (production or absorption) of heat with the *surroundings* due to the flow of an electric current through a conductor in the presence of a temperature gradient. Note: the Thomson effect, which is different from the *Joule effect*, also has nothing to do with the *Joule–Thomson effect*! The Thomson coefficient of a conducting material is defined by:

$$\tau = \frac{1}{I} \frac{dQ}{dx} \frac{dx}{dT}.$$

Titre—A solution is said to be titred when its composition is known. Titre is a dimensionless number, the ratio of the quantity characterizing a constituent to the same quantity characterizing the solution. However, the nature of the titre: mass, molar or volumetric, must be specified (see *Fraction*).

Tonometry—From Greek $\tau o \nu o \varsigma$, tension. Tonometry is the measurement of the lowering of vapour tension of a *solvent* in which a *solute* is dissolved. Tonometry may give access to the molar mass of the solute.

Let x_2 be the *mole fraction* of a solute, p_1° the vapour tension of the pure solvent, and p its vapour pressure above the solution. The *activity* of the solvent is: $a_1 = p_1/p_1^\circ$. When it is possible to apply *Raoult's law* to the solvent ($x_2 \approx 0$): $1 - x_2 = p_1/p_1^\circ$, then $x_2 = \Delta p/p_1^\circ$.

Δp is the tonometric lowering, the decrease in the vapour pressure of the solvent due to the presence of a solute. With a knowledge of the mass concentration of the solute, the measurement of Δp gives access to the molar mass.

Transfer variable—A name sometimes given to physical quantities which appear when they cross the boundary separating two systems. *Work* and *heat* are two examples of transfer variables.

A system may be characterized by its *energy E*, but not by its work *W* or by its heat *Q*, because neither d*W* nor d*Q* is generally an *exact* *differential*. The boundary must necessarily be well defined. Consider for instance a generator providing an electric current to a resistance used to heat a thermally isolated enclosure. If the system considered is the generator, it exchanges only work with its surroundings; if it is formed by the assembly (generator + resistance), it exchanges only heat with its surroundings; finally, if it is formed by the whole (generator + resistance + enclosure), it remains isolated and exchanges nothing with its surroundings.

Transformation—A transformation is defined by an initial state and a final state, independently of the path followed. For simplicity, a transformation is said to be reversible or irreversible according to whether the *process* is reversible or irreversible.

Transient state—An *open* system is in a transient state when not every transformation which occurs, or not every elementary step of the same overall reaction, takes place at the same rate, the reaction rates also being a function of the time. Such a state is called transient because it is not stable.

If the *constraints* and *fluxes* imposed on the system do not vary with time, the system evolves towards a *stationary state*. Prigogine showed that during a transient state, the rate of *entropy production* decreases with time: $d\dot{\sigma}/dt = d^2\sigma/dt^2 < 0$. The minimum of $\dot{\sigma}$ corresponds to a stationary state or an equilibrium state when the minimum equals zero.

Transition—From the Latin *transitio*, passing over. The word is used for a slow evolution between two states which may coexist at equilibrium. An allotropic transition for elements, or a polymorphic transition for compounds, may be observed only between two *enantiomorphic* crystal varieties.

Tricritical point—A point, characterized by its coordinates, at which three phases become identical. The application of the *phase rule* shows that three constituents are needed for the appearance of a tricritical point. In a ternary system, a tricritical point is necessarily *invariant*.

Triple point—The name given on a diagram to the meeting point of three equilibrium curves. On a *pressure–temperature diagram* relating to a pure substance, the triple point, where three phases coexist at equilibrium, corresponds to the meeting of three *monovariant* equilibrium curves; it is then *invariant*.

U

Uncompensated heat—The *entropy* created during an irreversible process is given, according to the *second law*, by:

$$d_i S \equiv d\sigma = dS - \frac{dQ}{T}.$$

De Donder and his school used, following Clausius, the term uncompensated heat, defined by:

$$dQ' = T d_i S.$$

The concept of *entropy production* is now used in preference to that of uncompensated heat. This latter expression is justified when considering an irreversible cycle:

$$\int_A^B T\,dS = \Delta Q_{AB} + \Delta Q'_{AB}, \qquad \int_B^A T\,dS = \Delta Q_{BA} + \Delta Q'_{BA},$$

with $\Delta Q'_{AB} > 0$ and $\Delta Q'_{BA} > 0$. Furthermore: $\oint T\,dS = 0$. Therefore:

$$\Delta Q_{AB} + \Delta Q_{BA} = -(\Delta Q'_{AB} + \Delta Q'_{BA}) < 0.$$

Over an irreversible cycle there is no compensation between the heat exchanged with the surroundings during the forward path AB and the return path BA. The overall thermal effect is to provide heat to the surroundings.

Undifferentiated equilibrium—A *system* is in an undifferentiated equilibrium state when it is possible to produce a finite modification of the state of the system by means of an infinitesimal modification of the state of the *surroundings*.

A sphere on a frictionless plane surface is a good example of a mechanical system in an undifferentiated equilibrium state. In thermodynamics, such states occur when it is possible to modify an extensive parameter while keeping intensive parameters unmodified.

At constant temperature, a liquid–vapour equilibrium is undifferentiated with respect to a change in volume of the system. In a more general way, an *invariant* equilibrium is undifferentiated with respect to any change in an extensive parameter. This is also true for a *monovariant* equilibrium, provided that an external constraint keeps $(n-1)$ intensities constant; the nth intensity will be then kept constant by the system.

Unfalsifiable—According to *Karl Popper*, a statement is said to be unfalsifiable when it cannot be proved, and whose incorrectness also cannot be proved. Example: 'There is a Greek sentence which, uttered in front of anybody who is averse to thermodynamics, will make them love the discipline' is an unfalsifiable statement. Thus, following Popper, it is not a scientific proposition. Actually, a large number of areas present unfalsifiable propositions which are justified by the exactness of their consequences.

Uniform—With intensive properties the same at each point of the space. 'Uniform' is more restrictive than 'homogeneous'. A *phase*, a homogeneous part of a mixture, is not necessarily uniform.

Units—With respect for the reader, the author of a scientific text keeps an eye on his language. For the same reason, since the SI system exists, he refrains from using units not in that system; he knows in fact that a force is expressed not in kg but in N, that an energy is expressed not in K, V, or m^{-1}, but simply in J!

The International System of Units, whose name and abbreviation SI were coined in 1960 at the 11th Conférence Générale des Poids et Mesures, consists of seven base units:

Length	metre	m
Mass	kilogram	kg
Time	second	s
Electric current	ampere	A
Thermodynamic temperature	kelvin	K
Amount of substance	mole	mol
Luminous intensity	candela	cd

To these seven basic physical quantities are added two supplementary units: plane angle, expressed in radians (symbol rad), and solid angle, which is expressed in steradians (symbol sr). All derived units are expressed in terms of the seven base units and two supplementary units.

Multiples and sub-multiples are expressed by prefixes:

U

exa-	E	10^{18}	atto-	a	10^{-18}
peta-	P	10^{15}	femto-	f	10^{-15}
tera-	T	10^{12}	pico-	p	10^{-12}
giga-	G	10^{9}	nano-	n	10^{-9}
mega-	M	10^{6}	micro-	μ	10^{-6}

kilo-	k	10^3	milli-	m	10^{-3}
hecto-	h	10^2	centi-	c	10^{-2}
deca-	da	10	deci-	d	10^{-1}

Universal constant—The name given to a physical quantity which retains the same value in all circumstances:

- Avogadro constant: $\mathcal{N} = 6.022\,136\,7 \times 10^{23}\,\text{mol}^{-1}$
- Boltzmann constant: $k = R/\mathcal{N} = 1.380\,658 \times 10^{-23}\,\text{J}\,\text{K}^{-1}$
- Faraday constant: $\mathscr{F} = \mathcal{N}\cdot\text{e} = 96\,485.309\,\text{C}\,\text{mol}^{-1}$
- Gravitational constant: $G = 6.672\,59 \times 10^{-11}\,\text{m}^3\text{kg}^{-1}\text{s}^{-2}$
- Planck constant: $h = 6.626\,075\,5 \times 10^{-34}\,\text{J}\,\text{s}$ (quantum of action)
  $\hbar = h/2\pi$ (quantum of angular momentum)
- Molar gas constant: $R = 8.314\,510\,\text{J}\,\text{mol}^{-1}\text{K}^{-1}$
- Stefan–Boltzmann constant: $\sigma = \pi^2 k^4/60\hbar^3 c^2$
  $= 5.670\,51 \times 10^{-8}\,\text{W}\,\text{m}^{-2}\text{K}^{-4}.$

Many other 'constants', such as the *equilibrium constant*, keep the same value in very particular circumstances; this is a misuse of the word, though tolerated by usage.

Universe—From the Latin *universus*, turned (*versus*) to form a whole (*unus*), the universe represents the entirety that is accessible to our experiment. For the thermodynamicist, the universe is made up of the *system* examined and its *surroundings* able to act on its evolution. By convention, the universe of the thermodynamicist is an *isolated* system:

'Universe' = 'System' + 'Surroundings'.

Whether the Universe (with a capital U) of the cosmologist is an isolated system still remains the subject of discussion.

Unmixing—Separation of a condensed solution into two condensed solutions having the same *aggregation state*.

Let there be a *mixing*: $(1-x)\langle A \rangle + x\langle B \rangle \rightarrow \langle A_{1-x}B_x \rangle$. The components A and B may be liquid or solid with the same structure as the solution $\langle A_{1-x}B_x \rangle$. The integral *Gibbs energy* of mixing is given by:

$$\Delta_{\text{mix}} G = RT[x \ln x + (1-x)\ln(1-x)] + \Delta_{\text{mix}} G^{\text{xs}}$$

with: $\Delta_{\text{mix}} G^{\text{xs}} = RT[x \ln \gamma_B + (1-x)\ln \gamma_A].$

The *activity coefficients* γ_A and γ_B being non-zero and finite, it is easy to verify (isn't it?) that $\Delta_{mix} G \to 0$ when $x \to 0$ or 1 with a vertical tangent. Unmixing is observed when the isothermal curve $\Delta_{mix} G = f(x)$ shows two inflection points, that is to say, when it is possible to draw a common tangent at two points of the curve. The compositions of the two solutions α and α' in equilibrium correspond to the abscissae of the contact points.

Let there be a solution M unmixing into two solutions N and P whose compositions are such that $x_N < x_M < x_P$. The Gibbs energy for the transformation:

$$\langle M \rangle \to y \langle P \rangle + (1-y) \langle N \rangle$$

is given by:

$$\Delta G = G_{\text{two-phase mixture}} - G_{\text{initial solution}} = y G_P + (1-y) G_N - G_M.$$

$\Delta G < 0$ when the chord NP is situated below the point M. The Gibbs energy of the two-phase mixture reaches a minimum together with ΔG when N and P correspond to the contact points α and α' of the common tangent to the curve $\Delta_{mix} G = f(x)$. The locus of the points α and α' on the temperature–composition diagram is called a 'binode'.

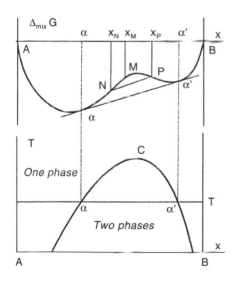

Unmixing: The isothermal curve *(above)* gives the integral Gibbs energy of mixing for the solution $A_{1-x} B_x$. The solutions whose compositions lie between α and α' are unstable with respect to the mixture $(\alpha + \alpha')$. The temperature–composition diagram *(below)* shows the unmixing zone of the solution $A_{1-x} B_x$ together with C, the critical point of the mixture.

U

Unmixing generally indicates repulsive interactions between constituents of the solution. At high temperatures, thermal agitation acts to favour mixing. Unmixing is often observed below a critical temperature T_c. Critical coordinates are obtained by expressing that the two inflection points meet, that is, by resolving the system:

$$\partial^2 \Delta_{mix} G / \partial x^2 = 0, \qquad \partial^3 \Delta_{mix} G / \partial x^3 = 0.$$

Unstable equilibrium—A system is in a state of unstable equilibrium when a spontaneous fluctuation, or an infinitesimal modification of a physical quantity characterizing it, has the effect of removing the system definitively from its equilibrium state and making it evolve towards a new equilibrium state. Consider for instance trinitroglycerol:

$$CH_2ONO_2\text{-}CHONO_2\text{-}CH_2ONO_2 \rightarrow 3\,CO_2 + \frac{5}{2}H_2O + \frac{1}{4}O_2 + \frac{3}{2}N_2.$$

Since this compound itself contains more oxygen than is necessary to burn all the carbon and all the hydrogen present, the slightest shock makes it explode, a result out of all proportion to the cause that provoked it.

In a calmer area, a supercooled liquid is also a good example of an unstable equilibrium: the presence of a seed in the medium is sufficient to trigger the crystallization.

The distinction between unstable and *metastable equilibrium* is sometimes unclear. Thus, when the temperature of a supercooled liquid becomes so low that the movements of the molecules begin to freeze, crystallization is no longer observed and we get a metastable glass.

V

Vacancy—The name given to an unoccupied crystal site. Vacancies are point defects which are thermodynamically stable. In fact, the formation of a vacancy at a site, an endothermic phenomenon ($\Delta_f H > 0$), correspondingly introduces disorder ($\Delta_f S > 0$). As a consequence, there is an equilibrium vacancy concentration which is calculated by seeking the minimum of the function $G = H - TS$.

Let $\Delta_f H$ be the molar enthalpy of formation of vacancies in a crystal and x the mole fraction of vacancies. If $G = 0$ is the *Gibbs energy* of a pure crystal without any vacancies, the Gibbs energy of an actual crystal will be given by:

$$G = x \Delta_f H + RT [x \ln x + (1-x) \ln(1-x)] - Tx \Delta_f S^{xs}.$$

The term $\Delta_f S^{xs}$ (excess molar entropy of formation of vacancies) is justified because the presence of a vacancy at a site modifies the state of the crystal by a relaxation effect on the surrounding atoms. The vacancy concentration at equilibrium is obtained by resolving the equation $dG/dx = 0$, whence:

$$\frac{x}{1-x} = \exp\left(\frac{\Delta_f S^{xs}}{R}\right) \cdot \exp\left(-\frac{\Delta_f H}{RT}\right),$$

an expression which may be written, x always being very small, as:

$$x = A \exp(-\Delta_f H / RT).$$

For the same reason (x very small), $\Delta_f S^{xs}$ and $\Delta_f H$ are independent of x. Vacancies may be considered as ordinary impurities and subject to the same thermodynamic treatment. $\Delta_f H$ then represents the partial molar enthalpy of mixing of a vacancy at infinite dilution.

However, the comparison must not be pursued too far: the experimenter can always control the amount of impurities introduced into a crystal, at least as long as saturation is not reached, whereas vacancies introduce themselves up to equilibrium. The experimenter can control the vacancy concentration only by using external *constraints*.

Van der Waals equation—Historically, this is the first equation of state of a fluid to account for the liquid–vapour transition, and it is also the simplest of the equations of state with two parameters which account, at least qualitatively, for the behaviour of a real gas. It was proposed in 1873 by Johannes Diderik Van der Waals, Dutch physicist (1837–1923), who gained the Nobel prize in physics in 1910:

$$\left(P + \frac{n^2 a}{V^2}\right)(V - nb) = nRT.$$

V

The term $n^2 a / V^2$ is the *internal pressure* representing the attraction exerted between molecules at short distances, by 'van der Waals forces'. The term nb, called the *covolume*, represents the incompressible volume occupied by the gaseous molecules.

The calculation of the *fugacity* of a van der Waals gas, a good exercise if somewhat irksome, does not present great difficulties:

$$\ln\frac{f}{P} = \int_0^P \left(V - \frac{RT}{P}\right)\mathrm{d}P$$

$$\ln f = \ln\frac{RT}{V-b} + \frac{b}{V-b} - \frac{2a}{RTV}.$$

The van der Waals equation accounts, at least qualitatively, for:

- the existence of a critical point [with $(\partial P/\partial V)_T = 0$ and $(\partial^2 P/\partial V^2)_T = 0$, it is easy to calculate $V_c = 3b$, $P_c = a/27b^2$, and $T_c = 8a/27bR$];
- the shape of locus of the minimum of the product PV on a diagram $PV = f(P)$ (*Amagat diagram*);
- the shape of the inversion curve of the *Joule–Thomson effect*.

From a quantitative point of view, the van der Waals equation, with only two parameters, is not accurate enough and the agreement between its predictions and actuality is poor. For instance, it does not forecast the maximum observed for the second virial coefficient and gives a critical compressibility factor $Z_c = P_c V_c/RT_c = 3/8 = 0.375$, whereas observed values generally lie between 0.29 and 0.31. Moreover, it predicts a ratio $T_b/T_c = 3.38$ [T_b: *Boyle temperature*, for which $\mathrm{d}(PV)/\mathrm{d}P = 0$ when $P = 0$], whereas experiment gives values around 2.75.

Van't Hoff equation—Sometimes wrongly called, for historical reasons, the van't Hoff isochore, it was proposed in 1886 by Jacobus Henricus van't Hoff, Dutch physicist (1852–1911), who developed the theory of *osmotic pressure* and earned the first Nobel prize in chemistry in 1901. The van't Hoff equation is now easily obtained by applying the *Gibbs–Helmholtz relation* to the *equilibrium constant*:

$$\frac{\mathrm{d}\ln K}{\mathrm{d}T} = +\frac{\Delta_r H^\circ}{RT^2}.$$

The integration of this equation gives the curve $K(T)$, provided that a point $K(T_0)$ be known.

Vaporization—Transition from the liquid state to the vapour state. Vaporization is an *endothermic* phenomenon. It is possible to cool a liquid by means of *adiabatic* vaporization.

Vapour—The term vapour is applied to gases when the temperature and pressure lie close to the liquid domain. A small decrease in temperature or a moderate increase in pressure are sufficient to condense a vapour. There is no well-defined borderline between gas and vapour. A vapour may be:

- saturated when it is in equilibrium with the liquid, the *saturated vapour pressure* being obtained by integration of the *Clapeyron equation*;
- dry when it is under a pressure lower than the saturated vapour pressure of the liquid, the liquid then being unstable;
- supersaturated when it is under a pressure higher than the saturated vapour pressure of the liquid, the vapour then being unstable.

Variance—The word has two different meanings:

IN STATISTICS, the variance σ_X^2 is defined by:

$$\sigma_X^2 = \langle(x - \langle x \rangle)^2\rangle = \langle x^2 \rangle - \langle x \rangle^2.$$

$\langle x \rangle$ is the *moment* of order 1, or mean, of the variable X.
$\langle x^2 \rangle$ is the moment of order 2, or mean of the squares, of X.
If x represents the different possible values taken by the variable X, the *distribution* of x presents a peak around x_{mp}, the most probable value for X. The square root of the variance, or *root mean square*, σ_X, gives the half-width of the peak. In the case of a *normal distribution*, the probability for the variable X to take a value included between $(x + \sigma_X)$ and $(x - \sigma_X)$ is 68%.

IN THERMODYNAMICS, the variance υ is the number of *intensive* parameters describing a system at equilibrium which is possible to modify without changing either the number φ or the nature of the phases at equilibrium; ς being the number of independent constituents, the variance is given by the Gibbs *phase rule*: $\upsilon = \varsigma + 2 - \varphi$.

Velocity—A velocity, distance covered per unit time, expressed in $m\,s^{-1}$, is a vector. A speed is a velocity which has lost its vector character.

V

Velocity of light—Since the 17th Conférence Générale des Poids et Mesures in 1983, the *metre* is defined from the velocity of the light in vacuum. By convention: $c = 299\,792\,458\,m\,s^{-1}$ exactly. In a material medium whose index of refraction is n, $\upsilon = c/n$.

Velocity of sound—The velocity of propagation of a vibration in a material with volumetric mass ρ and *Young's modulus E* is given by the general relationship:

$$v^2 = E/\rho.$$

The pressure variations which cause the sonic wave are, by hypothesis, *adiabatic*. From the definition of Young's modulus:

$$\frac{\delta l}{l} = \frac{\delta V}{V} = \frac{1}{E} \cdot \frac{\delta \mathscr{F}}{\mathscr{S}} = -\frac{1}{E} \cdot \delta P$$

$$\frac{1}{E} = -\frac{1}{V}\frac{\delta V}{\delta P} = -\frac{1}{V}\left(\frac{\partial V}{\partial P}\right)_S = +\frac{1}{\rho}\left(\frac{\partial \rho}{\partial P}\right)_S$$

$$v^2 = \left(\frac{\partial P}{\partial \rho}\right)_S.$$

Virial—The virial equation of state (from the Latin *vis, viris*: force), proposed by Kamerlingh Onnes in 1901, appears in the form of a series expansion of the product PV, as a function of $1/V$ or P:

$$PV = RT + \frac{B}{V} + \frac{C}{V^2} + \frac{D}{V^3} + \cdots$$

$$PV = RT + B'P + C'P^2 + D'P^3 + \cdots$$

Coefficients B, C, ..., called second, third, ... virial coefficients, are temperature-dependent. The second virial coefficient is negative at low temperatures, becomes zero at the *Boyle temperature*, is positive above that temperature, goes through a maximum, and tends towards zero when $T \to \infty$. The curve $B = f(T)$ allows easy access to T_i, the inversion temperature of the *Joule–Thomson effect* when $P \to 0$:

$$\mu_J = \left(\frac{\partial T}{\partial P}\right)_H = \frac{1}{c_P}\left[T\left(\frac{\partial V}{\partial T}\right)_P - V\right].$$

When $P \to 0$, the brackets become zero for $\mathrm{d}B/\mathrm{d}T = B/T$. The inversion temperature thus corresponds to the abscissa of the point of contact of the tangent drawn from the origin to the curve $B = f(T)$.

There are naturally relationships between the two series of coefficients, easily obtained by extracting P from the first equation to insert it into the second one:

$$B' = B/RT, \qquad C' = (C - B^2)/R^2T^2.$$

Near the critical point, the series expansion in $(1/V)$ is more realistic than that in P, because the actual behaviour of the fluid is better represented:

$$(\partial PV/\partial P)_T \to \infty \quad \text{when} \quad P \to P_c \text{ for } T = T_c.$$

The expansion as a function of P cannot rigorously give a vertical tangent with a finite number of terms.

The virial expansion allows easy calculation of the *fugacity*:

$$\ln \frac{f}{P} = \int_0^P \left(V - \frac{RT}{P} \right) dP$$

$$\ln f = \ln P + \frac{BP}{RT} + \frac{1}{2}(C - B^2)\left(\frac{P}{RT} \right)^2 + \cdots$$

Viscosity—The viscosity of a fluid (from Low Latin *viscosus*, sticky) appears through a resistance to the gliding of molecules or parallel molecular planes. There are two physical quantities characterizing viscosity: the dynamic viscosity η and the kinematic viscosity v, linked by the volumetric mass ρ: $v = \eta/\rho$. When the viscosity is quoted without any other indication, it is in principle η, the dynamic viscosity, or 'absolute viscosity', of the fluid.

THE DYNAMIC VISCOSITY is the *conductivity* of *momentum*. Consider the laminar flow of a fluid in a capillary whose diameter is d, under the influence of a pressure gradient $\Delta P/l$; the flow of the fluid (volume per unit time) through the capillary is given by Poiseuille's law:

$$q = \frac{\pi}{128} \cdot \frac{\Delta P}{l} \cdot \frac{d^4}{\eta}.$$

The dynamic viscosity η is expressed in Pa s, a unit still known under the old designation of poiseuille (symbol Pl). The poise is a cgs unit whose use must be discouraged ($1\,P = 0.1\,Pl = 0.1\,Pa\,s$; conveniently, $1\,cP = 1\,mPa\,s$). The dynamic viscosity of water at $20°C$ is $10^{-3}\,Pa\,s$ whereas that of the fused glass around $600°C$, the temperature at which it can be worked, varies between 10^3 and $10^6\,Pa\,s$. Above $10^{12}\,Pa\,s$, a liquid becomes glassy and the motion of the molecules is no longer possible. The viscosity of a liquid decreases with increasing temperature, whereas that of a gas increases.

THE KINEMATIC VISCOSITY, defined by $v = \eta/\rho$, represents the *diffusivity* of momentum. Its dimensions are those of a *diffusion*

V

coefficient and it is expressed in $m^2 s^{-1}$. The Stoke is a unit of kinematic viscosity whose use must be discouraged ($1\,St = 1\,cm^2 s^{-1}$).

Volt—The unit of electric potential in the SI system (symbol V), named from Alessandro Volta, Italian physicist (1745–1827), who discovered the galvanic cell in 1800:

$$1\,V = 1\,A\,\Omega = 1\,m^2\,kg\,s^{-3}\,A^{-1}.$$

Volta potential—The difference in electric potential between a point in the gaseous phase near the surface of a metal and a point in the gaseous phase near the solution in contact with the metal. This potential, also called the outer potential, linked to the *work function*, is measurable.

The outer potential ψ is directly related to the inner potential (or *Galvani potential*) φ by the relationship: $\varphi = \psi + \chi$, where χ represents the surface potential. With a galvanic cell:

$$(\psi_2 - \psi_1) = (\varphi_2 - \varphi_1) - (\chi_2 - \chi_1).$$

The difference $(\chi_2 - \chi_1)$ is often small compared with $(\varphi_2 - \varphi_1)$.

Volume—Volume is an *extensive* quantity whose conjugate tension is pressure. In the SI system, the volume is expressed in m^3. The elementary *work* exchanged with the *surroundings* by pressure forces is given by:

$$dW = -P\,dV.$$

Volume of mixing—The integral volume of mixing $\Delta_{mix}V$ is the volume change relating to the transformation:

$$\sum n_i \langle A_i \rangle \rightarrow \langle Mixture \rangle.$$

The integral molar volume of mixing is the integral volume of mixing reduced to one mole:

$$\sum x_i \langle A_i \rangle \rightarrow \langle Mixture \rangle \quad (\text{with } \sum x_i = 1).$$

The partial molar volume of mixing of i, or 'partial molar volume of dissolution' of i, represented by Δv_i and defined by:

$$\Delta v_i = (\partial \Delta_{mix}V/\partial n_i)_{T, P, n_{j \neq i}}$$

represents the difference between v_i, the partial molar volume of i into the mixture, and v_i°, the molar volume of i out of the mixture. δV, the volume change during the transformation:

$$\langle \text{Mixture} \rangle + \delta n_i \langle A_i \rangle \rightarrow \langle \text{Mixture} \rangle,$$

is given by:

$$\delta V = (v_i - v_i^\circ)\delta n_i = \Delta v_i\, \delta n_i.$$

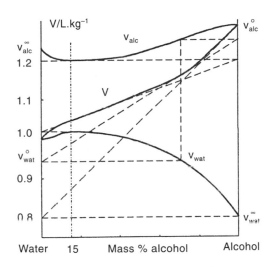

Volume of mixing: *Here, the integral volume and the partial volumes in the water–alcohol solution are not molar, but mass volumes. The integral volume shows an inflection at 15 mass% alcohol, whence the extrema observed on the partial volumes. The dashed lines show the construction of the partial mass volumes from the integral mass volume.*

The relationships between partial molar volumes of dissolution and integral volume of mixing are the following:

- the definition equation: $\Delta v_i = (\partial \Delta_{\text{mix}} V / \partial n_i)_{T,P,n_{j \neq i}}$;
- the *Gibbs–Duhem equation*: $\sum n_i\, d\Delta v_i = 0$ (*T* and *P* constant);
- the *Euler identity*: $\Delta_{\text{mix}} V = \sum n_i \Delta v_i$.

Volumes of mixing derive from *Gibbs energies of mixing*:

$$\Delta v_i = (\partial \Delta \mu_i / \partial P)_T = RT(\partial \ln a_i / \partial P)_T$$

$$\Delta_{\text{mix}} V = (\partial \Delta_{\text{mix}} G / \partial P)_T = RT \sum n_i (\partial \ln a_i / \partial P)_T.$$

V

In practice, the accuracy obtained by direct measurement of volumes is greater than that given by the above derivatives.

For *ideal solutions*: $a_i = kx_i$. Since activities are then independent of pressure, the volumes of mixing equal zero:

$$\Delta v_i = 0 \text{ and } \Delta_{mix} V = \sum n_i \Delta v_i = 0.$$

Volume–temperature diagram—The slope of *isobaric* curves is positive: $(\partial V / \partial T)_P = \alpha_P V$. The *critical* isobar shows at the critical temperature an inflection with a vertical tangent. The loci of points L and V characterizing the liquid and the vapour at equilibrium meet at the critical temperature with a vertical tangent. Along the saturated vapour curve $(\partial V / \partial T)_x$ tends towards $+\infty$ for the liquid and $-\infty$ for the vapour.

The law of rectilinear diameter, proposed by Cailletet and Mathias, then experimentally verified on various substances, allows the critical volume to be determined with good accuracy. By drawing, versus T, curves giving the density of liquid and vapour at equilibrium (or, equivalently, by drawing the curves $1/V$ vs T), the locus of the centre of the line LV is a straight line whose slope is negative. This line goes through the critical point where both L and V points meet with a vertical tangent.

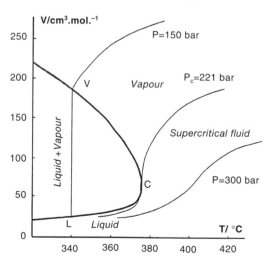

Volume–temperature diagram: *Water near the critical point C* ($56 \, cm^3 mol^{-1}$, $374°C$, $221 \, bar$).

W

Waage, Peter (1833–1900)—Norwegian chemist who, together with Cato Guldberg, expressed the *mass action law* in a mathematical form in 1864.

Wagner, Carl (1901–1977)—German physical chemist and metallurgist, who made an important contribution to the study of defects in crystals and in dilute solutions. He introduced *interaction parameters*.

Wagner reciprocity—Consider a mixture of solvent ($i=1$) and solute ($i=2, 3, \ldots$), in which the *reference state* for the solute i is the Henrian ideal solution ($a_i = \gamma_i x_i$, a_i being the *Henrian activity*). If there is only one solute i in the solvent 1, $\gamma_i \rightarrow 1$ when $x_i \rightarrow 0$. In the presence of different solutes j, k, $\ldots$, the *activity coefficient* γ_i may be expressed, in the neighbourhood of the solvent ($x_1 \approx 1$), by a series expansion:

$$\ln \gamma_i - \sum_{j=2}^{s} \left(\frac{\partial \ln \gamma_i}{\partial x_j} \right)_{x_{k \neq j}} x_j + \frac{1}{2} \sum_{j=2}^{s} \sum_{k=2}^{s} \left(\frac{\partial^2 \ln \gamma_i}{\partial x_j \partial x_k} \right) x_j x_k + \cdots$$

The first terms are the first-order *interaction parameters*:

$$\varepsilon_i^{(j)} - \left(\frac{\partial \ln \gamma_i}{\partial x_j} \right)_{x_{k \neq j}}$$

The Wagner reciprocity relationship expresses identity between the interaction of the solute j on the activity coefficient of i and the interaction of the solute i on the activity coefficient of j:

$$\varepsilon_i^{(j)} = \varepsilon_j^{(i)}.$$

An elegant proof of this result consists in applying the general relationship (see *Partial molar quantity*) linking the excess partial molar Gibbs energy of mixing, $\Delta \mu_i^{xs} = RT \ln \gamma_i$, and the excess integral Gibbs energy of mixing. Omitting the subscript 'mix' for simplicity:

$$\Delta G^{xs} = RT \sum x_i \ln \gamma_i$$

$$\Delta \mu_i^{xs} = \Delta G^{xs} + \left(\frac{\partial \Delta G^{xs}}{\partial x_i} \right) - \sum_{j=2}^{r} \left(\frac{\partial \Delta G^{xs}}{\partial x_j} \right) x_j$$

324 Watt

$$\Delta\mu_j^{xs}=\Delta G^{xs}+\left(\frac{\partial\Delta G^{xs}}{\partial x_j}\right)-\sum_{i=2}^{r}\left(\frac{\partial\Delta G^{xs}}{\partial x_i}\right)x_i.$$

Remember that in the above two expressions, ΔG^{xs} is given as a function of x_2, x_3, ..., which implies $(\partial\Delta G^{xs}/\partial x_1)=0$. The differentiation of both expressions with respect to x_j and x_i yields:

$$\frac{\partial\Delta\mu_i^{xs}}{\partial x_j}=\frac{\partial\Delta G^{xs}}{\partial x_j}+\frac{\partial^2\Delta G^{xs}}{\partial x_i\partial x_j}-\frac{\partial\Delta G^{xs}}{\partial x_j}-\sum_{i=2}^{r}x_i\left(\frac{\partial^2\Delta G^{xs}}{\partial x_i\partial x_j}\right)$$

$$\frac{\partial\Delta\mu_j^{xs}}{\partial x_i}=\frac{\partial\Delta G^{xs}}{\partial x_i}+\frac{\partial^2\Delta G^{xs}}{\partial x_j\partial x_i}-\frac{\partial\Delta G^{xs}}{\partial x_i}-\sum_{j=2}^{r}x_j\left(\frac{\partial^2\Delta G^{xs}}{\partial x_j\partial x_i}\right).$$

Owing to the independence of the order of differentiation, the two expressions on the right-hand sides are identical. Thus the left hand sides must also be identical. Now:

$$(\partial\Delta\mu_i^{xs}/\partial x_j)=RT\varepsilon_i^{(j)} \text{ and } (\partial\Delta\mu_j^{xs}/\partial x_i)=RT\varepsilon_j^{(i)},$$

whence:
$$\varepsilon_i^{(j)}=\varepsilon_j^{(i)}.$$

Remark: In the preceding proof, the partial derivatives with respect to x_i are valid, because ΔG^{xs} is expressed as a function of the mole fractions x_2, x_3, ..., which are independent variables. An erroneous demonstration of the Wagner reciprocity relationship would be:

$$\varepsilon_i^{(j)}=\frac{\partial\ln\gamma_i}{\partial x_j}=\frac{1}{RT}\frac{\partial^2\Delta G^{xs}}{\partial x_i\partial x_j}=\frac{1}{RT}\frac{\partial^2\Delta G^{xs}}{\partial x_j\partial x_i}=\frac{\partial\ln\gamma_j}{\partial x_i}=\varepsilon_j^{(i)}.$$

The mistake made in this proof, often presented, resides in writing $\Delta\mu_i^{xs}=RT\ln\gamma_i=(\partial\Delta G^{xs}/\partial x_i)$, whereas it would be correct to write $\Delta\mu_i^{xs}=RT\ln\gamma_i=(\partial\Delta G^{xs}/\partial n_i)_{n_{j\neq i}}$. A partial derivative with respect to x_i with $x_{j\neq i}$ constant has no physical meaning, given that the x_i are linked by a constraint!

Watt—The unit of power in the SI system (symbol W), from James Watt, Scots engineer (1736–1819):

$$1\,W=1\,J\,s^{-1}=1\,m^2\,kg\,s^{-3}.$$

Wave—A disturbance which propagates in a medium, characterized by its wavelength λ, its velocity of propagation c, and its frequency $v=c/\lambda$. The energy of a photon is given by $E=hv=hc/\lambda$, h being the Planck constant.

Weber—The unit of magnetic *flux* in the SI system (symbol Wb), from Wilhelm Weber, German physicist (1804–1891). The weber is the magnetic flux which, crossing a circuit consisting of only one loop, and reduced to zero at a uniform rate in one second, generates in the circuit an electromotive force of one volt:

$$1\,\text{Wb} = 1\,\text{V\,s} = 1\,\text{m}^2\,\text{kg\,s}^{-2}\,\text{A}^{-1}.$$

Weight—The name given to the gravitational force mg exerted on a mass m at the earth's surface, giving it an acceleration of free fall g. Weight has the dimensions of force, and in the SI system is expressed in newtons. Although in the current practice the acceleration of free fall may be constant, those who pay attention to their language will prefer to use mass rather than weight.

Wien's law—In 1895, the German physicist Wilhelm Wien (1864–1928, Nobel prize for physics in 1911), stated the law according to which the product of the temperature of a black body by the wavelength corresponding to the maximum energy emission is constant: $\lambda_{\max} T =$ const. This result may be deduced from the more general distribution law given by Planck in 1900:

$$\rho_\lambda(\lambda, T) = \frac{8\pi hc}{\lambda^5} \times \frac{1}{e^{hc/\lambda kT} - 1},$$

where $\rho_\lambda(\lambda, T)$, which is expressed in $\text{J\,m}^{-4}$, is the spectral energy density.

$\rho_\lambda(\lambda, T)\,d\lambda$ represents the energy density emitted between the wavelengths λ and $\lambda + d\lambda$ by a black body at a temperature T. By introducing the radiation frequency $v = c/\lambda$, we get:

$$\rho_v(v, T) = \frac{8\pi h v^5}{c^4} \times \frac{1}{e^{hv/kT} - 1}.$$

Work—An interaction between two *systems* is called work when the resulting *transformation* may be reproduced independently for each system with, as the only external effect, the displacement of a mass in a gravitational field.

RECOGNITION CRITERION: This results at once from the above definition. Two systems exchange work if it is possible for each system taken separately to undergo a transformation which brings it from the same initial state to the same final state with, as the only external effect,

W

the raising or lowering of a mass in a gravitational field. A *process* during which a system exchanges only work with the *surroundings* is called *adiabatic*.

An exchange of radiation between two lasers is work, whereas an exchange of radiation between a warm body and a cold one is not work. Indeed, although it is possible to increase the temperature of a system by lowering a mass in a gravitational field, it is impossible on the other hand to lower a temperature by raising a mass in a gravitational field.

EXPRESSIONS OF WORK: Work is classically defined as the product of a force by a displacement: $dW = \mathcal{F}\,dl$. This expression allows the work needed to displace a mass in a gravitational field, to lengthen a wire, or to move an electric charge in an electric field to be obtained directly. The unit of work in the SI system is the joule (symbol J):

$$1\,J = 1\,kg\,m^2\,s^{-2} = 1\,N\,m = 1\,Pa\,m^3 = 1\,C\,V.$$

It is possible to obtain various expressions of work by dividing a force by a physical quantity and multiplying the displacement by the same quantity. For instance:

- $dW = \Gamma\,d\mathcal{S}$, work of forces of *surface tension* (Γ: surface tension, in $N\,m^{-1}$; $\mathcal{S}$: surface, in m^2).
- $dW = -P\,dV$, work of forces of pressure (P: pressure, in $N\,m^{-2}$, or Pa; V: volume, in m^3). The minus sign is the consequence of the *sign convention* ($dW > 0$ if $dV < 0$).
- $dW = v\,d(mv)$, work of kinetic forces (mv: *momentum*, in $kg\,m\,s^{-1}$; v: velocity, in $m\,s^{-1}$).
- $dW = \mathcal{H}\,d(\mathcal{B}V)$, work of magnetic forces ($\mathcal{H}$: *magnetic excitation*, in $A\,m^{-1}$; $\mathcal{B}$: magnetic field, in T or $kg\,s^{-2}\,A^{-1}$; V: volume, in m^3).
- $dW = \mathcal{E}\,d\mathcal{Q}$, work of electric forces ($\mathcal{E}$: electric potential, in V or $kg\,m^2\,s^{-3}\,A^{-1}$; $\mathcal{Q}$: amount of electricity, in C or A s). A more general expression is: $dW = E\,d(DV)$ (E: *electric field*, in $V\,m^{-1}$ or $kg\,m\,s^{-3}\,A^{-1}$; D: *electric excitation*, in $A\,s\,m^{-2}$; V: volume, in m^3).
- $dW = \mu_i\,dn_i$, work of chemical forces (μ_i: *chemical potential* of the constituent i, in $J\,mol^{-1}$; n_i: amount of substance, in mol).

When a system exchanges work under different forms with the surroundings:

$$dW = \mathcal{F}\,dl + \Gamma\,d\mathcal{S} - P\,dV + v\,d(mv)$$
$$+ \mathcal{H}\,d(\mathcal{B}V) + E\,d(DV) + \sum \mu_i\,dn_i + \cdots$$

Remark 1: It is important to make a difference between work and *energy*. Work is not generally a state function (the *differential* dW is not *exact*). It is meaningless to speak of the work contained in a piece of wood! On the other hand, it is quite possible to define its energy. Work is a *transfer variable* which appears only at the boundary separating the system from its surroundings.

Remark 2: Whatever the nature of the work exchanged, the transformation may always be made with 100% efficiency. A system A may provide work to a system B which will transform it entirely into mechanical energy. This observation justifies the definition adopted for work which reduces it to the displacement of a mass in a gravitational field. In practice, efficiencies do not reach 100% owing to friction. However, friction may be reduced by approaching the conditions of *reversibility*.

Remark 3: It is possible to express the element of work by:

$$dW = \sum X_i \, dx_i,$$

in which x_i represents an *extensive* variable and X_i the conjugate *intensive* variable. The intensity X_i is a variable of *tension* only if the work exchanged, $X_i \, dx_i$, changes only the *internal energy* of the system. For instance, kinetic energy is *external energy*; electric and magnetic energies have internal and external contributions. Hence velocity v, magnetic excitation $\mathscr{H}$ and electric field E cannot be variables of tension. When electric energy may be expressed by $dW = \mathscr{E} \, d\mathcal{Q}$, the work exchanged d$W$ is only internal energy, so the electric potential $\mathscr{E}$ is a variable of tension. The elemental work $\mathscr{F} d\ell$, $\Gamma \, d\mathcal{S}$, $-P \, dV$, and $\mu_i \, dn_i$ exchanged by a system with its surroundings is all internal energy. As a consequence, $\mathscr{F}$, Γ, P, and μ_i are variables of tension.

Work function—The name given to the energy needed to extract an electron from the surface of a metal. It is related to the external potential, or *Volta potential*, by the relationship $w = -\mathscr{F}\psi$. If an electron is transferred from a phase α to a phase β, its energy change is:

$$w_\beta - w_\alpha = -\mathscr{F}(\psi_\beta - \psi_\alpha).$$

When two metals are in contact, electrons tend to leave the metal characterized by the smaller work function to enter in the other metal, where electrons are more strongly bonded. That explains the minus sign in the relation between external potential and work function.

W

Y

Young's modulus—The ratio of the stress exerted on a wire to its strain, named from Thomas Young (1773–1829), English doctor and physicist, sometime Egyptologist. He is well known for having provided, through his experiments on interference, an experimental basis for the wave theory of light. The author, with Helmholtz, of the trichromatic theory of colour vision, he was also involved in studies on surface tension and elasticity:

$$E = \frac{\sigma}{\varepsilon} = \frac{\mathscr{F}}{\mathscr{S}} \cdot \frac{l}{\delta l}.$$

$\sigma = \mathscr{F}/\mathscr{S}$ is the stress, that is, the force exerted in the direction of its axis on a cylindrical bar whose cross-section is $\mathscr{S}$.
$\varepsilon = \delta l/l$ is the strain, or lengthening of the bar.

Z

Zeotrope—From the Greek ζεω, to boil, and τροπη, evolution. Antonym of *azeotrope*, introduced by Świętosławski in 1945 to describe a solution whose boiling temperature, under constant pressure, evolves between an initial boiling point and a final boiling point. By extension, a mixture is called zeotropic when it is possible to separate its constituents by distillation. As a consequence, the temperature–composition diagram does not show a composition corresponding to an extremum on the *dew-point curve* or on the *boiling* curve.

Zeroth law—This was stated for the first time by Maxwell in 1872, after the *first* and *second laws*, given by Clausius in 1850.

Two systems A and B in thermal equilibrium with a third system C are in thermal equilibrium with one another.

Corollary: This law leads to the concept of temperature and makes the construction of a thermometer possible. Pressure P and volume V

are two independent parameters which can be modified separately. Two systems A and B, separated by a *diathermanous* boundary, the assembly (A + B) being *isolated*, are said to be in 'thermal contact'. At equilibrium, there is a relationship $F(P_A, V_A, P_B, V_B) = 0$. For instance, for two *ideal gases* at thermal equilibrium, this relationship is:

$$(P_A V_A / n_A) - (P_B V_B / n_B) = 0.$$

Consider now a system C in thermal equilibrium with A and B:

$$F(P_A, V_A, P_C, V_C) = 0$$

$$F(P_B, V_B, P_C, V_C) = 0.$$

The zeroth law states that the systems A and B are in thermal equilibrium. Mathematically, this is expressed by writing that the two above relationships imply:

$$F(P_A, V_A, P_B, V_B) = 0.$$

The function F must then take the form:

$$F(P_A, V_A, P_B, V_B) = t(P_A, V_A) - t(P_B, V_B).$$

Naturally, the function t is not unique, because any arbitrary function $T(t)$ obeys the condition. Each function t defines an arbitrary scale of temperature. Each equation $t = t(P, V)$, called an 'equation of state', defines on the plane (P, V) a family of *isotherms*. Once the scale of temperature has been chosen, it is possible to take any two of the three variables (t, P, V) as independent variables.

Remark: The zeroth law is an easy means of introducing the concept of temperature. However, the present tendency is not to consider the zeroth law as a true *principle*, but to deduce the notion of temperature from the *second law, corollary 7*. On inserting into the expression of the *internal energy* for a *reversible* transformation: $dQ = T\,dS$ (*Second law, corollary 11*), we obtain: $dU = T\,dS + \cdots$. The temperature T is the *tension* conjugate of the *extensity* entropy. Thermal equilibrium at the interface separating two systems implies equality of temperatures, a result which constitutes the statement of the zeroth law.

Z